AF801486

Behaviormetrics: Quantitative Approaches to Human Behavior

Volume 4

Series Editor

Akinori Okada, Professor Emeritus, Rikkyo University,
Tokyo, Japan

This series covers in their entirety the elements of behaviormetrics, a term that encompasses all quantitative approaches of research to disclose and understand human behavior in the broadest sense. The term includes the concept, theory, model, algorithm, method, and application of quantitative approaches from theoretical or conceptual studies to empirical or practical application studies to comprehend human behavior. The Behaviormetrics series deals with a wide range of topics of data analysis and of developing new models, algorithms, and methods to analyze these data.

The characteristics featured in the series have four aspects. The first is the variety of the methods utilized in data analysis and a newly developed method that includes not only standard or general statistical methods or psychometric methods traditionally used in data analysis, but also includes cluster analysis, multidimensional scaling, machine learning, corresponding analysis, biplot, network analysis and graph theory, conjoint measurement, biclustering, visualization, and data and web mining. The second aspect is the variety of types of data including ranking, categorical, preference, functional, angle, contextual, nominal, multi-mode multi-way, contextual, continuous, discrete, high-dimensional, and sparse data. The third comprises the varied procedures by which the data are collected: by survey, experiment, sensor devices, and purchase records, and other means. The fourth aspect of the Behaviormetrics series is the diversity of fields from which the data are derived, including marketing and consumer behavior, sociology, psychology, education, archaeology, medicine, economics, political and policy science, cognitive science, public administration, pharmacy, engineering, urban planning, agriculture and forestry science, and brain science.

In essence, the purpose of this series is to describe the new horizons opening up in behaviormetrics — approaches to understanding and disclosing human behaviors both in the analyses of diverse data by a wide range of methods and in the development of new methods to analyze these data.

Editor in Chief

Akinori Okada (Rikkyo University)

Managing Editors

Daniel Baier (University of Bayreuth)
Giuseppe Bove (Roma Tre University)
Takahiro Hoshino (Keio University)

Akinori Okada · Kazuo Shigemasu ·
Ryozo Yoshino · Satoru Yokoyama
Editors

Facets of Behaviormetrics

The 50th Anniversary of the Behaviormetric Society

Springer

Editors
Akinori Okada
Rikkyo University
Kanagawa, Japan

Ryozo Yoshino
The Institute of Statistical Mathematics
Tokyo, Japan

Kazuo Shigemasu
Graduate School of Human Relations
Keio University
Tokyo, Japan

Satoru Yokoyama
School of Business
Aoyama Gakuin University
Tokyo, Japan

ISSN 2524-4027 ISSN 2524-4035 (electronic)
Behaviormetrics: Quantitative Approaches to Human Behavior
ISBN 978-981-99-2239-0 ISBN 978-981-99-2240-6 (eBook)
https://doi.org/10.1007/978-981-99-2240-6

© The Editor(s) (if applicable) and The Author(s), under exclusive license to Springer Nature Singapore Pte Ltd. 2023
This work is subject to copyright. All rights are solely and exclusively licensed by the Publisher, whether the whole or part of the material is concerned, specifically the rights of translation, reprinting, reuse of illustrations, recitation, broadcasting, reproduction on microfilms or in any other physical way, and transmission or information storage and retrieval, electronic adaptation, computer software, or by similar or dissimilar methodology now known or hereafter developed.
The use of general descriptive names, registered names, trademarks, service marks, etc. in this publication does not imply, even in the absence of a specific statement, that such names are exempt from the relevant protective laws and regulations and therefore free for general use.
The publisher, the authors, and the editors are safe to assume that the advice and information in this book are believed to be true and accurate at the date of publication. Neither the publisher nor the authors or the editors give a warranty, expressed or implied, with respect to the material contained herein or for any errors or omissions that may have been made. The publisher remains neutral with regard to jurisdictional claims in published maps and institutional affiliations.

This Springer imprint is published by the registered company Springer Nature Singapore Pte Ltd.
The registered company address is: 152 Beach Road, #21-01/04 Gateway East, Singapore 189721, Singapore

Foreword

It is a great privilege to have been invited by Akinori Okada to write this foreword to the Volume "Facets of Behaviormetrics", celebrating the 50th Anniversary of the Behaviormetric Society.

It is natural to view the Behaviormetric Society as a sister society to the Psychometric Society that was founded in 1935. For a long time, the Psychometric Society was basically a North American Society, to become more international when it started to organize meetings outside North America, the first being in Uppsala in 1978.

The Behaviormetric Society, established in 1973, was originally called the Behaviormetric Society of Japan and was renamed the Behaviormetric Society in 2016. The Psychometric Society has its flagship journal Psychometrika, while the Behaviormetric Society has (beside Kodo keiryogaku—Journal of Behaviormetrics, in Japanese) the journal Behaviormetrika. The two journals exchange their table of content for every issue.

One of the significant events in the history of the Psychometric Society was the organization of its first meeting outside of North America and Europe. Not surprisingly, this first International Meeting of the Psychometric Society (IMPS) was held in Japan, and this was in 2001 in Osaka. The Behaviormetric Society played a key organizational role (next to the Japan Statistical Society and various other Japanese societies). With over 313 participants from 19 countries, the meeting was very successful (see also the Volume of Proceedings of this meeting). It was a great pleasure working with Haruo Yanai, Akinori Okada, Kazuo Shigemazu, and Yutaka Kano to start the preparation of the first IMPS in 1998.

As it turns out, I was able to meet my Japanese friends in person because of the following event. The Behaviormetric Society celebrated its 25th anniversary in September 1998 in Tokyo, and I was very honored to give a keynote address at that occasion, entitled "Relational data analysis, optimal scaling, and quantification methods".

The term "quantification" in the title honors the close similarity between data analytic research in The Netherlands, particularly at the Leiden Data Theory Department, and in Japan. Here visionary Chikio Hayashi was the originator of "Quantification Methods", which he developed since the 1950s. We may consider Louis Guttman and Chikio Hayashi the godfathers of quantification (of qualitative data). Research in this area turned into a global activity, with Jean-Paul Benzécri in France, Shizuhiko Nishisato in Canada, and the Albert Gifi—Data Theory group in the Netherlands. Chikio Hayashi's leadership continued, which can be seen from the fact that he is writing about "Data Science" as early as in the middle 1990s.

Moreover, in the meantime Chikio Hayashi was the Founding President of the Behaviormetric Society in 1973. He was succeeded by an impressive sequence of successors, Tadashi Hidano, Haruo Yanai, Kinji Mizuno, Meiko Sugiyama, Hiroshi Akuto, and Akinori Okada. I copied this list from the article "From the President" by Kazuo Shigemasu in 2015.

The areas of expertise of the members of the Behaviormetric Society cover a wide spectrum. From the beginning, they carried out interdisciplinary research on the development of data analytic methods and their application in relation to human behavior in a wide sense. Multivariate analysis and multidimensional scaling have always been in the center of interest, and the analysis of asymmetric data can be considered a Japanese speciality. Approaches such as network models, Bayesian analysis, machine learning, and big data analysis are embraced as well. Facets of Behaviormetrics offers a wonderful insight into the richness and breadth of behaviormetric research.

During my stay in Japan in 1998, I not only gave four (different) lectures in Tokyo and Osaka, but my hosts also organized visits to very special events. Chikio Hayashi arranged for me to watch the famous Yabusame (Horseback Archery) Festival on September 16 in Kamakura. Akinori Okada took me to a Sumo match in Tokyo and to a Sumo restaurant in Osaka, owned by one of his friends, a former Sumo wrestler. These were unforgettable experiences. Both Japanese culture and data analysis are close to my heart.

I wish to congratulate the Behaviormetric Society and its members with their 50th Anniversary. I am convinced that the future is bright for Behaviormetrics because of its great adaptability to new demands, both from a data analysis perspective and from modern society at large.

Leiden, The Netherlands
January 2023

Jacqueline Meulman

Preface

The Behaviormetric Society (BMS) was established in September 1973 to foster behaviormetrics, a field which covers a variety of quantitative approaches in order to reveal and comprehend human behavior in a broadest sense. The parent organization of the Society was the "Symposium of Behaviormetrics" that held the annual events from 1969 through 1972. The Society launched in September 1973 as the "Behaviormetric Society of Japan" with over 400 society members. The name of the society was altered in September 2016 when the publishing house of the official English journal *Behaviormetrica* was changed from Sasaki Printing & Publishing to Springer. The energetic and active research activity of the society brought the status of a registered organization of the Science Council of Japan to the society in August 1984. The society will celebrate its 50th anniversary in September 2023.

Behaviormetrics includes the entire range of quantitative approaches, from theoretical or conceptual to empirical or practical studies and researches in order to reveal, disclose, understand, and predict human behavior in a broadest sense. The purpose of the present volume is twofold: one is to look back on the traditional aspects of behaviormetrics exhibiting conventional features thereof: the other is to look into the future of behaviormetrics which exhibits features shaping the future of behaviormetrics.

The present volume contains 14 articles. Some of them are invited, and some of them are submitted by researchers including non-members of the Behaviormetric Society. Part I contains 10 invited articles and Part II contains four submitted articles. Articles are in alphabetical order of the first author of an article within each part. All articles are peer reviewed. The content of this volume is briefly described below for the sake of the convenience of the readers.

Daniel Baier and Sascha Vökler apply a cluster-based genetic algorithm and tabu search to the problem of selecting appropriate attribute levels in marketing. They analyze small- to large-size problems with promising results. Giuseppe Bove and Donatella Vicari present a procedure for dealing with asymmetric data which analyze symmetric and skew symmetric components separately through multidimensional scaling and cluster analysis, and then combine them based on the drift vector model. Chadjipadelis Theodore presents a timely study entitled "Citizens and The Pandemic: Values, Attitudes, and Impacts". He examines the attitudes of Greek citizens toward the pandemics and its effects, detects distinct groups of behaviors, and relates these profiles to other variables regarding information sources, views on democracy, moral values, and even the emotional impact of the overall situation. Mark de Rooij and Patrick J. F. Groenen developed a generalized logistic model which handles multiple outcome variables within a distance framework, which was given a charming name MELODIC. They also developed an efficient algorithm to estimate the parameters of this model. Applying the model to two data sets, they demonstrate that the results of the analysis can be interpreted by using a graphical presentation and using tables. Wolfgang Jagodzinski, Kazufumi Manabe, Hermann Dülmer, Carola Hommerich, and Eldad Davidov present their research on human values in Germany and Japan. Focusing on the cross-cultural comparability of social values, they examine whether the German and Japanese versions of Shalom Schwartz's Portrait Value Questionnaire are composed of scalar-equivalent and reliable indicators. P. M. Kroonenberg and M. Stoltenborgh apply multidimensional scaling and correspondence analysis to the data gathered from two handbooks in a particular research area in psychology to depict the interrelations among researchers and their research topics. The particular area the authors chose is the attachment research and they succeeded in visualizing the world of attachment research and its changes over the decades. Francisco Javier Ortín Cervera, Honorio Ros Múgica, and Innar Liiv present a new protocol based on blockchain so that the feedback from customers or consumers of goods or services can easily be feedbacked, evaluated, and rewarded for their feedback. Shizuhiko Nishisato gives a fitting commemoration of the 50th anniversary of the Behaviormetric Society, and presents 11 propositions to drive the mission of the society for the next 50 years. This may be said to be a historic proposition, comparable to the 23 problems of the great mathematician Hilbert, which appeared in 1900. Nishisato believes that each person participating in an interdisciplinary project needs to be fully committed to the project, not just on an ad hoc basis. Fulvia Pennoni and Miki Nakai use data from the Preference Parameters Study to investigate subjective well-being among Japanese using a hidden Markov model. They report that initial happiness plays an important role in inequality in the life course of subjective well-being in a substantial and sustained manner. Yoshio Takane and Peter G. M. van der Heijden investigate the asymptotic behavior of the log-likelihood ratio (LR) statistics when regularity conditions were not satisfied, and pointed out that they were not guaranteed to follow chi-square distributions. Violations of the regularity conditions typically occur when we compare the goodness of fit of a true T dimensional model against that of $(T+1)$-dimensional model. To avoid the difficulties, they recommend comparing of specific dimensionalities against saturated models, in which the problem of violating regularity conditions does not occur.

Syou Maki, Yasutoshi Hatsuda, Toshihiko Ishizaka, Sachiko Omotani, Naonori Koizumi, Yukako Yasui, Takako Saito, Michiaki Myotoku, and Tadashi Imaizumi reveal asymmetric relationships between antibiotic drugs by analyzing cross-resistance rates using asymmetric multidimensional scaling. They found that antibiotic drugs can be classified according to the strain groups of the antibiotics. Haruhiko Ogasawara introduces the multivariate power-gamma distribution, which is given by power transformations of the multivariate gamma distribution and derived its moments and other properties in the situation corresponding to the one-factor model of exploratory factor analysis. Yoosung Park and Tadahiko Maeda analyze time-series data from the Survey of Japanese National Character, which has been conducted by the Institute of Statistical Mathematics for over 60 years since 1953. This survey is said to have been the precursor of the GSS, WVS, and Eurobarometer. In this paper, they focus in particular on interpersonal relationships, which they call "giri-ninjo", a concept said to be unique to the Japanese. Kojiro Shojima investigates the total score distribution of a test, regarding it as the distribution of the sum of rolling unfair dice with a different number of faces, and derived a closed form of the distribution's probability mass function (PMF) and provide a Lord-Wingersky (LW) algorithm to estimate the PMF efficiently.

We thank the authors of the contributions in this book and referees who carefully reviewed the contributions and gave the authors valuable suggestions which are useful for improving their papers. We would like to express our special thanks to Jacqueline Meulman for her thoughtful Foreword. We are largely indebted to Reginald Williams and Yasuko Hase for their help with English in preparing the present volume. We appreciate Yutaka Hirachi and Rammohan Krishnamurthy at Springer Nature for their help in publishing this volume.

Tokyo, Japan — Akinori Okada
Tokyo, Japan — Kazuo Shigemasu
Kyoto, Japan — Ryozo Yoshino
Tokyo, Japan — Satoru Yokoyama
January 2023

Contents

Invited Articles

Product-Line Design Using Cluster-Based Genetic Algorithms and Tabu Search

Daniel Baier and Sascha Vökler

Abstract Selecting adequate attribute-levels (e.g., components, ingredients, materials, prices, qualities) for new and/or existing products is an important task for marketeers. The goal is to maximize a focal firm's overall revenue or profit. The typical knowledge base consists of customers' attribute-level partworths, marginal contributions, and descriptions of own and competing status quo products. However, since these so-called product-line design problems are known to be NP-hard, they often cannot be solved exactly. Instead, heuristics have to be applied. In this paper, we give an overview on proposed solution methods. Moreover, we apply two recent propositions—Cluster-Based Genetic Algorithms (CGA) and Tabu Search (TS)—to a sample of 460 small (up to about 10^6 possible solutions)-to-large-size problems (more than 10^{12} possible solutions). The results are promising: Especially CGA solves small-size problems accurately and in acceptable computing time (within seconds), the latter even when applied to medium- and large-size problems.

1 Introduction

Collecting and analyzing rankings, ratings, or choices for attribute-level combinations (fictive products) has a long tradition in marketing research (see, e.g., [26, 34, 36]). The approach is widespread since it allows to estimate attribute-level partworths and helps marketeers to understand customers' trade-off between product quality and price. So, e.g., [38] found out that each year more than 27,000 commercial conjoint analysis applications are conducted worldwide. Roberts et al. [41] support these findings by concluding from their citation analysis and surveys among

D. Baier (✉)
University of Bayreuth, Chair of Marketing & Innovation, Universitätsstraße 30, 95444 Bayreuth, Germany
e-mail: daniel.baier@uni-bayreuth.de

S. Vökler
Brandenburg University of Technology, Chair of Automation Technology, Platz der Deutschen Einheit 1, 03046, Cottbus, Germany
e-mail: sascha.voekler@b-tu.de

© The Author(s), under exclusive license to Springer Nature Singapore Pte Ltd. 2023
A. Okada et al. (eds.), *Facets of Behaviormetrics*, Behaviormetrics: Quantitative Approaches to Human Behavior 4, https://doi.org/10.1007/978-981-99-2240-6_1

marketing researchers, mediators, and practitioners that related tasks like new product development and introduction as well as pricing benefited the most from academic marketing research. Journal articles on conjoint analysis (including discrete choice analysis) had and have highest impact to marketing theory and practice [41].

So, for conjoint and cost-based product-line design, there exists a large number of well-known and often-cited journal articles (e.g., [13, 17, 21, 22, 30]). Problems are discussed from a practical point of view, solution methods are proposed and applied. A typical problem can be stated as follows: A focal firm wants to introduce new products (e.g., goods, services, or hybrid offers) into a market where own and/or foreign status quo products are already bought. The new products as well as the already available status quo products can be described as attribute-level combinations with respect to relevant attributes (e.g., components, ingredients, materials, prices, qualities). As a result of a conducted conjoint analysis application, estimated individual partworths for these attribute-levels are available from a sample of customers that allow to predict choices of each customer among sets of offered products. The product-line design problem now consists in finding attribute-level combinations for the new products that maximize an objective function, e.g., overall revenue, market share, or profit of the focal firm's new and status quo products.

However, in practice, the number of possible attribute-level combinations often is high and a so-called Complete Enumeration by evaluating all possible product-lines not possible. So, e.g., [44] demonstrates in his summary of 2,089 commercial conjoint analysis applications that 3–9 attributes with 2–7 levels are widespread, resulting in up to $7^9 = 40{,}353{,}607$ possible attribute-level combinations. If 3 or 9 new products have to be looked for, $(7^9)^3 = 6.57 \cdot 10^{22}$ or $(7^9)^9 = 2.84 \cdot 10^{68}$ possible product-lines would have to be evaluated (identical new products with respect to the attribute-levels allowed). Consequently, from the beginning, heuristics as solution methods have been proposed (see, e.g., [7, 22, 32]). In this paper, we provide an overview on these solution methods and discuss and apply two recently proposed heuristics—cluster-based genetic algorithms and tabu search. A sample of 460 generated problems is used for comparisons with respect to accuracy and computing time.

The paper is structured as follows: In Sect. 2, we discuss typical product-line problems formally. Section 3 gives an overview on proposed and applied solution methods. In Sect. 4, we discuss the two selected heuristics in more detail and apply them in Sect. 5. The paper closes in Section 6 with conclusions and an outlook.

2 Product-Line Design Problems

Based on the available input, a product-line design problem can be formalized by describing the assumed choice rule and the objective function to be maximized. We discuss well-known choice rules and objective functions but also demonstrate how product-line problems can be solved using a small example from the literature [17, 18].

2.1 Choice Rules

After conducting a conjoint analysis or discrete choice analysis application under the widespread linear-additive utility assumption [38, 55], the marketing researcher receives estimated partworths $\beta_{ik\ell} \in \mathbb{R}$ for a sample of I customers (i = 1,...,I) with respect to K attributes (k = 1,...,K) and L_k levels (ℓ = 1,...,L_k). The partworths reflect the customers' preferences for attribute-levels and can be used to evaluate J attribute-level combinations (products, candidates) from the customers' point of view. So, with $x_{jk\ell}$ as an indicator whether combination j (j = 1,...,J) has level ℓ of attribute k (=1) or not (=0), the utility u_{ij} of combination j for customer i can be estimated as

$$u_{ij} = \sum_{k=1}^{K} \sum_{\ell=1}^{L_k} \beta_{ik\ell} x_{jk\ell} \; \forall i = 1, \ldots, I; \; j = 1, \ldots, J. \tag{1}$$

Now, when a set of combinations is offered to a customer, the straightforward assumption would be that he/she prefers and/or buys the utility maximizing combination. This so-called first-choice rule is especially widespread in product-line design (see, e.g., [24, 25]): Customer i selects combination j among the available J combinations with probability

$$p_{ij} = \begin{cases} 1, & \text{if } u_{ij} \geq u_{ij'} \; \forall j' = 1, \ldots, J, \\ 0, & \text{else,} \end{cases} \quad \text{(First-choice)}, \; \forall i = 1, \ldots, I; \; j = 1, \ldots, J. \tag{2}$$

If two or more combinations have maximum utility (e.g., when having identical levels across all attributes), it is usually assumed that one of them is bought randomly. Alternatively, it can be assumed that they are bought with equally distributed probabilities (1 divided by the number of combinations with maximum utility).

However, the assumption that customers always buy the combination with maximum utility has often been criticized in the literature (see, e.g., [24, 33]). Even small utility differences lead to the prediction that the maximum utility combination is always bought, a contradiction to a complex data collection and estimation process with inaccurate and faulty partworths as results. Also, the focus on maximum utility could be wrong in a market where customers buy combinations more or less randomly (e.g., when customers have to choose among low involvement products or among more or less identical alternatives).

Probabilistic choice rules mitigate these problems: The Bradley–Terry–Luce (BTL) choice rule [8, 35] assumes that probabilities p_{ij} are proportional to the utilities u_{ij}, the logit choice rule [37, 40] assumes proportionality to exponentiated utilities:

$$p_{ij} = \frac{u_{ij}^{\alpha}}{\sum_{j'=1}^{J} u_{ij'}^{\alpha}} \text{ (BTL)}, \quad p_{ij} = \frac{\exp(\alpha u_{ij})}{\sum_{j'=1}^{J} \exp(\alpha u_{ij'})} \text{ (Logit)} \quad \forall \begin{array}{c} i = 1, \ldots, I; \\ j = 1, \ldots, J, \\ \alpha \geq 0. \end{array} \tag{3}$$

In both cases, the parameter α can be used to calibrate the choice rules. Values of α near or equal to zero reflect an equal distribution of probabilities among available combinations, large values reflect the first-choice rule. Often, past purchase behavior of the customers is used to estimate α. It should be mentioned that the BTL rule assumes non-negative utilities which are usually received by normalizing estimated individual partworths so that the minimum utility among all possible combinations receives a utility of 0 and the maximum utility among all possible combinations receives a utility of 1.

2.2 *Objective Functions*

Basing on estimated partworths for attribute-levels and an assumed choice rule, now the product-line design problem consists in selecting attribute-levels for R new products of a focal firm that maximize a given objective function (e.g., overall revenue, market share, profit) in a market with O own and F foreign status quo—already offered—products. Again, $x_{jk\ell}$ is the indicator whether combination j ($J = R+O+F$; j=1,...,J) has level ℓ of attribute k (=1) or not (=0). The $x_{jk\ell}$ values for the first R products are unknown. The case of modifying own established products is included in this formulation by increasing R and decreasing O.

Among many proposals in the literature, here, we first discuss profit maximization based on the BTL choice rule as discussed by [17, 18, 42]. In contrast to other proposals for profit maximization by, e.g., [10–12, 21, 25, 47, 49], this proposal is a very flexible model with contribution margins per sold product and fixed costs per period at an attribute-level. So, we have to describe customer i's attribute-level contribution margins $d_{ik\ell}$ and partworths $\beta_{ik\ell}$ as well as a weight ω_i (number of products per period bought by customer i). The periodical fixed costs are estimated by summing up the relevant attribute-level specific periodical fixed costs $f_{k\ell}$ (k = 1,...,K; ℓ = 1,...,L_k). This approach also allows to block certain attribute-levels for the new products by defining large $f_{k\ell}$ values for specific (k,ℓ) combinations. With these assumptions, the model can be formulated as follows (see [17, 18, 42]):

$$\sum_{i=1}^{I} \sum_{j=1}^{R+O} \sum_{k=1}^{K} \sum_{\ell=1}^{L_k} \omega_i d_{ik\ell} x_{jk\ell} p_{ij} - \sum_{j=1}^{R} \sum_{k=1}^{K} \sum_{\ell=1}^{L_k} x_{jk\ell} f_{k\ell} \longrightarrow \max! \tag{4}$$

subject to

$$\sum_{\ell=1}^{L_k} x_{jk\ell} \leq 1 \quad \forall\ j = 1, \ldots, R; k = 1, \ldots, K \tag{5}$$

$$p_{ij} = \frac{\left(\sum_{k=1}^{K} \sum_{\ell=1}^{L_k} \beta_{ik\ell} x_{jk\ell}\right)^{\alpha}}{\sum_{j'=1}^{J} \left(\sum_{k=1}^{K} \sum_{\ell=1}^{L_k} \beta_{ik\ell} x_{j'k\ell}\right)^{\alpha}} \quad \forall\ i = 1, \ldots, I;\ j = 1, \ldots, R{+}O, \tag{6}$$

$$\sum_{\ell=1}^{L_k} x_{jk\ell} = \sum_{\ell=1}^{L_{k+1}} x_{j(k+1)\ell} \quad \forall\ j = 1, \ldots, R; k = 1, \ldots, K-1, \tag{7}$$

$$x_{jk\ell} \in \{0, 1\} \quad \forall\ j = 1, \ldots, R; k = 1, \ldots, K; \ell = 1, \ldots, L_k. \tag{8}$$

The objective function (4) maximizes the weighted contribution margins from the new and the own status quo products minus the fixed costs for the new products. Constraint (5) ensures that a new product has at most one level per attribute. Constraint (6) specifies the BTL choice rule with p_{ij} as the choice/buying probability of customer i among the new and all status quo products. Constraint (7) ensures that products are offered complete with all attributes. Finally, constraint (8) reflects the binary restriction on the decision variables. It should be mentioned that this formulation allows identical new products (with respect to the attribute-levels). Since—when applying a probabilistic choice rule—these new products can generate additional choice probabilities and profit, this formulation could have an advantage. However, if we want to restrict the solutions to product-lines without identical new products, an additional restriction or—alternatively—a modification of the objective function (e.g., setting to a negative value in this case) is needed.

Other objective functions are closely related to the above formulation: If we ignore the monetary information—e.g., by setting all $d_{ik\ell}$ to a constant value and all $f_{k\ell}$ to 0, the profit maximization problems turn into a revenue, market share, or shares of choices maximization problem. Other choice rule assumptions are also included. If we set α to a large value (e.g., =10, or =100), a first-choice rule is assumed.

2.3 A Sample Product-Line Problem and Its Solution

In the following, we shortly discuss profit maximization using a small-size hypothetical yoghurt market. The example was also used in [17, 18] for demonstration, but here we extend the customer sample to $I = 400$ (instead of $I = 4$). A focal firm wants to introduce new products into this market where yoghurts that can be described by the $K = 3$ attributes "cup" material (P = plastic or G = glass), yoghurt "taste" (S = sour or M = mild), and "price" per cup (0.4 = €0.4, 0.5 = €0.5, 0.7 = €0.7). Up to now, $F = 5$ foreign status quo products are offered: (G;M;0,7) and (P;M;0,4) by company A, (P;M;0,4) and (P;S;0,5) by company B as well as (P;M;0,5) by company

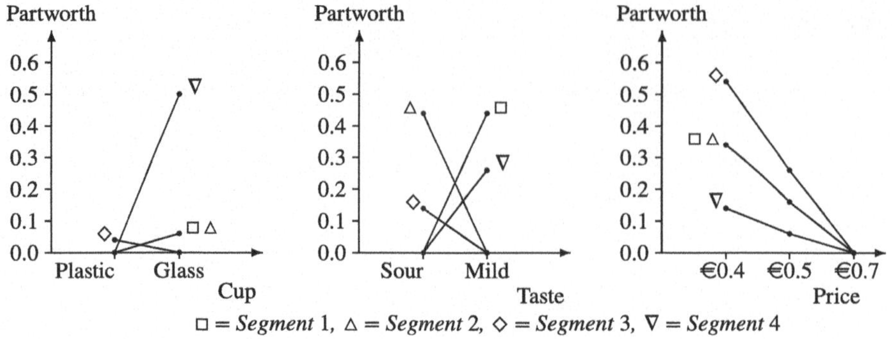

Fig. 1 Partworths across segments in our example (each segment consists of 100 customers)

Table 1 Partworths $\beta_{ik\ell}$ and mean number of products bought per customer and period ω_i across segments as well as attribute-level contribution margins per product $d_{ik\ell}$ (identical across segments) and attribute-level fixed costs per period $f_{k\ell}$ (each segment consists of 100 customers)

	Cup (k = 1)		Taste (k = 2)		Price (k = 3)		
	Plastic	Glass	Sour	Mild	€0.4	€0.5	€0.7
	$\ell = 1$	$\ell = 2$	$\ell = 1$	$\ell = 2$	$\ell = 1$	$\ell = 2$	$\ell = 3$
Segment 1 (ω_i = 200): Partworths $\beta_{ik\ell}$	0.00	0.06	0.00	0.44	0.33	0.17	0.00
Segment 2 (ω_i = 100): Partworths $\beta_{ik\ell}$	0.00	0.06	0.44	0.00	0.33	0.17	0.00
Segment 3 (ω_i = 50): Partworths β_{ikl}	0.05	0.00	0.14	0.00	0.55	0.27	0.00
Segment 4 (ω_i = 100): Partworths $\beta_{ik\ell}$	0.00	0.53	0.00	0.27	0.13	0.07	0.00
Contribution margins $d_{ik\ell}$ (€ per cup)	0.20	0.00	0.10	0.00	0.00	0.10	0.30
Fixed costs $f_{k\ell}$ (€ per period)	0	0	0	1000	0	0	0

C. The focal firm has—up to now—not introduced any products ($O = 0$) but knows that customers in this market have segment-specific partworths as reflected in Fig. 1 and Table 1.

The customers in segment 1 like a mild taste and a low price. Instead, segment 2 customers prefer a sour taste but also a low price. Segment 3 summarizes customers for whom price is more or less the only important attribute. For segment 4 customers, price is not important at all, but they are environmentally conscious and clearly prefer cups made of glass. Segment 1 contains heavy buyers of yoghurts (200 cups per customer and period) whereas segment 3 contains light buyers (50 cups per customer and period). Table 1 shows additional information for the focal firm: Contribution margins are directly related to the price of a cup but using plastic as material generates additional attribute-level contribution margins per cup (in contrast to glass), similar to the sour taste (in contrast to the mild taste). Moreover, producing mild yoghurts generates fixed costs of €1000 per period.

Now, we assume a linear-additive utility model and the BTL model with $\alpha = 1$ for transforming utilities to choice probabilities as in the previous subsection. In

Table 2 Market shares m_j before and after introducing a new product ($j = 1$) with predicted profit

		Market share					Profit
New product		(G;M;0,7)	(P;M;0,4)	(P;M;0,4)	(P;S;0,5)	(P;M;0,5)	(€ per
$j = 1$		$j = 2$	$j = 3$	$j = 4$	j=5	$j = 6$	period)
		0.176	0.248	0.248	0.151	0.177	
(G;M;0.4)	0.241	0.128	0.189	0.189	0.118	0.134	–1000.00
(G;M;0.5)	0.195	0.133	0.201	0.201	0.128	0.142	–121.01
(P;M;0.4)	0.198	0.143	0.198	0.198	0.121	0.141	783.10
(G;M;0.7)	0.139	0.139	0.215	0.215	0.141	0.151	873.09
(G;S;0.4)	0.217	0.142	0.196	0.196	0.107	0.142	975.62
(P;M;0.7)	0.089	0.155	0.226	0.226	0.146	0.159	1003.38
(P;M;0.5)	0.149	0.148	0.210	0.210	0.132	0.149	1015.18
(G;S;0.5)	0.172	0.147	0.208	0.208	0.115	0.150	1550.47
(P;S;0.7)	0.060	0.174	0.234	0.234	0.128	0.170	1623.50
(P;S;0.5)	0.118	0.166	0.219	0.219	0.118	0.159	2122.91
(G;S;0.7)	0.119	0.154	0.222	0.222	0.124	0.160	2133.39
(P;S;0.4)	0.166	0.159	0.207	0.207	0.110	0.151	2236.52

Table 2, the first row contains the market shares for the status quo products before an introduction of an additional product. Please note that market shares are mean choice probabilities across the $I = 400$ customers weighted by ω_i, the customer's number of cups bought per period. We see from the first row that status quo products 2 and 3 (with identical attribute-level combinations) have high market shares.

The other rows discuss possible introductions of a product-line with one new product ($R = 1$) together with the corresponding market shares of the new and the status quo products as well as the profit per period for the focal firm. Since there are $2{\cdot}2{\cdot}3 = 12$ possible attribute-level combinations for a new product we have 12 additional rows in Table 2. It is obvious that introducing (P;S;0.4) solves the product-line problem since it generates the highest profit for the focal firm (€2236.52 per period) whereas (G;M;0.4) generates the highest loss (€–1000.00). The latter results from generating a contribution margin of €0.00 from each sold cup of yoghurt whereas fixed costs of €1000.00 per period are generated (e.g., costs for enabling the production to manufacture yoghurts with a mild taste).

The demonstrated problem-solving can also be applied for product-lines with more than one new product: So, e.g., for $R = 2$ the product-line (P;S;0.4) and (G;S;0.7) generates the highest profit (€3748.40 per period), for $R = 3$ the product-line (P;S;0.4), (P;S;0.4), and (G;S;0.7) generates €4854.45, for $R = 4$ the product-line (P;S;0.4), (P;S;0.4), (G;S;0.7), and (G;S;0.7) with €5759.09 wins. Please note that the latter two product-lines contain identical attribute-level combinations since our product-line problem formulation has no corresponding restriction and due to few attributes this could be useful. If we restrict our solving method to generate product-lines without identical new products, the best product-lines would be for $R = 3$ the product-line (P;S;0.4), (P;S;0.5), and (G;S;0.7) with €4854.45 profit and for $R = 4$ the product-line (P;S;0.4), (P;S;0.5), (P;S;0.7), and (G;S;0.7) with €5427.05 profit.

3 Heuristics to Solve Product-Line Design Problems

In the previous sample problem, the number of possible solutions $\binom{2 \cdot 2 \cdot 3}{4} = 495$ with $K = 3$, $L_1 = L_2 = 2$, $L_3 = 3$, and $R = 4$ was very small (if identical new products are not allowed). Comparing all possible solutions needed 1.71 s on a standard notebook (Lenovo ThinkPad P53s with an 8th Generation Intel Core i7). However, in general, this so-called Complete Enumeration cannot be applied, since 3 to 9 attributes, each with 2–7 levels, are widespread in commercial conjoint analysis applications (see, e.g., [44]). So, if five new products are looked for, up to $\binom{7^9}{5} = 8.9 \cdot 10^{35}$ solutions would have to be checked. The mentioned standard notebook would need about $1{,}71/495 \cdot 8.9 \cdot 10^{35}$ seconds or about $9.8 \cdot 10^{25}$ years to solve the profit maximization product-line problem. In the following, we reflect heuristics as fast alternatives and discuss two recently proposed promising ones.

3.1 *Overview*

From the first discussions of product-line design problems in the 1970 and 1980s (e.g., [21–23, 30, 46, 56]), it became obvious that Complete Enumeration is too time-consuming. Kohli and Krishnamurti [31] showed that the problem in general is NP-hard, [22] mentioned that solving their discussed problems by Complete Enumeration with $\binom{32}{1} = 32$ up to $\binom{32}{5} = 201{,}376$ possible solutions needed up to 35 min on a mainframe DEC 10 computer whereas their proposed heuristics solved all problems in a few seconds.

Typical proposals in these early years were so-called two-stage heuristics, where in a first stage, a small number of promising attribute-level combinations were selected as so-called candidates. Then, in a second stage, all sets of candidates were tested as possible product-lines. So, e.g., [23] proposed to select in the first stage the utility maximizing attribute-level combination of each customer as candidates (the so-called best-in heuristic). Then, in the second stage, the algorithm starts with a random set of these candidates and iteratively improves the objective function by interchanging candidates not in and already in the set (the so-called interchange or product-swapping heuristic).

However, it soon became clear that constraining the solution space to sets of few pre-selected candidates is too restrictive, often leading not to the global maximum of the objective functions across all sets of attribute-level combinations [7, 32]. Therefore, later proposals (e.g., [1, 5, 7, 48, 51]) omitted this first stage and iteratively improved the attribute-levels of all new products without a candidate restriction.

Table 3 gives an overview on these proposals and applications of one- and two-stage heuristics by discussing which product-line problems were tackled (objective function, number of new products, choice rules assumed), which problem sizes were analyzed (number of possible solutions), which solution methods were applied, and which solution methods performed best during applications with respect to accuracy and computation time. Table 3 updates former overviews given by [25] as well as [3] where two-stage heuristics dominated. Nowadays, one-stage heuristics with adapted algorithms from combinatorial optimization and artificial life are widespread.

- Genetic algorithms [27] mimic biological evolution by natural selection. The algorithms start with an initial population consisting of random solutions. A subsequent population emerges from a previous one by preserving, recombining, and/or mutating selected "best" solutions. This transition of populations repeats until a stopping criterion is met. For product-line design problems, the random and improved solutions are attribute-level combinations for all new products (see, e.g., the proposals by [5, 48]).
- Simulated annealing [29] mimics annealing in metallurgy where heating and controlled cooling is used to improve material quality. The main idea is to iteratively modify components of an existing solution in the search for best ones, but to accept during this search with a certain probability deteriorations. The reduction of this probability over time (the cooling process) makes it more and more difficult for worse solutions to be selected and ends without early stagnation—hopefully—at the global maximum instead of local ones (see, e.g., [7], for a first application in product-line design).
- Ant colony optimization [15] reflects the foraging behavior of real ants. Artificial (real) ants repeatedly construct a valid solution (a path to the food) via components (sections) and mark the components with pheromones that reflect the overall solution quality (the shortness of the path). The amount of pheromones at the single components helps other ants to select good components in a probabilistic manner and to construct better solutions (shorter paths). For product-line design, the components are alternative levels for all new products and attributes (see, e.g., [1]).
- Particle swarm optimization [28] also has its roots in artificial life, but now in the behavior of animal swarms. Algorithmic basis is a population of particles (e.g., artificial birds, fishes) that iteratively adjust their positions in space according to their own past positions and evaluations as well as the past positions and evaluations of other particles. When it comes to solve product-line problems, the positions in space represent attribute-level combinations for the new products that can be evaluated and improved (see, e.g., [51]).

Obviously, the proposed heuristics can deal with large-size product-line design problems. So, e.g., [4] demonstrate that their "best" methods—cluster-based genetic algorithm as well as two variants of ant colony optimization: max-min ant system without and with local search—can successfully deal with up to $7.33 \cdot 10^{200}$ possible solutions in short computing time (up to 600 seconds on a standard notebook). In the following,

Table 3 Solving product-line design problems: An overview of method proposals and applications

References	Obj. func.	Choi. rule	Sing., mult.	#Potential solutions	#Prob. N, #cust. I, PW&SQP generation	Methods applied (best methods)
[56, 57]	M	FC	Sing., mult.	–	–	–
[21]	M,P	FC, BTL	Sing.	6581	N = 1, I = 486, PW, SQP: real	CE
[22]	BW, M,P	FC	Mult.	32–201,376	N = 1, I = 117, PW, SQP: real	CE,GH,PSH,LRH
[30]	M	FC	Sing.	$16–3.9{\cdot}10^{6}$	N = 48, $I\in$[100...400], PW: unif., SQP: rand.	CE,DPH,LRH
[32]	BW, M, P	FC	Mult.	$256–2.8{\cdot}10^{14}$	N = 81, $I\in$[50...], PW: unif., SQP: rand.	CE,DPH,LRH
[5]	BW, M	FC	Sing.	16–390625	N = 192, PW: unif., SQP: random	CE,DPH,GA
[2]	P	FC	Sing., line	$2016–3.66{\cdot}10^{15}$	N = 300, PW: unif., SQP: random	GA,BS, GA/BS hybrid
[45]	M	FC	Mult.	$3125–5.04{\cdot}10^{102}$	N = 217, I = 400, PW: beta, unif., SQP: rand.	BS, DCH, DPH, GH, GA,NPA
[50]	M	FC	Sing.	1024–59049	N = 120, PW: unif., norm., SQP: random	CE,GA,TS, pruning
[48]	P	Log.	Mult.	$28–4.6{\cdot}10^{10}$	N = 276, PW, SQP: n.a.	CE,GA,GH
[6]	M	FC	Mult.	$6.12{\cdot}10^{26}–1.88{\cdot}10^{38}$	N = 800, I = 200, PW: unif., SQP: random	GA,BS, GA/BS hybrid
[1]	M	FC	Sing.	$1.0{\cdot}10^{5}–1.0{\cdot}10^{9}$	N = 4200, $I\in${200, 250,300}, PW: beta	CE,ACO
[7]	P	FC	Mult.	$56–8.1{\cdot}10^{23}$	N = 1+120, $I\in${50, 100}, PW: real+unif., SQP: random	BS, CoA, DCH, DPH, GA, NPA, PSH, GH, DQH, LRH,SA
[51]	BW, M	BTL	Mult.	$1.58{\cdot}10^{15}–3.90{\cdot}10^{67}$	N = 160,$I\in${200,700}, PW: unif., SQP: rand.	GA, PSO
[54]	M	FC	Sing.	$625–1.0{\cdot}10^{20}$	N = 450, $I\in${250,500, 1000}, PW: unif.	ACO, BCO
[53]	P	Log.	Mult.	$70–8.06{\cdot}10^{23}$	N = 1+120,$I\in${50,100}, PW: real+unif.	FSTDE,GA,SA
[52]	P	FC	Mult.	$70–8.06{\cdot}10^{23}$	N = 1+120,$I\in${50,100}, PW: real+unif.	CoA, GA, DCH, DPH, GH, NPA, SA,TS
[4]	M,P	FC, BTL	Mult.	$6.96{\cdot}10^{5}–7.33{\cdot}10^{200}$	N = 78, $I\in$[200,1110], PW: unif., SQP: rand.	CGA,GA,PSO, MM,MML, SA
[39]	P	FC	Mult.	$70–8.06{\cdot}10^{23}$	N = 1+120, $I\in${20,75, 100}, PW: real+unif.	CSA,GA,SA

Abbreviations: BW: buyers welfare; BTL: BTL choice rule; #cust.: number of customers; FC: first-choice rule; log.: Logit choice rule; M: market share; mult.: multiple new products; n.a.: not available; obj. func.: objective function; P: profit; PW: partworth; #prob.: number of problems; sing.: single new products; SQP: status quo products; Exact methods: Complete Enumeration (CE); One-stage heuristics: Ant Colony Optimization (ACO); Bee Colony Optimization (BCO); Beam Search (BS); Clonal Selection Algorithm (CSA); Cluster-based GA (CGA); Coordinate Ascent (CoA); Dynamic Programming Heuristic (DPH); Fuzzy Self-Tuning Differential Evolution (FSTDE); Genetic Algorithm (GA); Lagrange Relaxation Heuristic (LRH); Max-Min Ant System (MM); MM with Local Search (MML); Nested Partition Algorithm (NPA); Particle Swarm Optimization (PSO); Simulated Annealing (SA); Tabu Search (TS); Two-stage heuristics: Divide-and-Conquer Heuristic (DCH); Greedy Heuristic (GH); Product-Swapping Heuristic (PSH) we discuss the cluster-based genetic algorithm more deeply together with the "best" heuristic as proposed by one of the most active authors in the field of product-line design, Tabu search by [52].

3.2 Cluster-Based Genetic Algorithm

Baier and Vökler [4] proposed to combine genetic algorithms with a modification of the best-in heuristic by [23]: First, cluster analysis (e.g., k-means) is applied to the customers' partworths in order to derive R customer segments (as many as new products are looked for). Then, according to their segment-specific utilities, for each segment the J_{best} attribute-level combinations with highest utilities are selected and—according to their ranks—combined to form J_{best} (initial) solutions for the product-line design problem. Together with other—randomly selected—solutions, these J_{best} solutions build the initial population for the genetic algorithm. In their comparisons, the cluster-based generic algorithm clearly outperformed the original genetic algorithm. The superiority was assumed to steam from—on one side—generating a diverse initial population as a good basis for later mutations and crossovers and—on the other side—starting from very good solutions as they reflect the preferences of larger shares of customers. Baier and Vökler [4] also demonstrated that $J_{best} = 50$, a population size of 500 solutions, a crossover rate of 0.8, and a mutation rate of 0.1 are promising parameter settings.

Baier and Vökler [4] developed an own implementation of genetic algorithms and cluster-based genetic algorithm in R (as well as of many other heuristics). Here, we rely on an alternative, the publicly available R package GA [43] that provides a flexible general-purpose toolbox for implementing applications of genetic algorithms. The package provides the basic operators needed but also allows to specify own operators and initial populations (as discussed above).

3.3 Tabu Search

Tabu search is a "higher level" heuristic that—like many local search heuristics that can be used as a basis—iteratively evaluates the neighborhood of a current solution (all solutions with slight differences from the current one) in order to find better solutions. However, during this iterative process, tabu search makes effective use of a so-called tabu list with solutions and neighborhoods that have already been evaluated and therefore shouldn't be evaluated again [16].

The algorithm was originally proposed by [19]. Tabu search makes—in contrast to other heuristics like genetic algorithms or particle swarm optimization—no use of a developing population of solutions but relies—like simulated annealing—on a single current solution. If the evaluation of the neighborhood (excluding the solutions in the tabu list) leads to an improvement, the better solution is stored as the best solution, serves as a current solution, but is also added to the tabu list. If there is no better solution in the neighborhood, also a slightly worse solution can become the current solution (in order to prevent from local maxima). The tabu list works as a first-in first-out list that helps to direct the search to not already evaluated neighborhoods, but its restricted size also allows—after some time—to forget the exclusion of formerly evaluated solutions [20].

Tsafarakis et al. [52] introduced tabu search to product-line design problems and demonstrated with their MATLAB implementation superior accuracy compared to other one-stage but also two-stage heuristics (e.g., coordinate ascent, genetic algorithms, divide-and-conquer heuristics, dynamic programming heuristics, greedy heuristics, and simulated annealing). They proposed to use small tabu list sizes (e.g., 60) and large numbers of iterations (e.g., 2, 200).

Nevertheless, we follow these proposals in our own realization of tabu search, but make use of the publicly available R package tabuSearch [14]. The package is based on the propositions by [20] as well as [16] but allows modifications if needed. So, e.g., we repeated the search with nine random solutions (as usual in tabu search) but also used the best of the above-discussed J_{best} solutions as a starting solution. For our yoghurt example from the last section, our application of tabu search as well as cluster-based genetic algorithm with the discussed parameter settings was able to find the global maxima for $R = 2$, 3, or 4 within seconds on a standard notebook.

4 Application to Simulated Problems and Comparisons

In the following, we demonstrate that the two proposed heuristics solve product-line design problems across a wide range of problem sizes. Up to about 10^6 possible solutions, we refer to small-size problems. From about 10^{12} possible solutions, we refer to large-size problems. The other problems are referred to as medium-size problems (see, e.g., [7], for similar size categorizations). In the first application with 190 small-size problems, we are able to compare the solutions by cluster-based genetic algorithm and tabu search with solutions by the accurate but slow complete enumeration and the inaccurate but fast two-stage product-swapping heuristic. In the second comparison, we discuss superiority among the three heuristics with respect to accuracy and speed using other 270 small- to large-size problems.

4.1 Comparison Using Small-Size Problems

For a first comparison, we generate small-size problems that allow to compare solutions from heuristics with solutions from complete enumeration. Similar to [7, 52], the number of attributes K is restricted to 4, 5, and 6; the number of levels for each attribute L_k to 2, 3, and 4; and the number of new products R to 2, 4, or 6. For each of the 27 combinations of K, L_k, and R, we generate data for 10 profit maximization product-line problems by drawing individual attribute-level partworths for $I = 100$ customers from iid [0, 1] uniform distributions and standardize them as described so that the BTL choice rule with $\alpha = 1$ can be applied. Moreover, a status quo market with $F = 3$ foreign products—with randomly drawn attribute-levels—is assumed. Attribute-level contribution margins are also drawn from iid [0, 1] uniform distributions, fixed costs are set to 0.

The number of possible solutions for these problems ranges from $\binom{2^4}{2} = 120$ to $\binom{4^6}{4} = 1.17 \cdot 10^{13}$. Since solving the problem with $K = 6$, $L_k = 4$, and $R = 2$ by complete enumeration already takes about 6 h on a standard notebook, we concentrate on the 190 problems with at most $\binom{4^6}{2} = 8{,}386{,}560$ possible solutions. Furthermore, we formed three blocks, a first one ($B = 1$) containing problems with up to 4,960 possible solutions ($B = 1$), a second ($B = 2$) with 29,403 up to 635,376, and a third block ($B = 3$) with more than 1,663,740 possible solutions.

Table 4 shows the results of applying Complete Enumeration (CE), Product-Swapping Heuristic (PSH), Cluster-based Genetic Algorithm (CGA), and Tabu Search (TS) to the 190 problems. We report means and standard deviations of the objective function (the profit for the focal company) divided by the global maximum (calculated via CE) as well as mean computation times on a standard notebook across all problems but also grouped by number of attributes K, of levels per attribute L_k, of new products R, and of possible solutions via three blocks B.

It is obvious that PSH outperforms the other solution methods with respect to computation time. Even within the block of largest of these small-size problems ($B = 3$), PSH needs on average 0.25 s to derive its solutions, in contrast to CGA with 4.46 s, TS with 13.26 s, and CE with 14,327.05 s (about 4 h).

In contrast to this, CGA clearly outperforms the other heuristics with respect to the objective function. Mean values of 1 and standard deviations near 0 demonstrate this superiority. The differences among the means are statistically significant according to a non-parametric Friedman test (see, e.g., [9]) with $\alpha = 0.05$. An additional non-parametric Friedman test with multiple comparisons demonstrates that this significance also holds for the pairwise comparisons between the heuristics across all problems and in all subgroups according to K, L_k, R, and B.

Some positive standard deviations in Table 4 with values 0.001 indicate that for some profit maximization product-line problems, CE found a better solution than CGA. A closer look into the results yields that this is the case for 15 of the 190 problems. CGA only was able to find the global maximum for the other 175 problems (92.10%), even if the objective function had near 1 values from 0.99687 up to 0.99985 in these cases. TS and PSH were less effective, also in this regard: TS only found in 17 cases the global maximum (8.95%) with values from 0.93795 up to 0.99975 in the other cases, PSH only found the global maximum in 2 cases (1.05%) with values of 0.89938–0.99948 in the other cases. The accuracy problems (not finding the global maximum) seem not to be dependent from the number of solutions (see, e.g., the standard deviation of 0.001 for the $R = 2$ subgroup of problems in contrast to the standard deviation of 0.000 of the $R = 3$ and $R = 4$ subgroups). In the next subsection,

Table 4 Results of Product-Swapping Heuristic (PSH), Cluster-based Genetic Algorithm (CGA), and Tabu Search (TS) applied to 190 small-size problems: Means (std. dev.) of the objective function divided by the global maximum (calculated through Complete Enumeration (CE)) and mean computation times; across all problems and grouped by number of attributes K, of levels per attribute L_k, of new products R, and of possible solutions via three blocks B

	Objective function			Computation time (in s)			
	PSH	CGA	TS	CE	PSH	CGA	TS
$K = 4$ ($n = 80$)	0.967 (0.024)	1.000 (0.000)	0.985 (0.013)	3579.00	0.15	2.73	7.61
$K = 5$ ($n = 60$)	0.974 (0.014)	1.000 (0.001)	0.982 (0.012)	1201.09	0.15	3.08	9.15
$K = 6$ ($n = 50$)	0.966 (0.017)	1.000 (0.001)	0.982 (0.012)	6466.81	0.21	3.65	11.50
$L_k = 2$ ($n = 90$)	0.969 (0.018)	1.000 (0.000)	0.988 (0.011)	1115.45	0.15	2.60	8.63
$L_k = 3$ ($n = 60$)	0.974 (0.021)	1.000 (0.001)	0.983 (0.011)	4588.03	0.15	3.34	9.59
$L_k = 4$ ($n = 40$)	0.961 (0.020)	1.000 (0.001)	0.974 (0.013)	7651.33	0.21	3.78	9.51
$R = 2$ ($n = 90$)	0.976 (0.016)	1.000 (0.001)	0.984 (0.014)	2592.19	0.14	2.72	6.49
$R = 3$ ($n = 60$)	0.966 (0.020)	1.000 (0.000)	0.983 (0.011)	2348.28	0.16	3.23	9.84
$R = 4$ ($n = 40$)	0.958 (0.021)	1.000 (0.000)	0.982 (0.012)	7688.29	0.22	3.67	13.96
$B = 1$ ($n = 70$)	0.975 (0.018)	1.000 (0.000)	0.992 (0.010)	7.30	0.12	2.06	5.46
$B = 2$ ($n = 80$)	0.967 (0.018)	1.000 (0.001)	0.980 (0.012)	1351.66	0.16	3.28	10.26
$B = 3$ ($n = 40$)	0.962 (0.024)	1.000 (0.001)	0.977 (0.012)	14327.05	0.25	4.46	13.26
All ($n = 190$)	0.969 (0.020)	1.000 (0.000)	0.984 (0.013)	3588.03	0.16	3.08	9.12

we want to analyze the performance of the heuristics across a larger sample with also medium- and large-size problems. However, for these large-size problems, CE is not applicable.

4.2 Comparison Using Small- to Large-Size Problems

The second comparison was intended to reflect the size range of real product-line design problems much better than the first comparison. Consequently, following [44]'s summary of 2,089 commercial conjoint analysis applications, our number of attributes K now has values 4, 5, and 6, the number of levels for each attribute L_k has values 2, 3, 4, and the number of new products R has values 2, 4, 6.

For each of these 27 combinations of numbers, we generate 10 profit maximizing product-line design problems by drawing individual attribute-level partworths for I = 200 customers from iid [0,1] uniform distributions and standardize them as described. Moreover, status quo markets with F = 3 foreign products—with randomly drawn attribute-levels—are assumed. Attribute-level contribution margins are also drawn from iid [0, 1] uniform distributions, fixed costs are set to 0.

Again, we grouped the problems into three blocks according to their number of possible solutions. This time the number of solutions in the first block ranges from 120 to $8.39 \cdot 10^6$ (B = 1, n = 90, small-size problems), in the second block from $7.50 \cdot 10^7$ to $1.141 \cdot 10^{12}$ (B = 2, n = 90, medium-size problems), and in the third block from $1.17 \cdot 10^{13}$ to $3.12 \cdot 10^{34}$ (B = 3, n = 90, large-size problems). Many problems are too large to be solved by CE exactly, so, we decided to compare the solutions with the best solution of any heuristic (the so-called assumed global maximum). Table 5 shows the results similar to Table 4 but now with values of the objective function divided by the assumed global maximum.

Obviously, now, the clear winner of the comparison is CGA: Faster than the other heuristics (at least for large-size problems) and always the best one (values of the objective function is 1, standard deviation is 0). The superiority with respect to the objective function is, again, statistically significant according to a non-parametric Friedman test (see, e.g., [9]) with $\alpha = 0.05$. Again, multiple comparisons support this significant superiority. Tabu search competes well in this field, being comparably fast as the two-stage heuristic PSH, but with significant superiority against PSH also in the objective values.

5 Conclusions and Outlook

In this paper, we gave an overview on well-known product-line design problems and fast heuristics as solution methods. We demonstrated that the new CGA (cluster-based genetic algorithm) outperforms other heuristics across all problem sizes and that CGA is able to find global maxima of the objective functions across a wide range of problems. Tabu search, also a recently proposed solution method for product-line design problems, competes well in this field, being superior to well-known two-stage heuristics with respect to accuracy and speed especially for large-size problems.

Of course, our comparison has some limitations. So, e.g., only some selected heuristics were compared (being the winners in former comparisons). Also, the comparison of the solutions achieved by the heuristics with the global maximum for the large-size problem is missing. However, again, the overview and comparisons in this paper have shown interesting starting points for further research since standard R packages can be used to solve real-world marketing problems very fast and accurate.

Table 5 Results of Product-Swapping Heuristic (PSH), Cluster-based Genetic Algorithm (CGA), and Tabu Search (TS) applied to 90 small- ($B = 1$), 90 medium- ($B = 2$), and 90 large-size problems ($B = 3$): Means (std. dev.) of the objective function divided by the assumed global maximum (best solution of an applied heuristic per problem) and mean computation times; across all problems and grouped by number of attributes K, of levels per attribute L_k, and of new products R

	Objective function			Computation time (in s)		
	PSH	CGA	TS	PSH	CGA	TS
$K = 4$ ($n =$ 90)	0.933 (0.026)	1.000 (0.000)	0.982 (0.011)	0.97	9.27	21.67
$K = 6$ ($n =$ 90)	0.950 (0.017)	1.000 (0.000)	0.981 (0.012)	5.16	15.39	43.74
$K = 8$ ($n =$ 90)	0.951 (0.016)	1.000 (0.000)	0.982 (0.010)	134.68	25.13	97.25
$L_k = 2$ ($n =$ 90)	0.943 (0.029)	1.000 (0.000)	0.991 (0.006)	0.88	6.95	22.74
$L_k = 4$ ($n =$ 90)	0.945 (0.018)	1.000 (0.000)	0.980 (0.009)	6.44	16.43	56.07
$L_k = 6$ ($n =$ 90)	0.946 (0.016)	1.000 (0.000)	0.974 (0.011)	133.49	26.41	83.86
$R = 2$ ($n =$ 90)	0.947 (0.022)	1.000 (0.000)	0.980 (0.014)	46.04	7.81	12.35
$R = 4$ ($n =$ 90)	0.945 (0.021)	1.000 (0.000)	0.982 (0.009)	46.93	16.95	45.79
$R = 6$ ($n =$ 90)	0.943 (0.022)	1.000 (0.000)	0.983 (0.009)	47.85	25.04	104.53
$B = 1$ ($n =$ 90)	0.940 (0.029)	1.000 (0.000)	0.986 (0.012)	0.69	4.93	9.6
$B = 2$ ($n =$ 90)	0.944 (0.019)	1.000 (0.000)	0.979 (0.011)	46.50	12.92	32.85
$B = 3$ ($n =$ 90)	0.950 (0.014)	1.000 (0.000)	0.980 (0.009)	93.63	31.95	120.22
All (n=270)	0.945 (0.022)	1.000 (0.000)	0.982 (0.011)	46.94	16.60	54.22

Acknowledgements The authors wish to thank the highly respected colleagues in the Behaviormetric Society for many fruitful discussions during conferences, workshops, and research stays as well as their inspiring insightful publications. Their convincing approach to connect theory, data science, and business practice formed the basis for our focus on collecting and analyzing customer data in order to solve marketing problems by self-developed or publicly available models, methods, and computer programs.

References

1. Albritton, M. D., & McMullen, P. R. (2007). Optimal product design using a colony of virtual ants. *European Journal of Operational Research, 176*(1), 498–520. https://doi.org/10.1016/j.ejor.2005.06.042
2. Alexouda, G., & Paparizzos, K. (2001). A genetic algorithm approach to the product line design problem using the seller's return criterion: An extensive comparative computional study. *European Journal of Operational Research, 134*(1), 165–178.
3. Baier, D., & Gaul, W. (1999). Optimal product positioning based on paired comparison data. *Journal of Econometrics, 89*(1–2), 365–392.
4. Baier, D., & Vökler, S. (2022) One-stage heuristics for product-line design: A comparison. Working paper, Fakultät für Rechts- und Wirtschaftswissenschaften, Universität Bayreuth.
5. Balakrishnan, P. V., & Jacob, V. S. (1996). Genetic algorithms for product design. *Management Science, 42*(8), 1105–1117.
6. Balakrishnan, P. V., Gupta, R., & Jacob, V. S. (2004). Development of hybrid genetic algorithms for product line designs. *IEEE Transactions on Systems, Man, and Cybernetics, 34*(1), 468–483.
7. Belloni, A., Freund, R., Selove, M., & Simester, D. (2008) Optimizing product line designs: Efficient methods and comparisons. *Management Science, 54*(9), 1544–1552. https://doi.org/10.1287/mnsc.1080.0864
8. Bradley, R. A., & Terry, M. E. (1952). Rank analysis of incomplete block designs: I. the method of paired comparisons. *Biometrika, 39*, 324–345.
9. Carrasco, J., García, S., Rueda, M., Das, S., & Herrera, F. (2020). Recent trends in the use of statistical tests for comparing swarm and evolutionary computing algorithms: Practical guidelines and a critical review. *Swarm and Evolutionary Computation, 54*, 100665.
10. Chen, K. D., & Hausman, W. H. (2000). Technical note: Mathematical properties of the optimal product line selection problem using choice-based conjoint analysis. *Management Science, 46*(2), 327–332.
11. Choi, C. S., & DeSarbo, W. S. (1993). Game theoretic derivations of competitive strategies in conjoint analysis. *Marketing Letters, 4*(4), 337–348.
12. Choi, C. S., & DeSarbo, W. S. (1994). A conjoint-based product designing procedure incorporating price competition. *Journal of Product Innovation Management, 11*(5), 451–459.
13. Dobson, G., & Kalish, S. (1988). Positioning and pricing a product line. *Marketing Science, 7*(2), 107–125.
14. Domijan, K. (2022). Package 'tabuSearch'—An R Package. CRAN—Comprehensive R Archive. https://cran.r-project.org/web/packages/tabuSearch/tabuSearch.pdf
15. Dorigo, M. (1992). Optimization, learning and natural algorithms. Ph.D. thesis, Politecnico di Milano, Dipartimento di Elettronica.
16. Fouskakis, D., & Draper, D. (2002). Stochastic optimization: a review. *International Statistical Review, 70*(3), 315–349.
17. Gaul, W., Aust, E., & Baier, D. (1995). Gewinnorientierte Produktliniengestaltung unter Berücksichtigung des Kundennutzens. *Zeitschrift für Betriebswirtschaftslehre, 65*(8), 835–855.
18. Gaul, W., & Baier, D. (2021). Simulation und Optimierung auf Basis der Conjointanalyse. In D. Baier & M. Brusch (Eds.), *Conjointanalyse: Methoden—Anwendungen—Praxisbeispiele* (2nd ed., pp. 275–306). Springer.
19. Glover, F. (1990). Tabu search: A tutorial. *Interfaces, 20*(4), 74–94.
20. Glover, F., & Laguna, M. (1998). Tabu search. *Handbook of combinatorial optimization* (pp. 2093–2229). Springer.
21. Green, P. E., Carroll, J. D., & Goldberg, S. M. (1981). A general approach to product design optimization via conjoint analysis. *Journal of Marketing Research, 45*(3), 17–37.
22. Green, P. E., & Krieger, A. M. (1985). Models and heuristics for product line selection. *Marketing Science, 4*(1), 1–19.
23. Green, P. E., & Krieger, A. M. (1987). A consumer-based approach to designing product line extensions. *Journal of Product Innovation Management, 4*, 21–32.

24. Green, P. E., & Krieger, A. M. (1988). Choice rules and sensitivity analysis in conjoint simulators. *Journal of the Academy of Marketing Science, 16*(1), 114–127.
25. Green, P. E., & Krieger, A. M. (1992). An application of a product positioning model to pharmaceutical products. *Marketing Science, 2*(1), 117–132.
26. Green, P. E., Rao, V. R. (1969). Nonmetric approaches to multivariate analysis in marketing. Working paper, Wharton School, University of Pennsylvania.
27. Holland, J. H. (1975). *Adaptation in natural and artificial systems*. University of Michigan Press.
28. Kennedy, J., & Eberhart, R. (1995). Particle swarm optimization. *Proceedings of IEEE International Conference on Neural Networks IV, Perth, Australia, 6*, 1942–1948.
29. Kirkpatrick, S., Gelatt, C. D., Jr., & Vecchi, M. P. (1983). Optimization by simulated annealing. *Science, 220*(4598), 671–680.
30. Kohli, R., & Krishnamurti, R. (1987). A heuristic approach to product design. *Management Science, 33*(12), 1523–1533.
31. Kohli, R., & Krishnamurti, R. (1989). Optimal product design using conjoint analysis: Computational complexity and algorithms. *Journal of Operational Research, 40*(2), 186–195.
32. Kohli, R., & Sukumar, R. (1990). Heuristics for product-line design using conjoint analysis. *Management Science, 36*(12), 1464–1478.
33. Louviere, J. J. (1988). Conjoint analysis modelling of stated preferences. *Journal of Transport Economics and Policy, 22*(1), 93–119.
34. Louviere, J. J., & Woodworth, G. (1983). Design and analysis of simulated consumer choice or allocation experiments: an approach based on aggregate data. *Journal of Marketing Research, 20*(4), 350–367.
35. Luce, R. D. (1959). *Individual choice behavior—a theoretical analysis*. Wiley.
36. Luce, R. D., & Tukey, J. W. (1964). Simultaneous conjoint measurement: A new type of fundamental measuring. *Journal of Mathematical Psychology, 1*, 1–27.
37. McFadden, D. (1976). Quantal choice analysis: A survey. *Annals of Economic and Social Measurement, 5*, 363–90.
38. Orme, B. K. (2019). *Getting Started with Conjoint Analysis* (4th ed.). Research Publishers LLC.
39. Pantourakis, M., Tsafarakis, S., Zervoudakis, K., Altsitsiadis, E., Andronikidis, A., & Ntamadaki, V. (2022). Clonal selection algorithms for optimal product line design: A comparative study. *European Journal of Operational Research, 298*(2), 585–595.
40. Punj, G. N., & Staelin, R. (1978). The choice process for graduate business schools. *Journal of Marketing Research, 15*, 588–598.
41. Roberts, J. H., Kayande, U., & Stremersch, S. (2014). From academic research to marketing practice: Exploring the marketing science value chain. *International Journal of Research in Marketing, 31*(2), 144–146.
42. Schön, C. (2010). On the optimal product selection problem with price discrimination. *Management Science, 56*(5), 896–902.
43. Scrucca, L. (2022). Package 'GA'—An R Package. CRAN—Comprehensive R Archive. https://cran.r-project.org/web/packages/GA/GA.pdf
44. Selka, S. (2013). *Validität computergestützer Verfahren der Präferenzmessung*. Dr.
45. Shi, L., Ólafsson, S., & Chen, Q. (2001). An optimization framework for product design. *Management Science, 47*(12), 1681–1692.
46. Shugan, S. M., Balachandran, V. (1977). A mathematical programming model for optimal product line structuring. Working paper, Series 7734 (October), Graduate School of Business, University of Rochester.
47. Steiner, W., & Hruschka, H. (2000). Conjoint-basierte Produkt(linien)gestaltung unter Berücksichtigung von Konkurrenzreaktionen. *OR Spektrum, 22*(1), 71–95.
48. Steiner, W., & Hruschka, H. (2002). Produktliniengestaltung mit genetischen Algorithmen. *Schmalenbachs Zeitschrift für betriebswirtschaftliche Forschung, 54*(7), 575–601.
49. Steiner, W., & Hruschka, H. (2003). Genetic algorithms for product design: How well do they really work? *International Journal of Market Research, 45*(2), 229–240.

50. Tarasewich, P., & McMullen, P. R. (2001). A pruning heuristic for use with multisource product design. *European Journal of Operational Research, 128*(1), 58–73.
51. Tsafarakis, S., Marinakis, Y., & Matsatsinis, N. (2011). Particle swarm optimization for optimal product line design. *International Journal of Research in Marketing, 28*(1), 13–22. https://doi.org/10.1016/j.ijresmar.2010.05.002
52. Tsafarakis, S., Zervoudakis, K., & Andronikidis, A. (2021). Optimal product line design using tabu search. *Journal of the Operational Research Society,* 1–12.
53. Tsafarakis, S., Zervoudakis, K., Andronikidis, A., & Altsitsiadis, E. (2020). Fuzzy self-tuning differential evolution for optimal product line design. *European Journal of Operational Research, 287*(3), 1161–1169. https://doi.org/10.1016/j.ejor.2020.05.018
54. Vökler, S., Krausche, D., & Baier, D. (2013). Product design optimization using ant colony and bee algorithms: A comparison. In B. Lausen, D. Van den Poel, & A. Ultsch (Eds.), *Algorithms from and for nature and life* (pp. 491–498). Springer.
55. Wittink, D. R., & Cattin, P. (1989). Commercial use of conjoint analysis: An update. *Journal of Marketing, 53*(3), 91–96.
56. Zufryden, F. S. (1977). A conjoint measurement-based approach for optimal new product design and market segmentation. In A. D. Shocker (Ed.), *Analytic approaches to product and market planning* (pp. 100–114). Marketing Science Institute.
57. Zufryden, F. S. (1982). Product line optimization by integer programming. *Proceedings* (pp. 100–114). Annual Meeting of ORSA/TIMS.

Graphical Analysis and Clustering of Asymmetric Proximities

Giuseppe Bove and Donatella Vicari

Abstract Proximity matrices are frequently asymmetric and analyzed by using the additive decomposition into symmetric and skew-symmetric components. A strategy for a preliminary graphical description and clustering of the two components based on some recent methods is presented. An empirical application allows to emphasize the advantages and the effectiveness of the proposal.

1 Introduction

In many fields, such as economics, sociology, marketing research and other behavioral sciences, asymmetric proximities between pairs of entities (e.g., import–export data, sociomatrices, brand switching, flows and migration data, etc.) are analyzed by using a variety of models and methods to detect meaningful relationships. The choice of a model depends on the purpose of the analysis, the conditions of data collection, and, more generally, the amount of information available on the phenomenon under investigation. In some cases, the theory suggests a particular explicit substantive model for the data: examples can be found both in experimental and non-experimental studies. When no substantive model is readily available and there is not a priori reason for preferring any particular model, we might look for graphical displays as preliminary investigation of the data that, similarly to the standard scatterplots, "allow the data to speak for themselves".

Furthermore, graphical representations may suggest the presence of clustering structures which can be identified from the application of methods of classifications.

From now on, the proximities ω_{ij} between pairs of entities (i,j) $(i, j = 1, 2, \ldots, n)$ are collected in a $(n \times n)$ data matrix $\mathbf{\Omega} = (\omega_{ij})$. In this contribution, we focus on the analysis of the symmetric and skew-symmetric components of $\mathbf{\Omega}$, given by

G. Bove (✉)
Dipartimento di Scienze della Formazione, Università degli Studi Roma Tre, Rome, Italy
e-mail: giuseppe.bove@uniroma3.it

D. Vicari
Dipartimento di Scienze Statistiche, Sapienza Università di Roma, Rome, Italy
e-mail: donatella.vicari@uniroma1.it

© The Author(s), under exclusive license to Springer Nature Singapore Pte Ltd. 2023
A. Okada et al. (eds.), *Facets of Behaviormetrics*, Behaviormetrics: Quantitative Approaches to Human Behavior 4, https://doi.org/10.1007/978-981-99-2240-6_2

matrices $\mathbf{M} = (m_{ij})$ and $\mathbf{N} = (n_{ij})$, respectively, with $m_{ij} = (\omega_{ij} + \omega_{ji})/2$ and $n_{ij} = (\omega_{ij} - \omega_{ji})/2$, where $\mathbf{M} = \mathbf{M}'$, that is, $m_{ij} = m_{ji}$, and $\mathbf{N} = -\mathbf{N}'$, that is, $n_{ij} = -n_{ji}$ $(i, j = 1, 2, \ldots, n)$. It is not difficult to see that the elements n_{ij} of matrix $\mathbf{N}$ are the deviations of the proximities from symmetry, because $n_{ij} = \omega_{ij} - m_{ij}$, and this implies $\omega_{ij} = m_{ij} + n_{ij}$, that is, in matrix notation is $\mathbf{\Omega} = \mathbf{M} + \mathbf{N}$, a unique additive decomposition widely studied (for a review, see [4]).

The two components can be modeled either independently or not, and can be represented either separately or jointly. Separate representations are usually more informative, in particular, for the skew-symmetric component, but they may be difficult to synthesize. Joint representations are easy to manage (e.g., spatial representations containing only n points) and suitable for detecting peculiar types of asymmetry.

In the following, an approach for the separate representation and classification of symmetry and skew symmetry will be described and applied to a real proximity matrix. The paper is organized as follows: in Sect. 2, a procedure is described for the separate graphical representation of symmetry and skew symmetry. The symmetric component is represented in a diagram by the Euclidean distance model, while the absolute value of the skew-symmetric component is represented by the Euclidean distance model in a separate diagram enriched by drift vectors [2] which display the signs of the skew symmetries. In Sect. 3, a method for clustering the asymmetric data is described which separately fits clustering structures to the symmetric and skew-symmetric components of the data.

In Sect. 4, an application to the Eurovision Song Contest data 2021 is presented to emphasize some advantages of the strategy proposed. Some final considerations on the methods and the results obtained are provided in the last section.

2 A Procedure for the Separate Graphical Representation of Symmetry and Skew-Symmetry

Constantine and Gower [5] remark that the symmetric component $\mathbf{M}$ and the skew-symmetric component $\mathbf{N}$ of any square matrix $\mathbf{\Omega}$ are uncorrelated and the following orthogonal breakdown of the Sum of Squares (SS) holds

$$\mathbf{SS}(\mathbf{\Omega}) = \mathbf{SS}(\mathbf{M}) + \mathbf{SS}(\mathbf{N}),$$

reflecting the uniqueness of the additive decomposition $\mathbf{\Omega} = \mathbf{M} + \mathbf{N}$. Thus, the two components can be viewed independently, and studied by separate models. An advantage of the separate analysis is that one can deal with the representation problem by adopting different kinds of models (e.g., spatial versus nonspatial) and different kinds of transformations (e.g., metric versus non-metric) for the two components. A drawback is that we lose the possibility to jointly analyze symmetry and skew symmetry in the same representation obtained by directly approximating matrix $\mathbf{\Omega}$. A procedure for a separate analysis is presented in the following parts of this section.

Representation of the symmetric component

The symmetric component **M** can be represented through a distance model using multidimensional scaling. When proximities are approximated in r dimensions, the most used choice for the distance model is

$$f(m_{ij}) = \hat{d}_{ij} = d_{ij}(\boldsymbol{x}_i, \boldsymbol{x}_j) + e_{ij} = \sqrt{\sum\nolimits_{s=1}^{r} (x_{is} - x_{js})^2} + e_{ij}, \tag{1}$$

where f is a monotone transformation, mapping the proximities ω_{ij} into a set of transformed values $\widehat{d}_{ij} = f(m_{ij})$, called *disparities* or *pseudo-distances*, x_{is} and x_{js} are the coordinates of row (column) i and row (column) j on dimension s, respectively, and the symmetric property holds ($d_{ij} = d_{ji}$). The estimation of the model provides a map where entries m_{ij} are approximated by distances in r dimensions: when the entries m_{ij} are similarities, the larger the values, the smaller the distances. The model can be easily estimated using routines available in standard statistical software for symmetric multidimensional scaling. An advantage of this model is that it is easy to incorporate both non-metric approaches and external information on the entities.

Representation of the skew-symmetric component

Gower [6] and Constantine and Gower [5] propose to represent the skew-symmetric component **N** by singular value decomposition that implies to analyze and compare triangle areas in (usually) two dimensions in order to detect asymmetric relationships. To facilitate the interpretation of the diagrams, methods based on distance models were also considered to represent skew symmetry (e.g., [2, 3]), the judgments on distances being cognitively less demanding than judgments on the areas of the triangles.

First of all, we note that any skew-symmetric matrix contains two types of information for each pair (i, j): size (or magnitude) and sign (or directionality) of the skew-symmetry, given by the absolute value $|n_{ij}|$ and $sign(n_{ij})$, respectively. Thus, each entry n_{ij} of matrix **N** can be written as

$$n_{ij} = sign(n_{ij})|n_{ij}|, \tag{2}$$

where $|n_{ij}|$ is the absolute value of n_{ij}, $sign(n_{ij}) = 1$ if $n_{ij} > 0$, $sign(n_{ij}) = -1$ if $n_{ij} < 0$, $sign(n_{ij}) = 0$ if $n_{ij} = 0$.

In this contribution, an easy-to-implement method is proposed that is based on the following two-step method.

STEP 1

The sizes of the skew symmetries $|n_{ij}|$ are symmetric ($|n_{ij}| = |n_{ji}|$), so matrix $\mathbf{T} = (t_{ij}) = (|n_{ij}|)$ can be considered a standard dissimilarity matrix that can be approximated in a *low-dimensional* (usually two-dimensional) Euclidean space by the distance model (1) where m_{ij} is replaced with t_{ij}. A map is obtained where the sizes of the skew symmetries t_{ij} are approximated by the distances d_{ij} (the larger the distance the larger the size).

STEP 2

The signs of the skew-symmetries $sign(n_{ij})$, collected into matrix $\mathbf{K}^{sign} = \left(k_{ij}^{sign}\right) = (sign(n_{ij}))$, can be represented by the drift vector method proposed in ([1, 2], par. 23.5). In the configuration obtained at STEP 1, an arrow is drawn from any point i to any other point j so that its length and direction correspond to the values in row i of the skew-symmetric matrix $\mathbf{K}^{sign}$ (the skew-symmetric matrix $\mathbf{N}$ in the original proposal of Borg and Groenen). For each pair (i, j), if n_{ij} is positive, the arrow goes from point i to point j, while, when n_{ij} is negative, the arrow points in the opposite direction. When the number of rows of the proximity matrix is large, the representation of all arrows may result into cluttered pictures, and it can be convenient to draw only the average vector (drift vector) of the arrow bundle attached to each point. The arrows representing the drift vectors go toward points having more frequently positive skew symmetries and they can exhibit systematic asymmetric trends.

We note that this two-step method allows to detect the general trend of the direction of the asymmetry in the proximity data matrix more easily than the two-step method proposed in [3]. We remark that in the present proposal, the length of the slide vector represents the homogeneity of the directions of the $(n–1)$ vectors emitting from the point.

3 A Clustering Model for Symmetry and Skew Symmetry

Methods for representing the asymmetries in low-dimensional spaces have been more widely studied than methods for clustering, especially the non-hierarchical ones (for a review, see [4]). For such kind of data, the main goal of a clustering method can be the identification of groups of entities which are similar not only for their average amounts, but also for the directions of the observed exchanges so that it is possible to discover "origin" and "destination" groups. The decomposition of matrix $\mathbf{\Omega} = \mathbf{M}+\mathbf{N}$ can be exploited in the clustering context because the symmetric and skew-symmetric components allow to obtain all the information on the intensities and directions of the exchanges, respectively. Following this line, several clustering methodologies have been proposed based on the independence of $\mathbf{M}$ and $\mathbf{N}$ which can be profitably used to get either separate partitions of the data or one single common partition which subsumes all the information from the entities.

Let us briefly recall here the non-hierarchical method proposed by [7], which aims at modeling the exchanges between unknown groups of entities by fitting two clustering structures to the components $\mathbf{M}$ and $\mathbf{N}$ of the observed data. Specifically, two partitions of the n entities in k groups are searched for to approximate $\mathbf{M}$ and $\mathbf{N}$,

$$\mathbf{M} = \mathbf{UC}(\mathbf{1}_{nk} - \mathbf{U})' + (\mathbf{1}_{nk} - \mathbf{U})\mathbf{CU}' + \mathbf{E}_{\mathrm{M}} \tag{3}$$

and

$$\mathbf{N} = \mathbf{VD}(\mathbf{1}_{nk} - \mathbf{V})' - (\mathbf{1}_{nk} - \mathbf{V})\mathbf{DV}' + \mathbf{E}_{\mathrm{N}}, \tag{4}$$

where $\mathbf{1}_{nk}$ denotes the $(n \times k)$ matrix of ones, **C** and **D** are diagonal weight matrices of size k, while **U** and **V** are the two $(n \times k)$ membership matrices which identify the two partitions. The membership matrix **U** is binary and subsumes a standard *complete* partition of the n entities into k groups (each entity is assigned to only one group and none of them are unassigned), while, in the binary membership matrix **V**, a number of rows is allowed to be a vector of zeros, which means that the corresponding entities remain unassigned to any group (*incomplete* partition).

In order to get a link between the two partitions, it may be assumed that $\mathbf{V} \subseteq \mathbf{U}$, i.e., any group $\mathcal{V}_j$ of the partition from the skew-symmetric component **N** is either equal or nested into the corresponding group $\mathcal{U}_j$ from the symmetric component **M** $(\mathcal{V}_j \subseteq \mathcal{U}_j, j = 1, \ldots, k)$.

When (3) and (4) are combined together, the clustering model for the observed asymmetric data is stated

$$\begin{aligned} \boldsymbol{\Omega} = \mathbf{M} + \mathbf{N} &= \left[\mathbf{UC}(\mathbf{1}_{nk} - \mathbf{U})' + (\mathbf{1}_{nk} - \mathbf{U})\mathbf{CU}'\right] \\ &+ \left[\mathbf{VD}(\mathbf{1}_{nk} - \mathbf{V})' - (\mathbf{1}_{nk} - \mathbf{V})\mathbf{DV}'\right] + b(\mathbf{1}_{nn} - \mathbf{I}_n) + \mathbf{E} \end{aligned}$$

subject to

$$\begin{aligned} &u_{ij} \in \{0, 1\}, (j = 1, \ldots, k); \sum_{j=1}^{k} u_{ij} = 1, (i = 1, \ldots, n) \\ &v_{ij} \in \{0, 1\}, (j = 1, \ldots, k); \sum_{j=1}^{k} v_{ij} \leq 1, (i = 1, \ldots, n) \\ &\mathbf{1}_n'\mathbf{VD1}_k = 0, \end{aligned} \tag{5}$$

where b is the constant term and **E** represents the part of $\boldsymbol{\Omega}$ not accounted for by the model. The first two sets of constraints in (5) identify the complete and incomplete partitions, respectively, while matrix **VD** is constrained to sum to zero for identifiability reasons. The model is fitted in a least-squares framework and an ALS algorithm is discussed in [7].

Several information may be derived from the estimates:

(a) constant b represents the baseline average amount of exchange regardless of any clustering;
(b) weight $c_j (j = 1, \ldots, k)$ in the diagonal matrix **C** measures the average amount of the exchanges between group $\mathcal{U}_j$ and any other group, while $d_j (j = 1, \ldots, k)$ in the main diagonal of **D** results to be the weighted average imbalance from all entities in group $\mathcal{V}_j$ directed to all other entities in different groups. This allows to qualify any group as either *origin* or *destination* according to the sign of its weight;

(c) interestingly, the entities which remain unassigned in $\mathbf{V}$ correspond to entities with (almost) symmetric values in the observed $\boldsymbol{\Omega}$.

4 Application to the Eurovision Song Contest (ESC) Data 2021

In order to illustrate the effectiveness of the methods sketched in Sects. 2 and 3, data from the Eurovision Song Contest 2021 have been analyzed.[1]

Thirty-nine countries initially participated in the contest. After two semi-finals, 20 countries (songs) qualified for the final, plus the six founder countries. Thus, 26 countries participated in the final with all 39 participating countries eligible to vote (professional jury and televoting).

With the aim of analyzing people's preferences and assessing whether similarities between countries in expressing their likings can be identified, only the detailed televoting results of the final have been considered for the analysis.

The countries participating in the final were Cyprus (CY), Albania (AL), Israel (IL), Belgium (BE), Russia (RU), Malta (MT), Portugal (PT), Serbia (CS), United Kingdom (GB), Greece (GR), Switzerland (CH), Iceland (IS), Spain (ES), Moldova (MD), Germany (DE), Finland (FI), Bulgaria (BG), Lithuania (LT), Ukraine (UA), France (FR), Azerbaijan (AZ), Norway (NO), Netherlands (NL), Italy (IT), Sweden (SE), and San Marino (SM).

The original similarity matrix contains the televoting score ω_{ij} for contestant i awarded by country j as a result of the televote; note that people cannot televote for their own country (diagonal entries are meaningless) and televoting scores from countries not qualified for the final have not been considered here, so the similarity matrix $\boldsymbol{\Omega}$ (Table 1) is square and of size (26×26). The average score is 4.46 (min $= 0$, max $= 24$) with very high dispersion (SD $= 5.83$). All columns have the same totals because all televotes from each country were divided up and converted into points so that each voting country awarded the same total score. Conversely, the row total scores determine the ranking of the contestants: Italy, France, Ukraine, Finland, Switzerland, Iceland, Lithuania, Russia, Greece, Sweden, Norway, Cyprus, Moldova, Serbia, Bulgaria, Malta, Portugal, Albania, Azerbaijan, Israel, San Marino, Belgium, United Kingdom, Spain, Germany, and Netherlands.

In order to convert the observed similarities into dissimilarities, the data have been preprocessed by using the following transformation:

$$\omega^*_{ij} = \left[1 + \frac{\omega_{ij}\left(\sum_{ij} \omega_{ij}\right)}{\left(\sum_i \omega_{ij}\right)\left(\sum_j \omega_{ij}\right)} \right]^{-1}, (i, j = 1, \ldots, 26)$$

[1] https://en.wikipedia.org/wiki/Eurovision_Song_Contest_2021#Detailed_voting_results.

Table 1 Eurovision Song Contest data: televoting scores of the final (2021)

	Televoting Country																									
Contestant	CY	AL	IL	BE	RU	MT	PT	CS	GB	GR	CH	IS	ES	MD	DE	FI	BG	LT	UA	FR	AZ	NO	NL	IT	SE	SM
CY		2			12	2		2		12																8
AL										7	7													10		
IL	2																			5	12					
BE																		2	1							
RU	3		10				1	6		1				12	5	1	7	4	4		6			7		1
MT	1		5	2					6	3		3	3									3	1		2	
PT											8	2	1							8			6			
CS						4					12				3		5			3				4	1	
GB																										
GR	12	8					2	3						7			2						10			7
CH		12	6	4	5		6	1	1	4		6	7	4	2	7	3	6	7	6	4	2	7		5	3
IS			1	5	1	5	3	5	10		5		5		6	12		3	6	4		10	8	6	10	
ES																										
MD		7			3		8									2				7	1			2		6
DE																										
FI	4	3	4	6	8	7	5	7	7	6	4	12	2	5	8		8	7	8	1	5	6	4	8	12	4
BG	7	5							8				8													2
LT	5		3	7	2	6			12		1	4	4	2	12	5			10			12	3	5	7	
UA	6	4	12	10	7	1	10	8	4	5	2	8	6	10	4	10	6	12		12	8	5	5	12	4	5
FR	8	6	8	12	6	3	12	10	5	8	6	7	12	6	10	6	10	8	5		2	4	12	1	6	10

(continued)

Table 1 (continued)

	Televoting Country																								
Contestant	CY	AL	IL	BE	RU	MT	PT	CS	GB	GR	CH	IS	ES	MD	DE	FI	BG	LT	UA	FR	AZ	NO	NL	IT	SE	SM
AZ			2		4			4		2				1			4		3			1				
NO				1		10			2			1			1	4		5			7				8	
NL																										
IT	10	10	7	8	10	12	7	12	3	10	10	5	10	8	7	8	12	10	12	10	10	7	2		3	12
SE		1		3		8	4				3	10		3		3	1	1	2	2		8				
SM																					3			3		

which actually allows to get dissimilarity measures assuming values equal to 0.5 in the case of statistical independence between contestant and voting countries, and maximum value equal to 1 when contestant i did not receive any score from country j. Values lower (greater) than 0.5 correspond to televoting scores higher (lower) than what expected by chance.

A graphical description of the transformed ESC data matrix can be obtained by the procedure for the separate graphical representation of symmetry and skew symmetry presented in Sect. 2. An ordinal multidimensional scaling was applied to the symmetric component of the transformed proximity matrix by using the *mds* routine of the SMACOF program in the R package. The diagram obtained from the solution in two dimensions is displayed in Fig. 1 (*Stress-I* = 0.19). Distances between points represent the average dissimilarities (transformed televote scores between countries), the higher the level the smaller the distance.

For instance, contestants which received very low levels of televote scores from the other countries, DE, GB, BE, ES, NL, AL, and SM, are positioned far from the center of the diagram and distant from the other countries. Conversely, contestants which received high levels of televote scores (IT, UA, FR, FI, and CH, which are the top five countries in the ranking) are positioned in the center of the diagram and seem to form a separate cluster of countries. Other groups of countries can be detected by looking at other areas of the diagram: CY, AZ, IL, RU, CS, BG in the first quadrant; the two groups: MT, LT, IS, SE and DE, GB, NO in the second and third quadrants, with BE and ES which remain isolated; NL and PT at the bottom of the diagram; and MD, AL, and GR in the fourth quadrant with SM isolated. This is just a first look at the presence of possible groups of countries that only a method of clustering is able to detect.

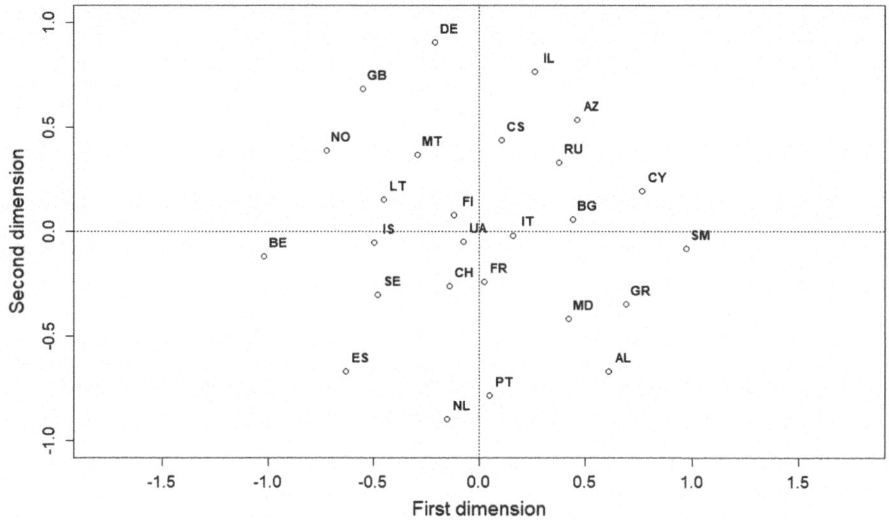

Fig. 1 Eurovision Song Contest data: multidimensional scaling of the symmetric component

The two-step method presented in Sect. 2 was then applied to the skew-symmetric component of the transformed proximity data to analyze the imbalances of the televotes exchanged. An interval multidimensional scaling was applied to the absolute value of the skew-symmetric component by using the *mds* routine of the SMACOF program, while the drift vectors were computed by an ad hoc routine in the R package. The diagram obtained from the solution in two dimensions is displayed in Fig. 2 (*Stress-I* = 0.33). Despite the low level of approximation of the solution, some imbalances between pairs of countries and an interesting trend can be observed in Fig. 2. For instance, the large distances between SM and AZ, BG and ES, and IL and MT represent large imbalances of the televote scores exchanged between such pairs of countries. Conversely, the small distances between GB and ES, BE and NL, and IT and LT represent null or small imbalances.

Moreover, the general direction of the asymmetry can be detected by looking at the trend of the drift vectors. Most of the drift vectors point to the right side of the diagram where the countries with high levels of televote scores, as IT, UA, FR, FI, IL, CH and LT, are positioned. This means that the skew symmetry from a country in the left side of the diagram to a country in the right side of the diagram is usually positive; the reverse occurs in the opposite direction.

In order to group the countries by taking into account both the symmetry and the skew symmetry of the data, model (5) by [7] was also fitted to the ESC data to better analyze the (lack of) reciprocity in the televoting preferences of people from different countries.

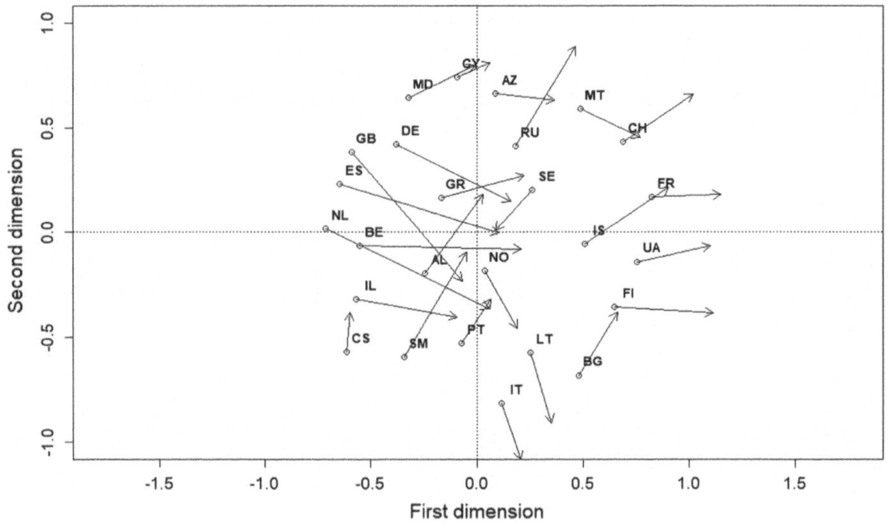

Fig. 2 Eurovision Song Contest data: multidimensional scaling of the absolute value of the skew-symmetric component. Drift vectors are attached to the country-points

The algorithm was applied to the same transformed data by setting the number of clusters from 1 to 7 and retaining the best solutions in 100 runs from different random starting partitions to prevent from falling into local optima.

Figure 3 reports the scree plot of the relative loss function values $\frac{\|\boldsymbol{\Omega} - \left(\widehat{\mathbf{M}} + \widehat{\mathbf{N}}\right)\|^2}{\|\boldsymbol{\Omega}\|^2}$ across the number of groups which exhibits a small decrease after the four-cluster solution for which only 9.7% of the score variability remains not accounted for.

From the symmetric matrix of the average exchanges of votes between countries, the following four groups of countries result (optimal *complete* partition $\widehat{\mathbf{U}}$):

$\mathcal{U}_1$= (CY, AL, IL, RU, CS, GR, MD, BG, AZ, SM),
$\mathcal{U}_2$= (CH, FI, UA, FR, IT),
$\mathcal{U}_3$= (PT, ES, DE, NL),
$\mathcal{U}_4$= (BE, MT, GB, IS, LT, NO, SE).

Moreover, the corresponding groups from the skew-symmetric component of the data (*incomplete* partition $\widehat{\mathbf{V}}$ nested into $\widehat{\mathbf{U}}$) differ for several countries:

$\mathcal{V}_1$= (AL, IL, BG, SM),
$\mathcal{V}_2$= (CH, FI, UA, FR, IT),
$\mathcal{V}_3$= (ES, DE, NL),

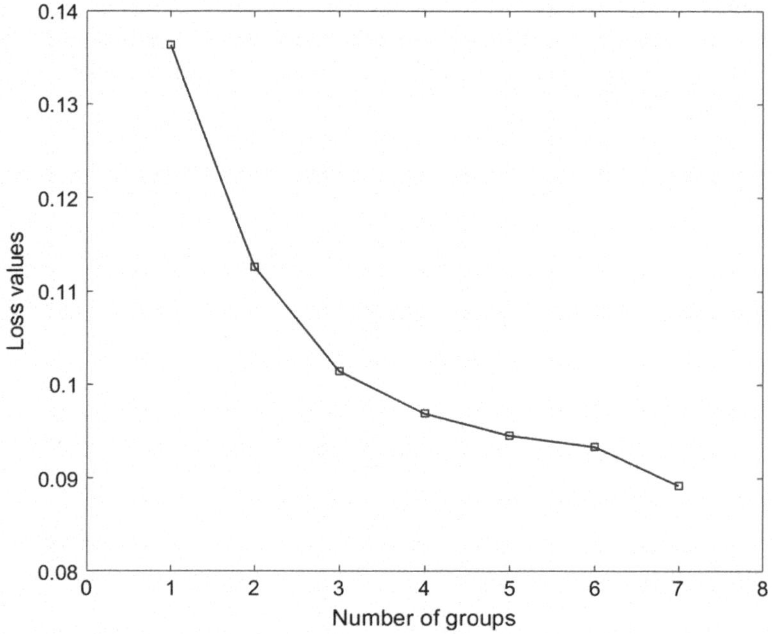

Fig. 3 Eurovision Song Contest data: scree plot of the loss function values versus number of groups

$\mathcal{V}_4$= (BE, GB).

Note that countries which remain unassigned in the groups resulting from the skew-symmetric component correspond to the countries with the smallest imbalances (least asymmetric) in expressing their televoting preferences.

The baseline average level is $\hat{b} = 0.6573$ which means that, regardless of any clustering, the amounts of the televoting scores are on average different from the case of independence (0.5).

The optimal weights from the symmetric component turn out to be $\hat{\mathbf{c}} = (0.1413, -0.1303, 0.1388, 0.1125)$, which provide a ranking of the groups $\{\mathcal{U}_1, \mathcal{U}_3, \mathcal{U}_4, \mathcal{U}_2\}$ in terms of increasing average amounts of votes exchanged between countries, regardless of their directions. In fact, the songs in group $\mathcal{U}_2$ obtained the highest scores from televoting and finished at the top of the final ranking. Moreover, from the estimated weights from the imbalances between groups $\hat{\mathbf{d}} = (0.0430, -0.1237, 0.0954, 0.0801)$ a different ranking results $\{\mathcal{V}_3, \mathcal{V}_4, \mathcal{V}_1, \mathcal{V}_2\}$, which indicates the directions of the televoting: i.e., the second group identifies the favorite songs ("destination" countries), because most people voted for songs of the countries in the second group much more than the other way around. Conversely, the third and fourth groups identify the countries where people cast the most votes in favor of countries not belonging to the same group.

The imbalances between groups are displayed in Fig. 4 where countries are represented as nodes and the estimated skew symmetries as arrowed arcs denoting the directions of the televoting. The arrangement of the nodes reflects the ordering of the groups obtained with the top-voted countries at the bottom. Note that the 12 countries which remain unassigned qualify countries with a low degree of asymmetry. All unassigned countries identify countries with small imbalances which denote reciprocity in the votes exchanged.

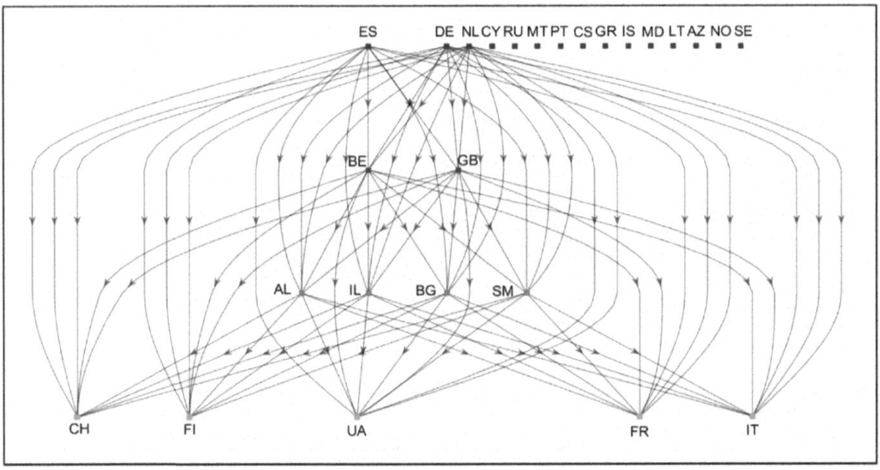

Fig. 4 Eurovision Song Contest data: estimated skew symmetries between groups of countries

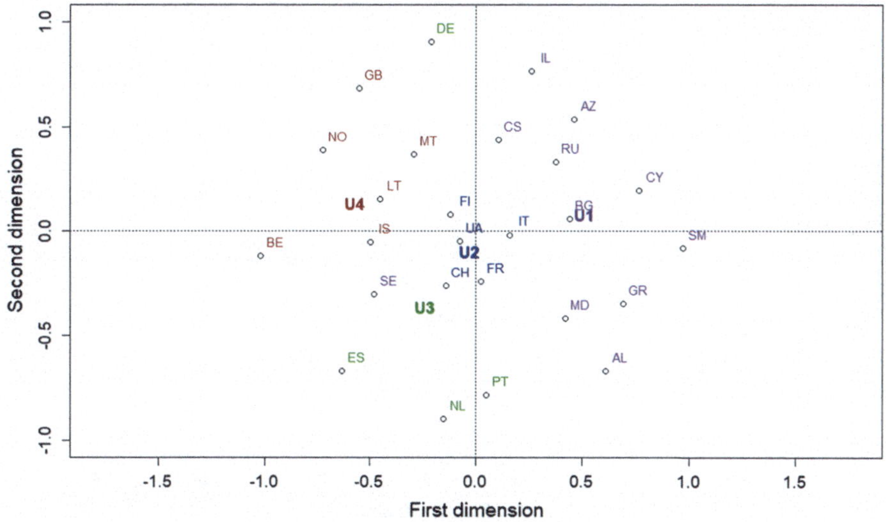

Fig. 5 Eurovision Song Contest data: multidimensional scaling and groups of countries with centroids obtained from the symmetric component

In order to link the clustering results with those obtained by the graphical methods based on multidimensional scaling, the centroids of the groups of the partitions $\widehat{\mathbf{U}}$ and $\widehat{\mathbf{V}}$ have been reported in Figs. 1 and 2. The four groups obtained from the symmetric and the skew-symmetric components are displayed in Figs. 5 and 6, respectively, where different colors indicate countries in different groups and their corresponding centroids. In Fig. 6, the vectors attached to the centroids of the four groups are computed as the averages of the drift vectors of the countries within each group. In Fig. 5, the groups are quite well separated, with only two exceptions (DE and SE). In Fig. 6, groups $\mathcal{V}_1$, $\mathcal{V}_2$ and groups $\mathcal{V}_3$, $\mathcal{V}_4$ partially overlap and are not clearly distinguished for the presence of countries not assigned to any group (countries with black labels). For the sake of clarity of the diagram, the unassigned countries could be removed. The global trend of the asymmetry is confirmed by the averages of the drift vectors attached to the centroids of the groups: they clearly point to the top countries of the final ranking including the winner (IT).

5 Conclusions

The analysis of symmetry and skew symmetry of a proximity matrix can be carried out by using different models and methods depending on the aims of the study and the available information. A preliminary graphical analysis and clustering help to detect the presence of structures in the data and to orient the study toward an accurate choice of the model. The strategy presented in this contribution allows to carry out

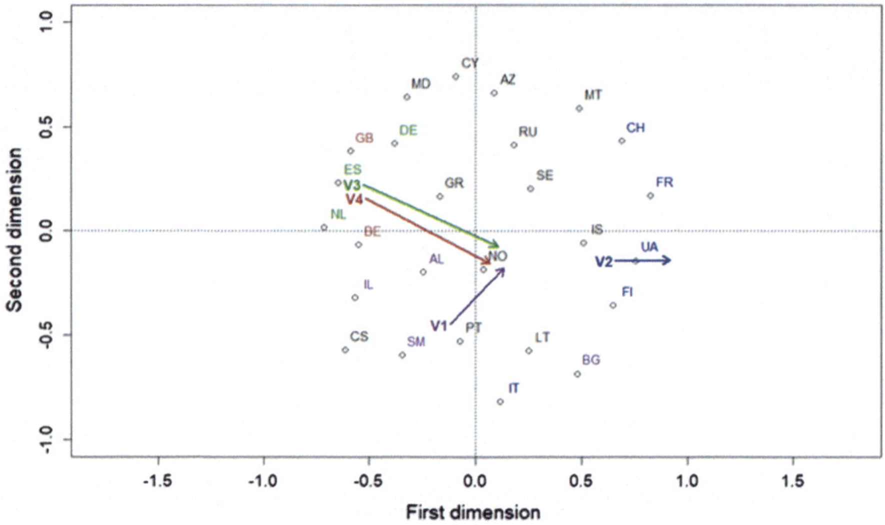

Fig. 6 Eurovision Song Contest data: multidimensional scaling and groups of countries with centroids obtained from the skew-symmetric component. Averages of the drift vectors are attached to the centroids of the groups. Unassigned countries are displayed in black

such preliminary analysis by using some recently proposed methods of asymmetric multidimensional scaling and cluster analysis. The application to empirical data shows a good level of coherence of the strategy, which allows an effective description of symmetry and skew symmetry. The graphical tools that characterize the methods presented make the description easy to understand.

References

1. Borg, I., & Groenen, P. J. F. (1996). Asymmetries in multidimensional scaling. In F. Faulbaum & W. Bandilla (Eds.), *Soft Stat '95: Advances in Statistical Software 5*. (pp. 31–35). Lucius & Lucius.
2. Borg, I., & Groenen, P. J. F. (2005). *Modern multidimensional scaling. Theory and applications* (2nd ed.). Springer.
3. Bove, G. (2012). Exploratory approaches to seriation by asymmetric multidimensional scaling. *Behaviormetrika, 39*, 63–73.
4. Bove, G., Okada, A., & Vicari, D. (2021). *Methods for the analysis of asymmetric proximity data*. Springer Nature.
5. Constantine, A. G., & Gower, J. C. (1978). Graphical representation of asymmetric matrices. *Applied Statistics, 3*, 297–304.
6. Gower, J. C. (1977). The analysis of asymmetry and orthogonality. In J. R. Barra, F. Brodeau, G. Romer, & B. van Cutsem (Eds.), *Recent developments in statistics* (pp. 109–123). North-Holland.
7. Vicari, D. (2020). Modelling asymmetric exchanges between clusters. In T. Imaizumi, A. Nakayama, & S. Yokoyama (Eds.), *Advanced studies in behaviormetrics and data science* (pp. 297–313). Springer Nature.

Citizens and The Pandemic: Values, Attitudes, Impact

Chadjipadelis Theodore

Abstract This study investigates in depth not only the attitudes and the feelings of individuals towards the Pandemic but also some important aspects of it such as vaccines, trust in experts, the fear of illness and also the compliance with the necessary measures that were imposed using a multi-variate method combining hierarchical cluster analysis, factorial correspondence analysis, and conjoint analysis. The aim is to explore profiles of respondents with reference to their stance towards the Pandemic and their political attitudes towards democracy, institutions, and moral values. The analysis is based on the use of a sample of almost 500 Greek citizens from whom data was collected using an electronic questionnaire in the period March–April 2022. The findings of the analysis reveal important correspondences between political behavior profiles and the way one behaves during the Pandemic period.

1 Introduction

This paper presents research findings on the attitudes of Greek citizens during the pandemic, capturing their views on the issue of vaccination, compliance with measures, trust in experts, and their sources of information about Covid-19 of during the pandemic crisis. At the same time, the impact of the pandemic on social and personal life during the restrictive measures is analyzed. In addition, an attempt is made to connect the above variables with the attitudes towards institutions and the political mobilization.

Many surveys and polls have been conducted in various countries regarding the attitudes of citizens towards government decisions during the pandemic. One of the first population-based studies carried out in Greece at the beginning of the pandemic focused on the region of Epirus and assessed the level of citizens' knowledge about the Covid-19 pandemic and their trust in official authorities based on age, gender, and period of report [7]. The results showed that the majority of participants expressed high levels of trust in Greek official authorities at the beginning of the pandemic,

C. Theodore (✉)
Aristotle University of Thessaloniki, Thessaloniki, Greece
e-mail: chadji@polsci.auth.gr

© The Author(s), under exclusive license to Springer Nature Singapore Pte Ltd. 2023
A. Okada et al. (eds.), *Facets of Behaviormetrics*, Behaviormetrics: Quantitative Approaches to Human Behavior 4, https://doi.org/10.1007/978-981-99-2240-6_3

with this trust decreasing over time. Furthermore, trust in the authorities was weaker among young people and among those who took part in the study after the restrictions were lifted. At the same time, in a survey conducted in various Western European countries, it appears that participants report higher political support, trust in the government, and satisfaction with democracy and its institutions than those who took part in the survey immediately before the virus appeared [2].

This specific research presented in this paper attempts to investigate thoroughly on a multiple level of analysis for various social, psychological, and political aspects of the citizens' attitudes towards COVID-19 pandemic, providing a methodological context for future analysis.

The analysis is based on the use of a sample of almost 500 citizens from whom data was collected using an electronic questionnaire in the period March–April 2022.

More specifically, to analyze the data, conjoint analysis is used, which is based on a scenario questionnaire [6], in order to analyze their: (a) attitude towards vaccination, (b) trust in experts and the state, (c) the risk of illness for themselves, (d) risk of illness for someone in their environment, and finally (e) compliance with the measures to protect against COVID-19.

Conjoint analysis, through experimental design, manages to highlight the factors and their contribution level in the formation of attitudes towards the pandemic. Moreover, the method is used to further analyze the individuals through their correlation with each factor.

In addition, this study explores another dimension of the pandemic, the effects on the emotional state of citizens. So, the research includes and analyzes questions that have to do with the feelings during the pandemic and the restrictive measures and also what the feelings were after the lifting of the measures.

The aim of the research and the analysis is to investigate in depth the attitudes of the Greek citizens with reference to the pandemic and its effect, detecting distinct groups of behavior and link these profiles to other variables about their sources of information, their views on democracy, and their moral values as well as the emotional impact of the overall situation on them.

The analysis develops combining the multi-variate methods of hierarchical cluster analysis and factorial correspondence analysis [1] in two steps, the joint analysis of all variables is attempted in order to identify all possible connections between groups of subjects and categories of the variables.

This analysis aims to identify and connect distinct discourses and individual behaviors [3] regarding the attitude towards critical issues of dealing with the pandemic, such as trust in experts and sources of information and the general attitude and trust towards institutions.

2 Methodology

To collect the data, an electronic questionnaire survey was carried out during March–April 2022 on a sample of 534 Greek citizens. Specific questions from the questionnaire were used to carry out the analysis. Eight questions were used for the analysis, which correspond to the research variables (Table 1). Of these, two demographic variables were used: gender and employment, two questions were about sources of information about the Pandemic and sources of information about politics, two questions concern symbolic representation by exploring views on democracy and personal moral values and one question corresponds to conjoint analysis with scenario analysis where the respondent is asked to put in ascending order of preference. In addition, the analysis includes not only the variable for the level of political interest but also the way in which someone chooses to mobilize for an issue that interests them.

To analyze the data, conjoint analysis method is used, which is based on a scenario questionnaire [6], in order to analyze their attitudes regarding:

(a) vaccination,
(b) trust in experts and the state,
(c) the risk of illness to themselves,
(d) risk of illness to a person in their environment, and
(e) compliance with the measures to protect against Covid-19.

The experimental design is based on the D-max criterion of parameter dispersion. The main effect of each level of the above five factors is estimated.

To investigate perceptions on democracy and moral values, we exploit symbolic representation [9]. Therefore, the respondents were asked to choose three pictures (among 2 different sets of 12 pictures) which represent themselves best (Figs. 1 and 2). Through this approach we investigate discourses of the democratic self and the moral self [12]

The variables are analyzed in the first step with HCA, producing clusters of pictures as well as clusters of the respondents. These new cluster membership variables are jointly analyzed with the rest of the political behavioral variables with AFC, which reorganizes all categories on the dimensions of the phenomenon, the total inertia [5]. These coordinates are clustered, producing groupings for the categories of the analysis. Then in a two-dimensional system all categories are positioned on the axes according to their coefficients [4]. The analysis was performed with the use of the software MAD, Méthodes d'Analyse des Données [8].

From the total of 534 questionnaires, the responses of 251 citizens with fully completed questionnaires, as far as the scenario questionnaire is concerned, were used for the analysis.

To check if there are differences in the characteristics of the 251 who answered the scenario questions (conjoint analysis) we used discriminant analysis, checking if the two groups K1 (did not answer) and K2 (answered) correspond to groups resulting from discrimination in terms of the characteristics: d3 occupation/age [pupils-students/employees-unemployed-other], d5 gender [male/female], e1 interest in politics [little-none/enough/very much], e7 attitude [not mobilized—leave it to those in charge—mobilized by social networks and media—address those in charge—collective mobilization], and e4 pandemic information sources [1−1042/ 2−1043 / 3−1051/ 4−1056/ 5−1057/ 6−1060/ 7−1061] (see Table 6).

Table 1 The variables of the analysis, coding, and categories

Code	Variable	Categories	Label
E3	Where do you mainly get informed about politics in general?	E3.1	TV-RADIO
		E3.2	NEWSPAPERS-PRESS
		E3.3	FAMILY
		E3.4	FRIENDS-SOCIAL CIRCLE
		E3.5	SOCIAL MEDIA
		E3.6	PARTY-MP'S I VOTE FOR
		E3.7	E-NEWSPAPER -INTERNET
E2	How close do you feel to each one of the following institutions?	e201	PARLIAMENT-PARTIES
		e202	CHURCH
		e203	MEDIA
		e204	MUNICIPALITY
		e205	REGIONAL AUTHORITY
		e206	ARMY
		e207	TRADE UNIONS
		e208	BUSINNESS
		e209	EUROPEAN UNION
		e210	SCIENTIFIC COMMUNITY
		e211	POLICE
		e212	JUSTICE
		E15.1	PROTEST
E15	What does "Democracy" mean to you?	E15.2	ANCIENT GREECE
		E15.3	DIRECT DEMOCRACY
		E15.4	E-DEMOCRACY
		E15.5	PARLIAMENT
		E15.6	RIOT
		E15.7	COUNCIL
		E15.8	VOLUNTEERING
		E15.9	MONEY-CORRUPTION
		E15.10	UPRISING
		E15.11	STRUGGLE
		E15.12	CHURCH
E16	Choose three (3) of the following pictures that represent you:	E16.1	ENTERTAINMENT
		E16.2	CAREER
		E16.3	CHRIST
		E16.4	LOVE AFFAIR
		E16.5	MEDITATION
		E16.6	MONEY
		E16.7	SUCCESS
		E16.8	VOLUNTEERING
		E16.9	NATURE
		E16.10	FAMILY
		E16.11	RIOT
		E16.12	ARMY

Fig. 1 The set of 12 pictures given to the respondents, representing various concepts of democracy

Fig. 2 The set of 12 pictures given to the respondent, representing various values in life

Two groups were formed using the five variables through discriminant analysis for which no distinction is made regarding their response to the scenario questions. The post hoc groups (resulting from the analysis) give 57%/43% placement to the ex-ante groups (responders/no responders). The percentage of respondents, according to the level of each variable, is given in Table 2. For example, we observe that 38.95% of men in the sample respond to the scenario questions and 51.45% of women respond. Slight variation is observed in terms of gender (the percentage of women is higher) and information sources (those in group 1061/TV-Radio-Internet have a higher percentage).

The analysis shows (Fig. 3) that the main factor (42.8% importance) is trust in experts, while in second place is the attitude towards vaccination (28.6% importance).

The weights of each factor variable for each participant in the analysis were also calculated and combined with the remaining variables. Table 3 summarizes descriptive statistics for the sample's weight on each factor.

In order to check the relationships for the five factors, we proceeded to analyze through factor analysis (Factor SPSS) the "positive" (marked in yellow in Fig. 3) utilities for each person, and also classify the five factors through cluster analysis.

Two factors emerge, the first with a weight of 33.9% and the second 24.3% (Table 4). These two factors are being formed by the polarization {fear-trust} and {compliance with measures} and the second by the polarization {vaccination-fear for others-trust-compliance with measures}.

Through the cluster (BAVERAGE) the distinction is made into two groups {Decided/concerned about the environment/observes measures}, {Afraid/Trusts experts} as seen in the dendrogram in Fig. 4.

3 Information Sources About COVID-19 Updates and Attitudes Towards the Pandemic

The respondents in their vast majority choose the Internet to get informed about the pandemic (60.3%). The second most popular source of information is the specialists and the traditional mass means like TV and radio while the least preferred option is the newspapers, friends, and family. Moreover, 3,7% of the respondents choose not to get informed at all, as the pandemic is of no interest to them.

Following the methodology of HCA and MCA in two steps, the analysis proceeds to detect specific profiles of behavior. Specifically, seven (7) clusters are detected which can be further merged into three major groups, as shown in Tables 6 and 7.

Clusters 1061 and 1042 are connected to TV-radio, Internet, and specialists; clusters 1060 and 1043 are connected to TV-radio, social media, and specialists with 1051 to be linked to no source specifically. Finally, clusters 1060 and 1043 are connected to friends, family, and the social circle in general.

Analyzing further with AFC we see that four (4) dimensions are derived from the phenomenon, the first dimension opposes selecting or not selecting TV, radio, and

Table 2 Results of the discriminant analysis, and percentage of the respondents for each category of the variables

Occupation/Age d3		Political interest e1		Gender d5		Mobilization e7		Information e4	
Pupils-students	39.71%	Little-none	44.83%	Male	38.95%	Not mobilized	45.00%	1042	44.12%
Employees-unemployed-other	48.07%	Enough	48.92%	Female	51.45%	Leave it to those in charge	46.31%	1043	40.38%
		Very much	46.51%			Mobilized by social networks and media	37.84%	1051	47.06%
						Address those in charge	46.61%	1056	41.03%
						Collective mobilization	53.26%	1057	34.48%
								1060	45.19%
								1061	56.69%

A	FEAR [ABOUT COVID 19]	.193	.241
	NO	-.193	.241
b	THINKING VACCINE	-.318	.241
	DECIDED	.318	.241
C	WORRY ABOUT FAMILY	.047	.241
	NO WORRIES	-.047	.241
D	TRUST SPECIALISTS ε.	.476	.241
	NO	-.476	.241
E	FOLLOW MEASOURS	.079	.241
	DO NOT	-.079	.241
(Constant)		4.500	.241

Importance Values

A	17.368
b	28.559
C	4.208
D	42.793
E	7.073

Averaged Importance Score

Fig. 3 Results of the conjoint analysis for the five factors. On the right, the importance values of each factor are given

Table 3 Descriptive statistics for the weight of each one of the five factors for the participants

	Fear (about Covid-19)	Decided (vaccine)	Worry about family	Trust specialists	Follow measures
mean	0.19	0.32	0.05	0.48	0.08
StDev	1.02	0.71	0.62	1.09	0.8
5%	-1.5	-0.85	-0.75	-2	-1.1
25%	-0.75	0	-0.5	-0.25	-0.5
50%	0.25	0.25	0	0.5	0
75%	1	0.75	0.25	1.5	0.75
95%	1.5	1.5	1	2	1.25

Internet as the main source of information, the second axis is created between the difference of those getting informed from family, friends, and social circle and those who are informed from TV and radio. The third axis depicts the antithesis between being informed from the social media and being informed from the newspapers and the last axis shows the opposition between being informed by the specialists and being informed by the newspapers and the Internet (Table 8).

These coordinates are clustered together again to produce three main profiles of information sources as shown in Table 9.

Proceeding to the incorporation of the conjoint analysis to the previous findings, the attitude towards the five factors is analyzed in relation to the sources of information about the pandemic:

- Fear-individual disease [A].
- Vaccine [YES] [B].
- Fear-social environment [C].
- Trust in experts [D].

Table 4 Results of factor analysis, and the rotated component matrix, showing the correlation of each "conjoint" factor to each one of the two components (factors)

Rotated Component Matrix[a,b]

	Component	
	1	2
Fear	0.859	0.086
Vaccine-decided	-0.116	0.713
Fear –social/family	0.221	0.689
Trust specialists	0.66	0.32
Follow measures	-0.677	0.35

Extraction Method: Principal Component Analysis.
Rotation Method: Varimax with Kaiser Normalization.[a,b]

a. Rotation converged in 3 iterations.

b. Only cases for which conjoint = 1 are used in the analysis phase.

- Compliance with measures [E].

For the analysis, the individual coefficients obtained from the Conjoint Analysis were transformed into array variables through the raw data program [10] and then into the form of a generalized 0–1 table through DIAS program [11].

Table 10 explains how each cluster which is created with reference to the sources of information is connected to the factors from A to E of the conjoint analysis. Cluster 1042 which is informed from the Internet and the specialists is afraid, is positive towards the vaccine, and is also afraid for their social environment. Cluster 1043, informed from TV, radio, and the social media, is not afraid to be sick with Covid-19, is positive towards the vaccine, is however afraid for their social environment, and shows no trust towards experts and complies with the preventive measures. Furthermore, cluster 1056 is generally not afraid, does not comply to the measures and does not trust the experts but is positive to the vaccine, while the close cluster

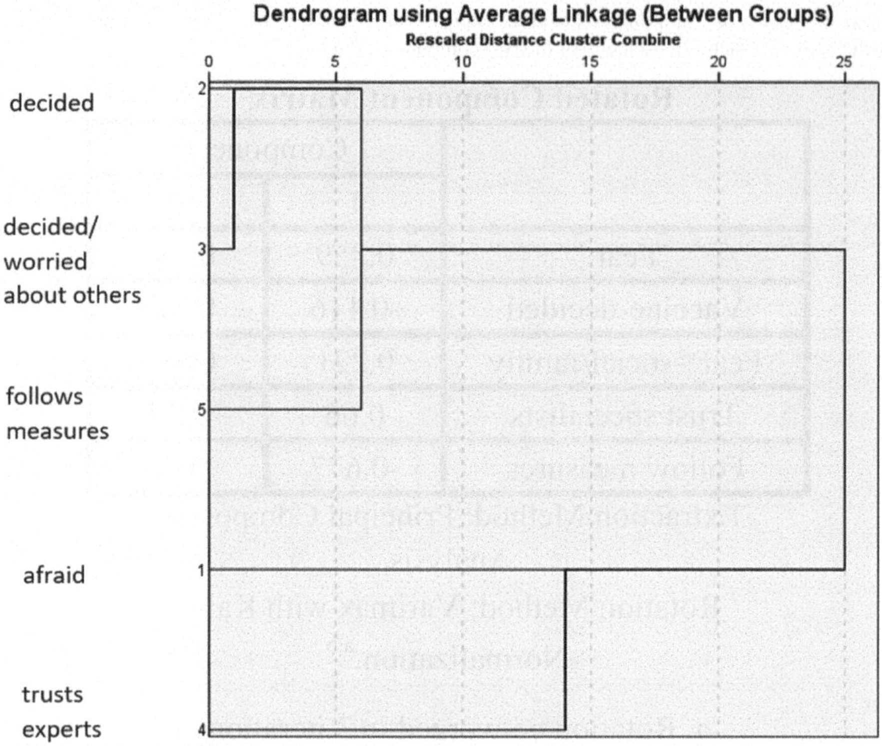

Fig. 4 Dendrogram produced by the hierarchical cluster analysis on the coefficients of the factors produced in the factor analysis

Table 5 Where you are mainly informed from about issues related to the COVID-19 pandemic? Frequencies of responses

TV-Radio	41.4%
Newspapers-press	3.2%
Family	5.4%
Friends-social circle	8.1%
Social media	21.3%
E-Newspaper-Internet	60.3%
Specialists	42.7%
No info about COVID-19/No interest	3.7%

1057 differentiates to the previous one on the vaccine, as these respondents is negative on this. These two clusters meet on their information sources which are family, friends, and social circle.

Similarly, Table 11 presents the coefficient of each group for each one of the five factors of the conjoint analysis. People who prefer to get informed from other

Table 6 Summary of the clusters for information sources about COVID-19

Cluster	N%		
1061	29.40%	tv-radio	e-newspaper -internet
1042	25.50%	e-newspaper -internet	specialists
1060	19.50%	tv-radio	specialists
1043	9.70%	tv-radio	social media
1056	7.30%	friends-social circle	
1057	5.40%	family	friends-social circle
1051	3.20%		

Table 7 Seven (7) clusters if information sources about COVID-19 and their connection to the sources

INT	**1061**	**1042**	**1060**	**1043**	**1056**	**1057**	**1051**
E4.10							
E4.11	X		X	X			
E4.20							
E4.21							
E4.30							
E4.31						X	
E4.40							
E4.41					X	X	
E4.50							
E4.51				X			
E4.60							
E4.61	X	X					
E4.70							
E4.71		X	X				
	157 29,40%	136 25,50%	104 19,50%	52 9,70%	39 7,30%	29 5,40%	17 3,20%
	TV-radio e-newspaper - internet	e-newspaper - internet specialists	TV-radio specialists	TV-radio social media	friends-social circle	family friends-social circle	

Table 8 AFC results, coordinates of the categories on each one of the four axes

IND	#F1	COR	CTR	#F2	COR	CTR	#F3	COR	CTR	#F4	COR	CTR
E4.10	−259	528	95	155	191	70	55	23	11	−89	61	31
E4.11	366	528	133	−221	191	98	−79	23	15	125	61	44
E4.20	−8	12	1	22	103	3	55	632	18	16	54	2
E4.21	232	12	5	−682	103	72	−1684	632	555	−493	54	52
E4.30	−32	112	3	−58	371	15	6	5	1	6	4	1
E4.31	546	112	38	994	371	263	−122	5	5	−111	4	5
E4.40	−16	18	1	−89	577	35	32	75	6	12	11	1
E4.41	178	18	7	1009	577	401	−366	75	66	−143	11	10
E4.50	−156	510	45	11	2	1	−115	274	63	26	15	4
E4.51	573	510	168	−44	2	2	420	274	232	−100	15	14
E4.60	405	589	156	−24	1	2	−7	0	1	−320	366	270
E4.61	−267	589	102	15	1	1	4	0	1	210	366	177
E4.70	277	566	105	75	42	16	−57	23	11	208	320	166
E4.71	−373	566	141	-102	42	21	75	23	15	−281	320	223

Table 9 Three profiles of how respondents choose to get informed on COVID-19 issues and news

22	**24**	**25**
E4.61	E4.31	E4.11
E4.71	E4.41	E4.51
		E4.21
e-newspaper -internet	family	TV-radio
specialists	friends-social circle	newspapers-press
		social media

Table 10 Connection between clusters of information and attitudes towards the pandemic

			A	B	C	D	E
1042	e-newspaper -internet	specialists	YES	YES	YES		YES
1043	tv-radio	social media	NO	YES	YES	NO	YES
1051			YES		NO		
1056	friends-social circle		NO	YES		NO	NO
1057	family	friends-social circle	YES	NO	YES	YES	NO
1060	tv-radio	specialists				YES	YES
1061	tv-radio	e-newspaper -internet					

Table 11 Connection between clusters of information and attitudes towards the pandemic

gr-e4			Afraid (for oneself)	Vaccine (YES)	Afraid for social circle	trusts experts
1042	e-newspaper - internet	specialists	0,23	0,21	0,04	0,52
1043	tv-radio	social media	0,13	0,48	0,08	0,52
1051			0,50	0,34	0,16	1,06
1056	friends-social circle		-0,03	0,44	-0,08	0,08
1057	family	friends-social circle	0,48	0,10	-0,23	0,60
1060	tv-radio	specialists	0,19	0,36	0,16	0,70
1061	tv-radio	e-newspaper - internet	0,17	0,33	0,02	0,32
Total	friends-social circle		0,19	0,32	0,05	0,48

people such as family or friends are generally more negative towards the measures, the vaccine, the experts, and show lower levels of fear towards the pandemic.

4 Emotional Impact During and After the Lockdown Measures

Analyzing feelings during the extended lockdown period as a prevention measure against the COVID-19 pandemic, HCA detects seven (7) clusters, further merged into three which show a specific emotional profile of how the respondents felt during this period. Clusters 1052 and 1053 comprise a 10.3% of the sample who had positive feelings during the lockdown. Cluster 1057 (14%) mentions to feel only anger and the rest of the clusters are spread among the rest of the negative feeling (Tables 12 and 13).

From the joint analysis of the HCA and the conjoint findings, it is shown that positive feeling during the lockdown is negative towards the vaccine (Tables 14 and 15).

The same analysis is applied for the feelings of the respondents in the end of the lockdown measures. Following a similar pattern, 5 clusters are found and a 14% of the respondents now have negative feelings with the end of the lockdown measures, a 17.4% has mixed feelings and the rest of the people feel mostly positive feelings of joy and relief (Tables 16 and 17).

The conjoint coefficients show that respondents with positive feelings don't feel fear for themselves or their environment and in general are positive towards the vaccine the measures and the specialists, while those with negative feelings upon the lifting of the measures for the pandemic are afraid of getting sick but not for their

Table 12 Feelings during the lockdown, seven clusters of emotional behavior

IND	1052	1053	1057	1060	1061	1062
A. Anxiety				X	X	
B. Worry				X	X	X
C. Joy	X					
E. Anger			X	X		
F. Sorrow				X	X	
G. Loneliness					X	
H. Carelessness	X	X				
	3,4%	6,9%	14%	20,2%	19,7%	35,8%

Table 13 Clusters of emotional behavior and frequency of replies

Joy/carelessness	3,4%
Carelessness	6,9%
Anger	14,0%
Anxiety/worry/anger/sorrow	20,2%
Anxiety/worry/sorrow/loneliness	19,7%
Worry	35,8%

social circle and are negative towards vaccination, specialists, and measures (with the exception of group 1059 who are positive about the vaccine) (Tables 18 and 19).

The most interesting point coming out from the analysis is that both emotional profiling (during and after the pandemic) are further connected between them (Table 20). More specifically the respondents who had positive feeling during the pandemic are the same who seem to feel bad about the end of the measures and the return to normal everyday life, while the rest of the respondents are these who initially felt in a negative way and upon the lifting of the measures started to feel well again.

Table 14 Connection between emotions during lockdown and attitudes towards the pandemic

		A	B	C	D	E
1052	joy/carelessness		NO	YES		NO
1053	carelessness	YES	NO	NO	YES	YES
1057	anger			YES	NO	
1060	anxiety/worry/anger/sorrow	NO		NO		YES
1061	anxiety/worry/sorrow/loneliness		NO	NO	YES	NO
1062	worry		YES	NO	YES	

Table 15 Connection between emotions during lockdown and attitudes towards the pandemic

gr-f1		Afraid (for oneself)	Vaccine (YES)	Afraid for social circle	trusts experts	compliance to measures
1052	joy/carelessness	0,21	0,33	0,13	0,33	-0,13
1053	carelessness	0,42	0,21	0,10	0,60	0,10
1057	anger	0,08	0,05	0,16	0,18	0,05
1060	anxiety/worry/anger/sorrow	0,09	0,42	-0,03	0,45	0,13
1061	anxiety/worry/sorrow/loneliness	0,24	0,31	-0,03	0,60	0,11
1062	worry	0,25	0,35	0,08	0,54	0,07
Total		0,19	0,32	0,05	0,48	0,08

5 Demographics, Political Interest, Mobilization, and Attitudes to the Pandemic

For the analysis, two demographic variables were used, gender and occupation status. A significant difference is observed between men and women when it comes to almost all the factors of the conjoint analysis for the attitudes towards the pandemic (Table 21). As a result, women are opposite to almost all factors (fear, vaccine, worry for others, trust in experts) but seem to comply with the measures, while men are positive to everything except from the measures.

In a same manner, a strong difference is found within occupation status, between being a student and being employed or unemployed. The basic group variation here is the age of the respondents as the first are younger in age from the second. Students and therefore younger respondents are negative towards all factors while the older respondents are afraid for themselves and for others and are positive towards the vaccination, the experts, and the measures (Tables 23 and 24).

A significant relationship is found between the degree of political interest and the attitudes towards the pandemic (Table 25). According to Table 26, high political interest is combined with the fear of the pandemic, the positive attitude towards the vaccine, and also the fear of those around us getting sick. On the contrary, those who

Table 16 Feelings after the lockdown, five clusters of emotional behavior

IND	1054	1059	1062	1058	1063
		14,2028			
A. Joy				8,5055	
	2,8559	5,8836			6,6183
B. Excitement			3,0292	36,6645	
				5,2322	5,2322
C. Anxiety	4,086	23,3458	29,226		
				3,3895	3,3895
D. Fear	3,8618	17,1871	24,3423		
	11,4772	18,0871	5,4049		
E. Relief				3,2975	3,3696
	1,8148				
G. Melancholy		80,822			
			2,6669	2,6669	2,6669
H. Inconvenience	65,0081	26,8586			
		4,4789			
I. Impatience				4,1321	
	6,9%	7,1%	17,4%	17,4%	51,1%

Table 17 Clusters of emotional behavior and frequency of replies

Anxiety/fear/inconvenience	6,9%
Anxiety/fear/melancholy/inconvenience	7,1%
Excitement/anxiety/fear	17,4%
Joy/excitement/impatience	17,4%
Relief	51,1%

have little to no political interest at all seem not to be afraid of the pandemic, neither for themselves nor for others, while they are also negative about the vaccine. Those who claim to have enough political interest are negative about vaccines claim not to worry about others but trust the experts.

Regarding political mobilization it is found that those who choose not to mobilize for a problem of theirs or let others do their work are positive in most factors (vaccine,

Table 18 Connection between emotions during lockdown and attitudes towards the pandemic

		A	B	C	D	E
1054	Anxiety/Fear/Inconvenience	Yes	No	No	No	No
1059	Anxiety/Fear/Melancholy/Inconvenience	Yes	Yes		No	No
1062	Excitement/Anxiety/Fear	No	Yes	Yes	Yes	Yes
1058	Joy/Excitement/Impatience		Yes	No		Yes
1063	Relief	No			Yes	

Table 19 Connection between emotions during lockdown and attitudes towards the pandemic

gr-f5		Afraid (for oneself)	Vaccine (YES)	Afraid for social circle	trusts experts	compliance to measures
1054	anxiety/ fear/ inconvenience	0,18	0,07	-0,40	0,58	-0,02
1058	anxiety/ fear/ melancholy/ inconvenience	0,22	0,28	0,03	0,37	0,05
1059	excitement/ anxiety / fear	0,31	0,48	0,27	0,90	0,38
1062	joy/ excitement/ impatience	0,02	0,39	-0,04	0,19	0,12
1063	relief	0,24	0,32	0,11	0,56	0,05
Total		0,19	0,32	0,05	0,48	0,08

Table 20 How feelings during and after the pandemic measures are related

			F5 feelings after				
			1054	1058	1059	1062	1063
			anxiety/ fear/ inconvenience	joy/ excitement/ impatience	anxiety/ fear/ melancholy/ inconvenience	excitement/ anxiety / fear	relief
F1 feeling during	1052	joy/carelessness	38,89%	5,56%	16,67%	22,22%	16,67%
	1053	carelessness	27,03%	5,41%	29,73%	13,51%	24,32%
	1057	anger	4,00%	26,67%	6,67%	6,67%	56,00%
	1060	anxiety/worry/ anger/sorrow	4,63%	22,22%	7,41%	20,37%	45,37%
	1061	anxiety/worry/ sorrow/loneliness	4,76%	21,90%	6,67%	31,43%	35,24%
	1062	worry	3,66%	12,04%	2,09%	12,57%	69,63%

Table 21 Connection between gender and attitudes towards the pandemic

d5c: gender	Afraid (for oneself)	Vaccine (YES)	Afraid for social circle	trusts experts	compliance to measures
Man	0.32	0.46	0.17	0.68	-0.08
Woman	0.14	0.26	0	0.39	0.15
Total	0.19	0.32	0.05	0.48	0.08

Table 22 Connection between gender and attitudes towards the pandemic

	Afraid	Vaccine	Worried for others	Experts	Measures
Man	Yes	Yes	Yes	Yes	No
Woman	To no	To no	To no	To no	To yes

Table 23 Connection between occupation status and attitudes towards the pandemic

d3c: occupation	Afraid (for oneself)	Vaccine (YES)	Afraid for social circle	trusts experts	compliance to measures
Student	0.24	0.1	-0.19	0.49	-0.09
Employed/ unemployed/else	0.19	0.34	0.07	0.47	0.1
Total	0.19	0.32	0.05	0.48	0.08

Table 24 Connection between occupation status and attitudes towards the pandemic

	Afraid	Vaccine	Worried for others	Experts	Measures
Student		No	No	No	No
Employed/unemployed/else	Yes	To yes	To yes	Yes	To yes

Table 25 Connection between political interest and attitudes towards the pandemic

e1c: political interest	N%	Afraid (for oneself)	Vaccine (YES)	Afraid for social circle	trusts experts	compliance to measures
little/none	32.4 %	0.12	0.30	0.05	0.30	0.10
enough	43.3 %	0.28	0.29	0.02	0.61	0.08
very much	24.2 %	0.13	0.38	0.09	0.45	0.06
Total		0.19	0.32	0.05	0.48	0.08

Table 26 Connection between political interest and attitudes towards the pandemic

	Afraid	Vaccine	Worried for others	Experts	Measures
Little/none	No	To no	To no		
Enough		To no	To no	Yes	
Very much	Yes	Yes	Yes	No	

Table 27 Connection between political mobilization and attitudes towards the pandemic

e7c: political mobilization	N%	Afraid (for oneself)	Vaccine (YES)	Afraid for social circle	trusts experts	compliance to measures
I do nothing	3.70%	0.31	0.19	0.31	0.75	-0.11
I let the people in charge do their job	27.90%	0.17	0.34	0.03	0.54	0.2
I am active through social networks (FB, Instagram, etc.)	6.90%	0.02	0.2	0.18	0.36	-0.23
I am addressing the authorities personally	44.20%	0.38	0.34	0	0.51	0.03
I take part with others in demonstrations	17.20%	-0.16	0.29	0.08	0.29	0.16
Total		0.19	0.32	0.05	0.48	0.08

Table 28 Connection between political mobilization and attitudes towards the pandemic

	Afraid	Vaccine	Worried for others	Experts	Measures
I do nothing			Yes		
I let the people in charge do their job		Yes	Yes	Yes	Yes
I am active through social networks (FB, Instagram, etc.)	No	No			
I am addressing the authorities personally	Yes	No	To no	To yes	
I take part with others in demonstrations	No		To no	No	No

worry for others, trust in experts, and compliance to measures). Those who answered that they participate in mobilizations are negative in the majority of the factors with the exception of the vaccine in which they do not show any correlation. Those who mobilize through social media declare that they are not afraid and are vaccine negative while those who are mobilized with their personal involvement, addressing those in charge, declare that they are afraid, trust the experts but are vaccine negative and are not concerned about their social environment (Tables 27 and 28).

6 Information Sources About Politics

Hierarchical cluster analysis of information sources for political issues reveals seven (7) distinct clusters, of which the most populous is the 1061 group to which 36.3% belong, which is informed by television, radio, electronic newspapers, and the Internet. The least popular media seems to be the party itself or the MPs I vote for only 2.2% while also only 8.8% are those who state that they are informed by the printed press (Table 29).

Analyzing these groups together with the factors of conjoint analysis (Tables 29 and 30), it turns out that the large group 1061 who are informed by TV, radio, and the Internet are afraid of the Pandemic, trust the experts and follow the measures. Those who are informed by newspapers and magazines report that they are afraid, they are positive about the vaccine, they are worried about those around them, but they do not trust experts and they are not positive about the measures. The exact same behavioral profile is maintained by those who choose to be informed only by television and radio.

Table 29 Connection between information source for politics and attitudes towards the pandemic

gr-e3	N%		Afraid (for oneself)	Vaccine (YES)	Afraid for social circle	trusts experts	compliance to measures
1041	20,8%	e-newspaper -internet	0.02	0.15	0.01	0.26	0.3
1051	2,2%	party-MP's I vote for	0.18	0.14	0.18	0.21	0.14
1056	11,6%	tv-radio	0.27	0.41	0.21	0.83	0
1058	9,2%	friends-social circle	0.1	0.46	-0.08	0.29	-0.06
1059	8,8%	newspapers-press	0.45	0.47	0.17	0.52	-0.1
1060	11,0%	friends-social circle	0.08	0.35	-0.06	0.19	0.08
1061	36,3%	tv-radio / e-newspaper -internet	0.25	0.3	0.05	0.62	0.07
Total			0.19	0.32	0.05	0.48	0.08

Table 30 Connection between information source for politics and attitudes towards the pandemic

	Afraid	Vaccine	Worried for others	Experts	Measures
E-newspaper-Internet	To no	To yes	To no	Yes	Yes
Party-MP's I vote for	To no	No	Yes	To no	Yes
Tv-radio	Yes	To yes	To yes	To no	To no
Friends-social circle	To yes		Yes	Yes	No
Newspapers-press	Yes	To yes	To yes	No	To no
Friends-social circle	No		No	No	Yes
TV-radio/ E-newspaper-Internet	To yes			Yes	To yes

Also of interest is the group informed by friends and their general social circle who state that they are afraid both for themselves and for others and that they trust the experts but do not follow the measures. Exactly the opposite behavioral profile is shown by those who are informed by friends and family, i.e., they are not afraid either for themselves or for those around them, they do not trust the experts but follow the measures.

7 Closeness and Trust in Institutions

The research proceeds with the analysis of the variable related to how close one feels to various democratic institutions (political, social, and ecumenical). The application of HCA results in seven (7) clusters which are described in Table 31. Among these groups there are those who declare that they feel close to all institutions (13.7%) and those who declare that they do not feel close to any of the institutions (18.4%). A total of 24% state that they feel close to the European Union, the scientific community, and justice. In addition the groups 1048 (10.7%) and 1054 (8.6%) also state that they feel closer to similar political institutions. 12.5% of the respondents feel close to ecumenical institutions such as the church, the army, and the police along with the municipal and regional authorities, while 12.2% report that they feel close to parliament and parties to trade unions and businesses.

From the analysis of these clusters with the factors of the conjoint analysis, an interesting finding derives for those who feel closer to ecumenical institutions, i.e.,

Table 31 Connection between closeness to institutions and attitudes towards the pandemic

gr-e2	N%		Afraid (for oneself)	Vaccine (YES)	Afraid for social circle	trusts experts	compliance to measures
1048	10,7%	municipality, regional authority	0.13	0.39	-0.04	0.6	-0.1
1054	8,6%	parliament-parties, media, European union, scientific community	0.3	0.27	0.23	0.77	0.02
1056	18,4%	none	0.02	0.38	0.13	0.35	0.18
1057	13,7%	all	0.52	0.52	0.06	0.76	-0.28
1058	12,5%	church, municipality, regional authority, army, police	-0.15	0.15	-0.17	0.17	0.26
1060	24,0%	European union, scientific community, justice	0.22	0.24	0.07	0.35	0.13
1061	12,2%	parliament-parties, media, trade unions, business	0.29	0.27	0.03	0.54	0.33
Total			0.19	0.32	0.05	0.48	0.08

Table 32 Connection between closeness to institutions and attitudes towards the pandemic

	Afraid	Vaccine	Worried for others	Experts	Measures
Municipality, regional authority		Yes	No	Yes	No
Parliament-parties, media, European union, scientific community	To yes	Yes		Yes	Yes
None	To no	Yes	No		Yes
All	Yes	Yes	To no	Yes	To no
Church, municipality, regional authority, army, police	To no	No	No	No	To yes
European union, scientific community, justice		No	To yes		To yes
Parliament-parties, media, trade unions, business	Yes	To no	Yes	No	Yes

church army and police, as they seem not to be afraid or worry about others, are negative towards the vaccine, don't trust the experts but do comply with the measures. Respondents who feel closer to political institution such as the parliament or the European union but also feel close to the scientific community say that they are afraid, they are positive about the vaccine, trust the experts, and comply with the measures.

Between the two groups who say all or none when it comes to closeness to the institutions, they seem to differentiate on the afraid factor and the compliance to the measures, where the "all" cluster seems to be afraid and to be against the measures and the "none" group to agree with the measures but does not feel afraid (Table 32).

8 Perceptions of Democracy and Personal Values

To analyze the next two variables, perceptions of democracy, and personal values, we use the answers of the respondents on the part of the symbolic representation including pictures which represent various concepts of these two thematic.

Regarding perception of democracy, HCA detects eight (8) clusters, that is today eight democratic profiles or "selves" (Table 33). Three of these profiles 1050, 1051, and 1053 perceive democracy in an antagonistic way including concepts of struggle, protest, and uprising, followed by elements of ancient Greece and direct democracy. Cluster 1053 consists of 10.9% of the sample, connecting democracy to church and ancient Greece. Clusters 1055 and 1056 conceptualize democracy in its modern representative form linking it to parliament, councils, ancient Greece and volunteering and e-democracy. A last cluster of 2.8% involves corruption and riot in its concept.

Table 33 Connection between perceptions on democracy and attitudes towards the pandemic

gr-e15	N%		Afraid (for oneself)	Vaccine (YES)	Afraid for social circle	trusts experts	compliance to measures
1050	10,5%	protest, ancient Greece, direct democracy	0.33	0.31	0.24	0.33	0.21
1051	10,7%	protest, e-democracy, uprising	-0.21	0.3	-0.16	0.29	0.23
1053	16,1%	ancient Greece, church,	0.22	0.37	0.06	0.42	0.07
1054	10,9%	direct democracy, uprising, struggle	0.32	0.17	0.33	0.53	0.32
1055	17,3%	parliament, council,	0.09	0.5	-0.08	0.49	0.03
1056	14,1%	ancient Greece, e-democracy, council, volunteering	0.3	0.25	-0.02	0.57	-0.11
1057	17,6%	ancient Greece, e-democracy, volunteering	0.34	0.23	0.12	0.67	0.03
1062	2,8%	ancient Greece, e-democracy, riot, money-corruption	0.05	0.2	0.1	-0.35	-0.35
(blank)			1.5	0.25	0.25	1.5	-0.25
Total			0.19	0.32	0.05	0.48	0.08

The attitudes of the groups in relation to this variable and their behavior towards the pandemic It does not seem to be characterized by any specific pattern. So, for example, the group linking democracy with ancient Greece and the church or those who think of democracy as ancient Greece, e-democracy, and volunteering appear to be positive on all conjoint factors, while those who perceive democracy in more contemporary terms of parliamentary democracy or councils are negative on all factors but remain positive on the vaccine. Specifically, negative to vaccine and experts appear to be those who associate democracy with protest or uprising and its direct form. A last significant finding is that those who connect democracy with money, corruption, and riot remain against the experts and the measures but are ok with the vaccine and are worried for others but not themselves (Table 34).

Following the same methodology for the analysis of moral values (Table 35), HCA gives seven (7) clusters of moral "selves", with the majority of the respondents to belong in cluster 1060 (44,2%) which is linked to Christ, volunteerism, nature, and family. The Greek traditional ethnocentric concept of "Christ, family, army" is joined by 10.1% of the respondents, while 14,8% sets as only value the love affair and 12,7% has entertainment, career, and money as basic moral values. Meditation and nature are followed by 9.4%, success as a sole oral value is connected to the 7.5% of the sample and the last moral concept includes money, success, volunteering but also riot, followed only by 1.3%

Table 34 Connection between perceptions on democracy and attitudes towards the pandemic

	Afraid	Vaccine	Worried for others	Experts	Measures
Protest, ancient Greece, direct democracy	Yes	No	Yes	No	Yes
Protest, e-democracy, uprising	No	Yes	No	To yes	No
Ancient Greece, church	Yes	Yes	Yes	Yes	Yes
Direct democracy, uprising, struggle	Yes	No	Yes	No	Yes
Parliament, council	To no	Yes	To no	To no	No
Ancient Greece, e-democracy, council, volunteering	No	To no	No	Yes	To no
Ancient Greece, e-democracy, volunteering	To yes	To yes	To yes	Yes	To yes
Ancient Greece, e-democracy, riot, money-corruption	To no	To yes	To yes	No	No

Table 35 Connection between personal moral values and attitudes towards the pandemic

gr-e16	N%		Afraid (for oneself)	Vaccine (YES)	Afraid for social circle	trusts experts	compliance to measures
1037	1,3%	money, success, volunteering, riot	0.25	0.42	-0.17	-0.33	-0.25
1048	7,5%	success	0.45	0.21	-0.11	0.98	-0.27
1050	10,1%	Christ, family, army	0.08	0.36	-0.06	0.34	0.16
1054	9,4%	meditation, nature	0.39	0.14	0.27	0.24	0
1056	14,8%	love affair	0.24	0.31	0.12	0.55	0.31
1060	44,2%	Christ, volunteering, nature, family	0.14	0.32	0.04	0.44	0.06
1062	12,7%	entertainment, carrier, money	0.15	0.5	-0.02	0.71	0.02
Total			0.19	0.32	0.05	0.48	0.08

The largest cluster, with major values those of Christ, volunteerism, nature, and family, is afraid of the pandemic, is pro-vaccine, worries over others, trusts the experts, and complies with the measures. The group connected to the ethnocentric values of "Christ, family, army" are not afraid but worry for others, are positive towards the vaccine but do not trust the experts and do not follow the measures. In general, all clusters are pro-vaccine except from the group connected to meditation,

Table 36 Connection between personal moral values and attitudes towards the pandemic

	Afraid	Vaccine	Worried for others	Experts	Measures
Money, success, volunteering, riot	Yes	Yes	To no	No	No
Success	Yes	Yes	No	Yes	No
Christ, family, army	No	Yes	Yes	To no	To no
Meditation, nature,	Yes	No	To yes	No	To no
Love affair	No	To no	To yes		Yes
Christ, volunteering, nature, family	To yes	Yes	To yes	To yes	To yes
Entertainment, career, money	No	Yes	No	Yes	

nature and the group connected to love affair (Table 36). When it comes to the experts, clusters connected to the ethnocentric value model; the cluster with the meditation and nature; and the cluster connected to money, success, volunteering, and riot are those who seem to be against. Finally, most of the clusters seem to be against the measures except the "love affair", "Christ, volunteering, nature, family" clusters who are positive to the measures and "entertainment, career, money" who are not connected with this factor at all.

9 Discussion

The research carried out and presented in this text aimed not only to investigate attitudes during the pandemic but also to connect them with various characteristics concerning the feelings experienced by citizens during the measures and after their removal, as well as with political characteristics and personal values.

The originality of the research lies in its methodology, which incorporates conjoint analysis in order to identify, through the analysis of scenarios, the factors that shape the behavior of citizens towards the Pandemic and highlights hidden attitudes such as being for or against vaccination, the trust towards the experts the compliance to the measures, but also the fear of someone getting sick and the concern for others in one's environment. In addition, the research methodology also includes symbolic representation where, with the use of images, the participants are investigated in terms of their perceptions of democracy as well as their personal framework of values.

The conjoint analysis identifies as the most important factors influencing the way one is positioned towards the pandemic, the attitude towards vaccination, and the trust towards experts.

The media preferred by citizens to be informed about the pandemic is mainly television and radio, followed by social media, the Internet, and the specialists. The analysis shows that those who prefer to be informed by their family and friends or their social environment are the ones who are more negative towards the vaccine but also do not trust the experts.

Regarding the feelings during the pandemic and after the lifting of the measures, two main groups are identified: the great majority declares that during the measures had negative feelings which became positive after the end of the measures, while there is also an approximately 15% who declares that during the measures they had positive feelings and with the return to normalcy they faced negative feelings. The first group fears less than the second and agrees in general with the vaccine, the measures, and the specialists. The second group is more afraid of getting sick and is negative towards vaccine, specialists, and measures.

A very interesting connection is also observed in the demographic characteristics of the sample with its attitude towards the pandemic. More specifically, the analysis shows how women are more compliant to the measures more, in general, they are not afraid, but they are negative towards the vaccine and specialists, while men report that they are afraid both for themselves and for those around them, they are positive about the vaccine and specialists, but they don't apply the measures. Therefore, we observe two diametrically opposed behaviors for the two genders. The exact same contrast is observed in relation to the variable of employment which essentially indicates the age of the respondent. So the youngest, i.e., the students, are negative towards the vaccine to the specialists and the measures and report that they are not worried about their social environment. This lack of trust to authorities is observed for the young also in previous research [7]; however, in this study, this behavior is analyzed further in connection to their behavior to the specialists. Meanwhile, those of older age, i.e., the workers, the unemployed, etc. state that they are afraid of themselves and for those around them and are positive both towards the vaccine and towards the experts and the measures.

The research then proceeds by analyzing the political characteristics of the participants and linking them to their attitudes towards the pandemic. In relation to political interest, it is observed how much they fill in that they are very interested in politics, they are afraid of getting sick themselves or those around them are positive to the vaccine but not to the specialists while as political interest decreases there is no fear of the pandemic and citizens are more negative regarding the vaccine. Political mobilization also seems to be linked to the attitude towards the pandemic with those who have low political mobilization, i.e., do nothing about a problem or let those in charge do their job being positive towards both vaccination and to experts and measures and to worry about those around them. Mobilization through social media appears to be more associated with negative attitudes towards the vaccine and fear of disease. Collective mobilization through protests seems to be negatively associated with all factors, while mobilization by individual initiative is associated with the fear of disease and trust in experts, but even in this case they remain vaccine negative and do not worry about those around them. Here the main distinction is mobilization versus non-mobilization where the former are associated with a positive attitude towards the factors, while the latter have a negative attitude towards almost all the factors with the exception of personal mobilization who state that they are afraid of the disease.

Preferred source of information for politics seems to have a more incoherent pattern when it comes to attitudes towards the pandemic. What stands out in this analysis is the group of those who choose to be informed by friends and their social circle as they are the most negative in most factors, they are not afraid for themselves or their social environment and do not trust experts. In addition, those who are informed by direct party sources are negative towards the vaccine and the experts, while those being negative towards measures and experts are those who are informed by traditional media such as television, radio, and also the printed press, i.e., newspapers and magazines.

The findings from the analysis of proximity to various democratic institutions are also of great interest. HCA combined with AFC highlights three main axes which reflect the respondents' feelings of closeness and trust to the institutions: a group that trusts in modern representative democratic institutions such as the Parliament, the municipality, the European Union, another group includes those who feel closer to the ecumenical institutions, i.e., in the church, the army, the police, and finally there is a group which declares that feels not only closer to the parliament and the parties but also to the unions and the businesses which expresses the more competitive working class. Moreover, two groups also are distinct: those who trust all institutions and those who trust none [2] try to interpret trust in democracy and institutions only in the form of government, while in this study institutions are analyzed in multiple levels such as political, ecumenical, and social. As a result, trust is not only measured single dimensionally but in a multidimensional way which reveals groups of people who show a specific attitude towards political institutions in general, which subsequently affects their attitude towards the pandemic. As a result, the group that feels closer to the ecumenical institutions is the one with the most negative attitude towards the factors. These respondents who feel closer to the church, the army, and the police state to be negative towards the vaccine and the experts have no fear of getting sick and are not worried for the others, but they do comply with the measures.

The last part of the analysis is focused on the symbolic representation of the democratic and the moral self. Regarding the perceptions on democracy, the group that understands democracy as ancient Greece and church is positive on all five factors. What stands out from the analysis is that those who connect democracy with protest and uprising and direct democracy are negative towards the vaccine and the experts and those who perceive democracy in its representative form are negative on all factors except the vaccine which they agree upon. When it comes to the moral framework of the respondents, Christ, volunteerism, family, and nature are the moral concept which is positive on all five factors, feeling afraid of getting sick, as well as worrying for others, they trust the experts, comply with the measures, and agree with the vaccination. Again, we see that riot as a value, followed by money, success, and volunteering in the same moral framework is negative to the measures and trusts not the experts, but also states not to be worried about others. An interesting moral discourse is the traditional Greek ethnocentric triptych "nation, religion family", supported by a 10% of the sample, who do not feel afraid of getting sick, do not trust the experts, and do not comply with the measures are pro-vaccine though. Against the vaccine are also those who prefer more naturalistic values connected to personal "well-being" such as nature, love, or meditation.

The extensive analysis manages to highlight the multilateral behavior of Greek citizens during the pandemic and to investigate any possible connection to characteristics related to their political behavior and attitudes, highlighting also the effect of the pandemic on their emotions. This has been managed with conjoint analysis which here was combined with the hierarchical cluster analysis and the factorial correspondence analysis, which enabled the multilevel and joint analysis of complicated variables, shedding lights to a complex phenomenon.

This has been a first attempt to investigate and analyze the phenomenon for the case of Greece, but nevertheless it can be the starting point for a series of analyses, using the specific methodology, in other countries as well with the possibility of a comparative approach between the different case studies.

References

1. Benzécri, J. P. (1973). *L'analyse des données* (Vol. 2, p. 1). Dunod.
2. Bol, D., Giani, M., Blais, A., & Loewen, P. J. (2021). The effect of COVID-19 lockdowns on political support: Some good news for democracy? *European Journal of Political Research, 60*(2), 497–505.
3. Chadjipadelis, T., & Panagiotidou, G. (2022). Semantic map: Bringing together groups and discourses. In *Data Clustering*. IntechOpen.
4. Greenacre, M. (1984). *Theory and applications of correspondence analysis*. Academic.
5. Greenacre, M. (2007). *Correspondence analysis in practice*. Chapman and Hall/CRC Press.
6. Hauser, J. R., & Rao, V. R. (2004). Conjoint analysis, related modeling, and applications. In *Marketing research and modeling: Progress and prospects* (pp. 141–168). Springer.
7. Kanellopoulou, A., Koskeridis, F., Markozannes, G., Bouras, E., Soutziou, C., & Chaliasos, K., et al. (2021). Awareness, knowledge and trust in the Greek authorities towards COVID-19 pandemic: results from the Epirus health study cohort. *BMC Public Health*, 21(1), 1–14.
8. Karapistolis, D. (2010). Software Method of Data Analysis MAD. *See further at* http://www.pylimad.gr
9. Marangudakis, M., & Chadjipadelis, T. (2019). *The greek crisis and its cultural origins*. Palgrave Macmillan.
10. Moschidis, O., Markos, A., & Chadjipadelis, T. (2022). Hierarchical clustering of mixed-type data based on barycentric coding. *Behaviormetrika*, 1–25.

11. Moschidis, O., & Chadjipadelis, T. (2017). A method for transforming ordinal variables. In *Data science: Innovative developments in data analysis and clustering* (pp.285–294). Springer International Publishing.
12. Taylor, C. (1989). *Sources of the self*. Harvard University Press.

The MELODIC Family for Simultaneous Binary Logistic Regression in a Reduced Space

Mark de Rooij and Patrick J. F. Groenen

Abstract Logistic regression is a commonly used method for binary classification. Researchers often have more than a single binary response variable and simultaneous analysis is beneficial because it provides insight into the dependencies among response variables as well as between the predictor variables and the responses. Moreover, in such a simultaneous analysis the equations can lend each other strength, which might increase predictive accuracy. In this paper, we propose the MELODIC family for simultaneous binary logistic regression modeling. In this family, the regression models are defined in a Euclidean space of reduced dimension, based on a distance rule. The model may be interpreted in terms of logistic regression coefficients or in terms of a biplot. We discuss a fast iterative majorization (or MM) algorithm for parameter estimation. Two applications are shown in detail: one relating personality characteristics to drug consumption profiles and one relating personality characteristics to depressive and anxiety disorders. We present a thorough comparison of our MELODIC family with alternative approaches for multivariate binary data.

1 Introduction

Logistic regression [1, 6, 14] is one of the most commonly used tools for binary classification. Although the logistic function has been known since the early 19th century, the logistic regression model was developed in the second half of the 20th century [15]. Adaptions of logistic regression models have been developed to make it more flexible, through basis expansion, or less flexible, by means of regularization. For an overview, see Friedman et al. [27].

M. de Rooij (✉)
Methodology and Statistics department, Leiden University, Leiden, The Netherlands
e-mail: rooijm@fsw.leidenuniv.nl

P. J. F. Groenen
Econometric Institute, Erasmus University, Rotterdam, The Netherlands
e-mail: groenen@ese.eur.nl

© The Author(s), under exclusive license to Springer Nature Singapore Pte Ltd. 2023

A. Okada et al. (eds.), *Facets of Behaviormetrics*, Behaviormetrics: Quantitative Approaches to Human Behavior 4, https://doi.org/10.1007/978-981-99-2240-6_4

Researchers regularly have more than a single response variable. That is, there are applications where several response variables can be predicted from a common set of predictor variables [8]. Examples include:

- The analysis of depressive and anxiety disorders. Mental disorders are highly prevalent in modern western societies and a high degree of comorbidity can often be observed among these disorders. In the Netherlands Study for Depression and Anxiety [51] data were collected on a large number of subjects about their personality and about mental disorders [55].
- The analysis of drug consumption profiles. Fehrman et al. [26] are interested in subjects' drug consumption profiles and how these relate to personality characteristics such as sensation seeking and impulsivity. They collected data about the consumption of 18 different drugs.
- In clinical trials, the effect of treatments is established where the outcome can be dichotomous, cured or not cured. Treatments, however, come with side effects and these can be coded as present or absent. It is important to study treatment and side effects together in order to obtain the whole picture. For an empirical example, see [49].
- Psychosocial problems frequently occur in young adults. To screen for these problems in community settings, for example, during large-scale general health checkups, the Strengths and Difficulties Questionnaire (SDQ) can be used as it is a relatively short instrument. The SDQ has two parts: a self-report and a parent report. To be useful as a screening tool it should have good validity properties, that is, it should be able to predict certain psychosocial problems. Vugteveen et al. [67] investigated the validity of the SDQ with respect to four diagnoses.

With multiple binary outcomes, it is possible to fit a logistic regression model separately for each outcome, but it is often wise to build a single multivariate model. In such a multivariate model, the dependencies between the various outcomes can be better understood and strength can be borrowed between the different outcomes. Second, such a multivariate model is more parsimonious in the sense that less parameters have to be estimated. Furthermore, estimated regression weights may be better in terms of mean squared error. Stein et al. [56], for example, showed that simple averages of a multivariate normal distribution are inferior to shrunken averages in terms of mean squared error, where the shrinkage tends toward the average of averages. Breiman and Friedman [8] show that shrinkage of coefficients of several multiple regression models toward each other is beneficial in terms of predictive accuracy. In a similar vein, building several logistic regressions in a reduced space might provide better estimates of the regression coefficients in terms of mean squared error.

There are basically two broad ways of analyzing multivariate data and performing dimension reduction: The first is based on inner products from which principal component analysis [41, 45, 50] and reduced rank regression [44, 60] are derived; the second is based on distances which have led to multidimensional scaling [32, 37, 62, 63] and multidimensional unfolding [11, 12, 38, 52]. This distance framework is conceptually easier than the inner product framework and leads to a more straightforward interpretation [22]. Distances, especially Euclidean and Manhattan, are all around us and can already be understood by very young children.

In multidimensional unfolding, we generally have a dissimilarity matrix between two sets of objects. The goal is to find a low-dimensional mapping including points for the row objects and the column objects such that the distances between the points of the two sets are as close as possible (often in the "least squares" sense) to the observed dissimilarities. We will develop a family of models based on similar ideas.

In this paper, we will develop a family of logistic models within a distance framework. We call it the MELODIC family, written out, the MultivariatE LOgistic DIstance to Categories family. More specifically, we will develop a framework of models in which both participants and the categories of the different response variables have a position in low-dimensional Euclidean space. The distances between the position of a participant and the positions of the two categories of a single response variable determine the probabilities for these two response options. The position of a participant will be parameterized as a linear combination of the predictor variables.

The family extends the recently proposed multivariate logistic distance models [69] which built on earlier logistic distance models [20, 57, 59] and can be considered as examples in the "Gifi goes logistic" framework as laid out by De Leeuw [17] and Evans [24].

In the next section, we will develop the general model and two constrained variants. We will discuss properties of the models and provide interpretational rules. Two types of these rules can be distinguished: the numerical and the graphical. These two modes of interpretation for a single model are beneficial because there are those people who say that "a graph is worth a thousand words", the so-called *graph people* [28], while others (the *table people*) firmly disagree [29]. In Sect. 3, we develop an Iterative Majorization or MM algorithm [19, 33, 39, 43] for estimating the parameters of our models by minimizing a deviance function. Section 4 describes two illustrative applications. In Sect. 5, we discuss related statistical models and provide some comparisons. We conclude, in Sect. 6, with a general discussion of our developments and some possibilities for further investigation.

2 MELODIC Family

2.1 Data and Notation

We consider a system with P explanatory, predictor, or independent variables X and R outcome, response, or dependent binary variables Y. That is, we have a sample of observations $\{\mathbf{x}_i, \mathbf{y}_i\}_1^n$ with $\mathbf{x}_i \in \mathbb{R}^P$ and $\mathbf{y}_i \in \{0, 1\}^R$. The response variables will be recoded into indicator vectors $\mathbf{g}_{ir}$ of length two, where the first element equals 1 if $y_{ir} = 0$ and the second element equals 1 if $y_{ir} = 1$.

We will use the following notation.

- $i = 1, \ldots, n$ for individuals (participants, subjects, objects).

- $p = 1, \ldots, P$ for predictor variables (explanatory or independent variables).
- $r = 1, \ldots, R$ for response variables (outcome or dependent variables).
- $m = 1, \ldots, M$ an indicator for the dimensions.
- There is a set of predictor variables $X = \{X_p\}_{p=1}^{P}$. The observed values of the predictor variables are collected in the $n \times P$ matrix $\mathbf{X}$ with elements x_{ip}. We assume, without loss of generality, that the predictor variables are centered, that is $\mathbf{1}^\top \mathbf{X} = \mathbf{0}$.
- There is a set of response variables $Y = \{Y_r\}_{r=1}^{R}$. Observed values of the response variables are collected in the $n \times R$ matrix $\mathbf{Y}$. The matrix has elements $y_{ir} \in \{0, 1\}$. We will code the responses in a super indicator matrix $\mathbf{G}$ having $C = 2R$ categories, that is,
$$\mathbf{G} = [\mathbf{G}_1 | \mathbf{G}_2 | \ldots | \mathbf{G}_R].$$
- $\mathbf{B}$ represents a $P \times M$ matrix with regression weights for the predictor variables.
- $\mathbf{u}_i$ is an M vector with coordinates for person i in M-dimensional Euclidean space. These coordinates will be collected in the $n \times M$ matrix $\mathbf{U}$ with elements u_{im}.
- $\mathbf{V}_r$ is a $2 \times M$ matrix having the coordinates of category 0 (i.e., $\mathbf{v}_{r0}$) in the first row and in the second row the coordinates of category 1 (i.e., $\mathbf{v}_{r1}$), both for response variable r. These matrices will be collected in the $2R \times M$ matrix $\mathbf{V} = \left[\mathbf{V}_1^\top, \ldots, \mathbf{V}_R^\top\right]^\top$ with elements v_{rcm}.
- The observations are $\{\mathbf{x}_i, \mathbf{y}_i\}_1^n$.
- We define a block diagonal matrix $\mathbf{J}$ with 2×2 diagonal blocks $\mathbf{I}_2 - \frac{1}{2}\mathbf{1}\mathbf{1}^\top$.
- We use tildes for current estimates in the iterative process, that is, $\widetilde{\mathbf{B}}$ represents the matrix with estimates in a given cycle of the algorithm.
- Diag() denotes the operator that takes the diagonal values of a matrix and places them in a vector.

2.2 General Model

We define the conditional probability that person i is in class c ($c = \{0, 1\}$) of response variable r, $\pi_{rc}(\mathbf{x}_i) = P(Y_{ir} = c|\mathbf{x}_i)$ as

$$\pi_{rc}(\mathbf{x}_i) = \frac{\exp(-\delta(\mathbf{u}_i, \mathbf{v}_{rc}))}{\exp(-\delta(\mathbf{u}_i, \mathbf{v}_{r0})) + \exp(-\delta(\mathbf{u}_i, \mathbf{v}_{r1}))}, \tag{1}$$

where $\delta(\cdot, \cdot)$ denotes half the squared Euclidean distance

$$\delta(\mathbf{u}_i, \mathbf{v}_{rc}) = \frac{1}{2}\sum_{m=1}^{M}(u_{im} - v_{rcm})^2 = \frac{1}{2}\sum_{m=1}^{M}\left(u_{im}^2 + v_{rcm}^2 - 2u_{im}v_{rcm}\right), \tag{2}$$

in M-dimensional Euclidean space. The dimensionality M has to be chosen by the researcher with possible values being between 1 and $\min(P, R)$. The coordinates of the subjects ($\mathbf{u}_i$) are assumed to be a linear combination of the predictor variables, that is, $\mathbf{u}_i = \mathbf{x}_i^\top \mathbf{B}$, where $\mathbf{B}$ is a $P \times M$ matrix with regression weights. The coordinates of category c of response variable r on dimension m are denoted by v_{rcm} and collected in the M-vector $\mathbf{v}_{rc}$.

Every subject i is thus represented in an M-dimensional Euclidean space. Moreover, this subject has a distance to a point representing category 1 of response variable r and to a point representing category 0 of response variable r. These two distances determine the probability for the subject to answer with either of these categories; the smaller the distance, the larger the probability. In other words, a subject is most likely to be in the closest class.

The log odds in favor of the 1 category and against category 0 for response variable r given the subject's position are given by

$$\log \frac{\pi_{r1}(\mathbf{x}_i)}{1 - \pi_{r1}(\mathbf{x}_i)} = \log \frac{\pi_{r1}(\mathbf{x}_i)}{\pi_{r0}(\mathbf{x}_i)} = \delta(\mathbf{u}_i, \mathbf{v}_{r0}) - \delta(\mathbf{u}_i, \mathbf{v}_{r1}), \tag{3}$$

a simple difference of squared Euclidean distances. This log odds can be further worked out as

$$\log \frac{\pi_{r1}(\mathbf{x}_i)}{\pi_{r0}(\mathbf{x}_i)} = \sum_{m=1}^{M} \left[\frac{1}{2}(v_{r0m}^2 - v_{r1m}^2) + \mathbf{x}_i^\top \mathbf{b}_m (v_{r1m} - v_{r0m}) \right], \tag{4}$$

where we see that the effect of predictor variable p on response variable r is determined by the regression coefficients b_{pm} and the distance between the two categories. In general, the further apart the two categories are, the better they are discriminated by the predictor variables. If the two categories fall in the same position in the Euclidean space, they are *indistinguishable* [2] based on this set of predictor variables.

Let us define $a_r^* = \frac{1}{2}\sum_{m=1}^{M}(v_{r0m}^2 - v_{r1m}^2)$ and $\mathbf{b}_r^* = \sum_{m=1}^{M} \mathbf{b}_m (v_{r1m} - v_{r0m})$. Then the log odds can be written as

$$\log \frac{\pi_{r1}(\mathbf{x}_i)}{\pi_{r0}(\mathbf{x}_i)} = a_r^* + \mathbf{x}_i^\top \mathbf{b}_r^*, \tag{5}$$

showing that the model can be interpreted as standard binary logistic regression models. We call the b^* the model *implied coefficients*.

Equation 5 also shows that our model is equivalent to a logistic reduced rank regression model. The matrix of regression coefficients with columns $\mathbf{b}_r^*$ has a rank M constraint and can be decomposed in a matrix with $\mathbf{b}_m$ and a matrix containing the differences $v_{r1m} - v_{r0m}$. The logistic reduced rank regression is a special case of reduced rank vector generalized linear models as has been proposed by Yee and Hastie [70].

2.3 Constrained Models

In the general model described above, the categories of the response variables lie freely somewhere in the M-dimensional space. Sometimes, however, researchers already have an idea about the underlying structure of the response variables. In the literature about depressive and anxiety disorders, for example, one theory says that fear and distress are its underlying dimensions, where each dimension comprises a subset of the disorders. In terms of our models, this means that the categories of the response variables pertaining to the distress dimension lie on a single dimension (i.e., the coordinates of the categories of these responses for the other dimensions equal zero). Similarly, for the categories of the response variables pertaining to the fear dimension, the coordinates on the distress dimension all equal zero.

If a specific response variable pertains to, say, dimension 1, the class coordinates on all other dimensions are set to zero, that is $v_{r1m} = v_{r0m} = 0, \forall m \neq 1$. Such a structure simplifies the model and its interpretation, because in the log odds definition (see Eq. 4) the last term becomes zero for several dimensions and only the regression weights of the dimension to which the response variable pertains are important for the discrimination of the categories of that response variable.

One further constraint is to let all response variables have the same discriminatory ability. In that case, $(v_{r1m} - v_{r0m}) = 1$ for the dimensions to which response variable r pertains. For this constrained model, Worku and De Rooij [69] showed that the parameters can be estimated using standard software for logistic regression by using a structured design matrix for the predictors. This paper describes them as members of a larger family of models, the MELODIC family.

2.4 Graphical Representation

When the dimensionality equals two ($M = 2$), the model can be easily represented graphically. This representation shows (1) the categories of the response variables as points, (2) a decision line for every response variable designating the predicted class at a specific point, (3) variable axes for the predictor variables, and 4) the subjects' positions as points. Many aspects of the interpretation of these graphical representations follow the theory of biplots as discussed in Gower and Hand [30] and Gower et al. [31].

Let us first look at a graphical representation of a single response variable and a set of subjects, of which three of them are highlighted. Figure 1 gives such a graph where A0 and A1 are the two categories of a response variables named A, and i, j, and k present three subjects. The line halfway between classes A0 and A1 represents the decision line, in other words, the line represents the points for which the odds are even. The log odds that subject i chooses A0 instead of A1 are clearly in favor of class A1, because that is the closest class.

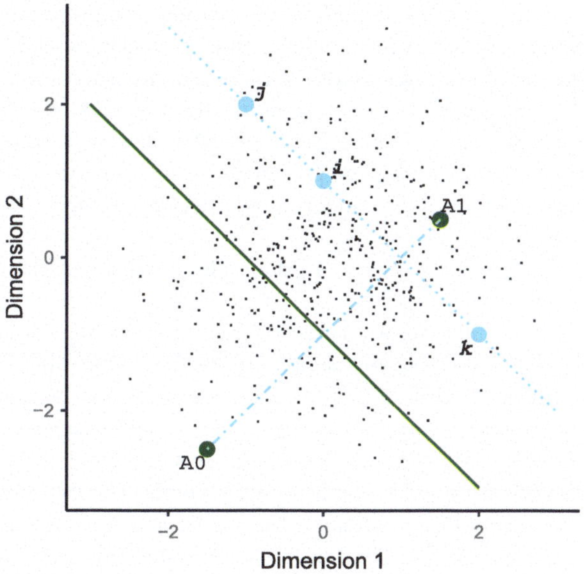

Fig. 1 Graphical representation with a single dichotomous response variable A (with categories A0 and A1) and three participants (i, j, and k). The solid green line represents the decision line where the probabilities for A0 and A1 are equal. The blue dotted line projects the points representing the three subjects onto the A01 line (blue dashed-dotted line). All points on this dotted line represent observations with the same log odds

The squared distances from Subject i to categories A0 and A1 additively decompose into one part toward the line through A0 and A1 (i.e., the A01 line) and one part along this line. Equation 3 shows that the log odds are defined in terms of a difference in squared distances, and therefore the part toward the A01 line drops out of the equation. In more detail, for this example, we have

$$\log \frac{\pi_{A0}(\mathbf{x}_i)}{\pi_{A1}(\mathbf{x}_i)} = \delta(\mathbf{u}_i, \mathbf{v}_{A1}) - \delta(\mathbf{u}_i, \mathbf{v}_{A0}).$$

According to the Pythagorean theorem, the squared distance $\delta(\mathbf{u}_i, \mathbf{v}_{A1})$ can be decomposed into $\delta(\mathbf{u}_i, \mathbf{v}_{A01}) + \delta(\mathbf{v}_{A01}, \mathbf{v}_{A1})$, where $\mathbf{v}_{A01}$ is the coordinate of the projection of $\mathbf{u}_i$ on the A01 line. Using this decomposition for both terms, we obtain

$$\begin{aligned}\log \frac{\pi_{A0}(\mathbf{x}_i)}{\pi_{A1}(\mathbf{x}_i)} &= (\delta(\mathbf{u}_i, \mathbf{v}_{A01}) + \delta(\mathbf{v}_{A01}, \mathbf{v}_{A1})) - (\delta(\mathbf{u}_i, \mathbf{v}_{A01}) + \delta(\mathbf{v}_{A01}, \mathbf{v}_{A0}))\\ &= \delta(\mathbf{v}_{A01}, \mathbf{v}_{A1}) - \delta(\mathbf{v}_{A01}, \mathbf{v}_{A0}).\end{aligned}$$

For person j or k, we can use the same decomposition. The projections for the three subjects are however equal, and therefore the log odds of category A0 against A1 for persons i, j, and k are equal (and for all three subjects in favor of category A1). As noted above, the decision line represents the set of positions where the odds are even, that is, the log odds are equal to zero. More generally, iso log odds curves, which are curves where the log odds equal any constant, are straight lines parallel to these decision lines and orthogonal to the A01 line. An example of such an iso log odds line is the one through the points representing the three subjects (blue dotted).

The variable axes can be understood as representations of subjects with varying scores on the corresponding predictor variables and an average score on all other predictor variables. In this way, we can interpret the variable axis by moving along the variable axes and computing the log odds for each response variable. More formally, let us denote by $\mathbf{d}_r$ the M-vector with differences $(v_{r1m} - v_{r0m})$ and let predictor variable p be presented by its regression weights $\mathbf{b}_p$ (with $\mathbf{b}_p^\top$ row p of matrix $\mathbf{B}$), then we can write the effect of predictor variable p on response variable r as

$$\mathbf{b}_p^\top \mathbf{d}_r = \|\mathbf{b}_p\| \cdot \|\mathbf{d}_r\| \cdot \cos(\mathbf{b}_p, \mathbf{d}_r),$$

showing that the log odds are largest when the direction of the variable axis for predictor variable p is parallel to the line connecting the two categories of response variable r, while it is zero if the variable axis is orthogonal to this line. Figure 2 illustrates this property. In Fig. 2, the variable axes are represented for four predictor variables. We use the convention that the labels attached to the variable axes are placed on the positive side of the variable. The two categories of a response variable (A0 and A1) are depicted by points. The variable axis for predictor variable X_3 is parallel to the A01 line, indicating that this variable discriminates this response variable well, while the variable axis for predictor X_2 is almost orthogonal to the

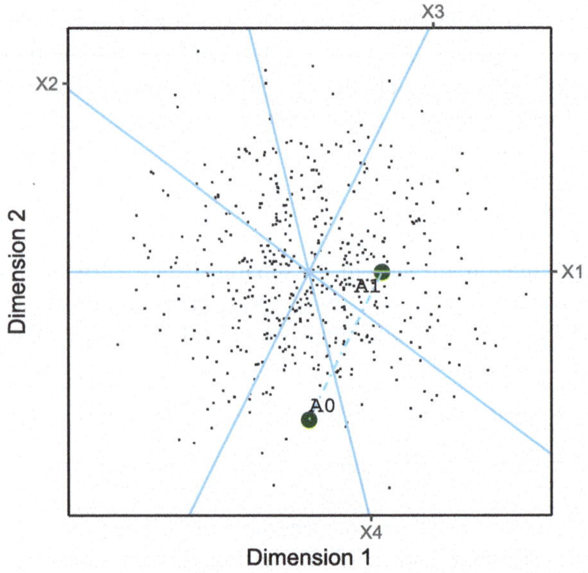

Fig. 2 Graphical representation with the two class points of a single response variable named A (the classes are represented by the points labeled A0 and A1), variable axes for four predictor variables, and subject points (small dots). The dashed-dotted line connecting A0 and A1 represents the vector $\mathbf{d}_A$. Predictor variable X_3 discriminates well between the two classes because the direction is parallel to the line joining the two classes, whereas predictor variable X_2 hardly discriminates between the two classes because the variable axis is almost orthogonal to the line joining the two classes

A01 line, indicating that X_2 does not discriminate between these two classes. We could draw the projections of A0 and A1 on each of the variable axes to see the discriminatory power: the further apart these projections are, the higher the power. For example, the projections of the two points onto variable X_2 are very close to each other, indicating that X_2 does not discriminate between these two classes well.

The discriminative power depends not only on the distance between the projections but also on the estimated value of the regression coefficients. Larger regression weights indicate more general discriminative power for the complete set of response variables. We will indicate the value of the regression weight by using markers along the variable axis in steps of 1 standard deviation. The further apart these markers are, the larger the regression weights and the higher the discriminative power will be.

With R binary response variables, the number of different possible response profiles is 2^R. When $M = R$, each of these profiles can be perfectly represented. In lower dimensional space ($M < R$), however, not all response profiles find a place in the solution. For example, in a one-dimensional space, only $R + 1$ different response profiles are represented. In two-dimensional space, the number of represented profiles is $\sum_{m=1}^{2} \binom{R}{m}$ [13].

In the constrained models, the number of represented response profiles is lower than in the general model because all decision lines are either horizontal or vertical.[1] With five response variables, of which three pertain to the first dimension and two to the second, the model represents $4 \times 3 = 12$ response profiles, which is even smaller than the 16 in the general unconstrained model, and much smaller than the $2^R = 32$ possible response profiles.

3 An MM Algorithm

In this section, we develop an MM algorithm, where the first M stands for "majorize" and the second M for "minimize". Such algorithms are also known as iterative majorization (IM) algorithms. MM algorithms have the property of guaranteed descent and, in MM algorithms, it is easy to use low-rank restrictions. The global idea of MM algorithms is that, instead of minimizing the original loss function, we seek an auxiliary function that (1) touches the original function at the current estimates, (2) lies above the original function, and (3) is easy to minimize. For a detailed treatment of the general principles of IM or MM, we refer to Heiser [39] and Hunter and Lange [43]. In the following subsections, we will majorize a deviance function with a least squares function. This results in a fast algorithm. Before developing the algorithm, we will discuss the identification of model parameters.

[1] Assuming we do not have a response variable pertaining to multiple dimensions.

3.1 Admissible Transformations

Before we develop an algorithm for the estimation of model parameters, we must discuss indeterminacies, that is, admissible transformations that change neither the estimated probabilities nor the loss value.

1. Multidimensional scaling and unfolding models, in general, have *translational freedom*. We center $\mathbf{X}$ so that the origin of the Euclidean space is fixed at the average value of the predictor variables.
2. The model has *rotational freedom*: any map can be rotated without changing the distances or the probabilities. We will require that $\frac{1}{n}\mathbf{B}^\top\mathbf{X}^\top\mathbf{X}\mathbf{B} = \mathbf{I}$ so that the rotational indeterminacy is removed. Reflection can be removed by requiring that the regression weights for the first predictor variable are positive.
3. The model

$$\pi_{rc}(\mathbf{x}_i) = \frac{\exp(\theta_{irc})}{\exp(\theta_{ir0}) + \exp(\theta_{ir1})}$$

is invariant under an additive constant in the "linear predictor" (θ_{irc}), that is

$$\frac{\exp(\theta_{irc})}{\exp(\theta_{ir0}) + \exp(\theta_{ir1})} = \frac{\exp(\theta_{irc} + \zeta_r)}{\exp(\theta_{ir0} + \zeta_r) + \exp(\theta_{ir1} + \zeta_r)}. \tag{6}$$

For our distance model, $\theta_{irc} = -\delta(\mathbf{u}_i, \mathbf{v}_{rc})$, we can add a constant to the squared distances *per response variable*, implying that the term $\sum_{m=1}^{M} u_{im}^2$ can be removed from the distance formulation. We will see a further simplification below.

3.2 Algorithm for the Unconstrained Model

The deviance function to be minimized is

$$\begin{aligned} L(\mathbf{B}, \mathbf{V}) &= -2\sum_{i=1}^{n}\sum_{r=1}^{R}\sum_{c=0}^{1} g_{irc} \log \pi_{rc}(\mathbf{x}_i) \\ &= -2\sum_{i=1}^{n}\sum_{r=1}^{R}\sum_{c=0}^{1} g_{irc} \log \left(\frac{\exp(\theta_{irc})}{\exp(\theta_{ir0}) + \exp(\theta_{ir1})} \right), \end{aligned} \tag{7}$$

where for our model $\theta_{irc} = -\frac{1}{2}\delta(\mathbf{u}_i, \mathbf{v}_{rc})$. Groenen et al. [34], De Leeuw [18] and Groenen and Josse [35] show that the function

$$f_{ir}(\boldsymbol{\theta}_i) = -2\sum_{c=0}^{1} g_{irc} \log \frac{\exp(\theta_{irc})}{\exp(\theta_{ir0}) + \exp(\theta_{ir1})},$$

is majorized by

$$g_{ir}(\boldsymbol{\theta}_i, \tilde{\boldsymbol{\theta}}_i) = f_{ir}(\tilde{\boldsymbol{\theta}}_i) + (\boldsymbol{\theta}_i - \tilde{\boldsymbol{\theta}}_i)^\top \nabla f_{ir}(\tilde{\boldsymbol{\theta}}_i) + \frac{1}{4}\|\boldsymbol{\theta}_i - \tilde{\boldsymbol{\theta}}_i\|^2,$$

with

$$\nabla f_{ir}(\tilde{\boldsymbol{\theta}}_i) = -(\mathbf{g}_{ir} - \tilde{\boldsymbol{\pi}}_{ir}),$$

and where the tilde means that it is evaluated using the current estimates. A proof of the majorizing property is given in the appendix of Groenen and Josse [35].

By analogy to Groenen and Josse, we therefore have that

$$\begin{aligned} L(\mathbf{B}, \mathbf{V}) &\leq \frac{1}{4}\|\mathbf{Z} - \boldsymbol{\Theta}\|^2 + L(\tilde{\mathbf{B}}, \tilde{\mathbf{V}}) - \frac{1}{4}\|\mathbf{Z}\|^2 + \left\|\mathbf{G} - \tilde{\boldsymbol{\Pi}}\right\|^2 \\ &= \frac{1}{4}\|\mathbf{Z} - \boldsymbol{\Theta}\|^2 + \text{constant} = g(\mathbf{B}, \mathbf{V}) + \text{constant}, \end{aligned} \tag{8}$$

where $\mathbf{Z} = \{z_{irc}\}$ with $z_{irc} = \tilde{\theta}_{irc} + 2(g_{irc} - \tilde{\pi}_{irc})$ and $\boldsymbol{\Theta} = \{\theta_{irc}\}$ with $\theta_{irc} = -\frac{1}{2}\sum_{m=1}^{M}(v_{rcm}^2 - 2u_{im}v_{rcm})$. In matrix terms, we can write $\boldsymbol{\Theta}$ as

$$\boldsymbol{\Theta} = -\frac{1}{2}\left(\mathbf{1}\mathbf{d}_v^\top - 2\mathbf{X}\mathbf{B}\mathbf{V}^\top\right),$$

and $\mathbf{Z}$ as

$$\mathbf{Z} = \tilde{\boldsymbol{\Theta}} + 2(\mathbf{G} - \tilde{\boldsymbol{\Pi}}).$$

Therefore, the objective function to be minimized in every iteration is

$$g(\mathbf{B}, \mathbf{V}) = \left\|\mathbf{Z} + \mathbf{1}\mathbf{d}_v^\top/2 - \mathbf{X}\mathbf{B}\mathbf{V}^\top\right\|^2.$$

The third indeterminacy, outlined above, allows us to rewrite the minimization function as

$$g(\mathbf{B}, \mathbf{V}) = \left\|\mathbf{Z}\mathbf{J} + \frac{1}{2}\mathbf{1}\mathbf{d}_v^\top\mathbf{J} - \mathbf{X}\mathbf{B}\mathbf{V}^\top\mathbf{J}\right\|^2, \tag{9}$$

with $\mathbf{J}$ a symmetric block diagonal matrix with 2×2 diagonal blocks $\mathbf{J}_r$, the usual centering matrices, $\mathbf{J}_r = \mathbf{I}_2 - \frac{1}{2}\mathbf{1}\mathbf{1}^\top$.

Let us define the matrices $\mathbf{A}_l = \mathbf{I}_R \otimes [1, 1]^\top$ and $\mathbf{A}_k = \mathbf{I}_R \otimes [1, -1]^\top$ with $\otimes$ the Kronecker product, such that $\mathbf{V}$ can be reparametrized as

$$\mathbf{V} = \mathbf{A}_l\mathbf{L} + \mathbf{A}_k\mathbf{K}$$

with $\mathbf{L}$ the $R \times M$ matrix with response variable *locations* and $\mathbf{K}$ the $R \times M$ matrix representing the *discriminatory power* for the response variables. As a numerical example with two response variables, consider

$$\mathbf{V} = \begin{bmatrix} 1 & 0 \\ 3 & 2 \\ \hdashline 0 & 0 \\ 4 & 6 \end{bmatrix} = \begin{bmatrix} 1 & 0 \\ 1 & 0 \\ 0 & 1 \\ 0 & 1 \end{bmatrix} \begin{bmatrix} 2 & 1 \\ 2 & 3 \end{bmatrix} + \begin{bmatrix} 1 & 0 \\ -1 & 0 \\ 0 & 1 \\ 0 & -1 \end{bmatrix} \begin{bmatrix} -1 & -1 \\ -2 & -3 \end{bmatrix},$$

where the second matrix on the right-hand side of the equation represents $\mathbf{L}$ (the locations or midpoints of the class coordinates) and the last matrix on the right-hand side of the equation represents $\mathbf{K}$. The larger the absolute values in row r of matrix $\mathbf{K}$, the better the two categories from this response variable (r) can be discriminated by the predictor variables; if the values equal zero, the categories cannot be discriminated and the positions of the two categories of a response variable fall in the same place. In the numerical example, the second response variable is better discriminated (that is, the class points are further apart).

Using this reparametrization in our loss function, the last term $\mathbf{XBV}^\top\mathbf{J}$ simplifies to $\mathbf{XBV}^\top\mathbf{J} = \mathbf{XBK}^\top\mathbf{A}_k^\top$ because the term $\mathbf{XBL}^\top\mathbf{A}_l^\top\mathbf{J} = \mathbf{0}$ as $\mathbf{A}_l^\top\mathbf{J} = \mathbf{0}$. Let us now have a closer look at the second term of the loss function (Eq. 9), that is, $\mathbf{1d}_v^\top\mathbf{J} = \mathbf{Jd}_v$. This term can be rewritten as

$$\begin{aligned} \mathbf{d}_v = \text{Diag}(\mathbf{VV}^\top) &= \text{Diag}\left((\mathbf{A}_l\mathbf{L} + \mathbf{A}_k\mathbf{K})(\mathbf{A}_l\mathbf{L} + \mathbf{A}_k\mathbf{K})^\top\right) \\ &= \text{Diag}\left(\mathbf{A}_l\mathbf{LL}^\top\mathbf{A}_l^\top + \mathbf{A}_k\mathbf{KK}^\top\mathbf{A}_k^\top + \mathbf{A}_l\mathbf{LK}^\top\mathbf{A}_k^\top + \mathbf{A}_k\mathbf{KL}^\top\mathbf{A}_l^\top\right), \end{aligned}$$

where $\text{Diag}(\mathbf{X})$ creates a column vector of the diagonal elements of $\mathbf{X}$. It can be verified that the terms $\mathbf{J}\text{Diag}\left(\mathbf{A}_l\mathbf{LL}^\top\mathbf{A}_l^\top\right)$ and $\mathbf{J}\text{Diag}\left(\mathbf{A}_k\mathbf{KK}^\top\mathbf{A}_k^\top\right)$ are both equal to zero. Therefore,

$$\text{Diag}(\mathbf{VV}^\top) = \mathbf{J}\text{Diag}\left(2\mathbf{A}_k\mathbf{KL}^\top\mathbf{A}_l^\top\right).$$

Focusing on a single response variable r, we can rewrite the corresponding part of the previous equation as

$$\mathbf{k}_r^\top\mathbf{l}_r \begin{bmatrix} 1 & 1 \\ -1 & -1 \end{bmatrix},$$

from which it follows that $\frac{1}{2}\mathbf{1d}_v\mathbf{J}$ can be written as

$$\frac{1}{2}\mathbf{1d}_v^\top\mathbf{J} = \mathbf{1}\text{Diag}(\mathbf{KL}^\top)^\top\mathbf{A}_k^\top.$$

Going back to our loss function and using the decomposition into $\mathbf{K}$ and $\mathbf{L}$, it becomes

$$g(\mathbf{B}, \mathbf{K}, \mathbf{L}) = \|\mathbf{ZJ} + \mathbf{1}\text{Diag}(\mathbf{KL}^\top)^\top\mathbf{A}_k^\top - \mathbf{XBK}^\top\mathbf{A}_k^\top\|^2,$$

that equals

$$g(\mathbf{B}, \mathbf{K}, \mathbf{L}) = 2\left\|\frac{1}{2}\mathbf{Z}\mathbf{J}\mathbf{A}_k + \mathbf{1}\mathrm{Diag}(\mathbf{K}\mathbf{L}^\top)^\top - \mathbf{X}\mathbf{B}\mathbf{K}^\top\right\|^2$$

and can be rewritten as

$$g(\mathbf{B}, \mathbf{K}, \mathbf{L}) = 2\left\|\frac{1}{2}\mathbf{Z}\mathbf{J}\mathbf{A}_k + \mathbf{1}\mathbf{a}^\top - \mathbf{X}\mathbf{B}\mathbf{K}^\top\right\|^2,$$

where $\mathbf{a}$ is a vector with elements $a_r = \sum_{m=1}^{M} k_{rm} l_{rm}$. The elements of $\mathbf{a}$ can be estimated independently of $\mathbf{K}$, because values of $\mathbf{L}$ always exist that together with the k_{rm} can reconstruct a_r (see below). Therefore, this latter loss function can be solved separately for 1) $\mathbf{a}$ and 2) $\mathbf{B}$, $\mathbf{K}$.

To update $\mathbf{a}$, define

$$\tilde{\mathbf{Z}}_1 = \frac{1}{2}\mathbf{Z}\mathbf{J}\mathbf{A}_k,$$

then the update is $\mathbf{a}^+ = -\tilde{\mathbf{Z}}_1^\top \mathbf{1}/n$.

To find the update for $\mathbf{B}$ and $\mathbf{K}$, let us define

$$\tilde{\mathbf{Z}}_2 = \frac{1}{2}\mathbf{Z}\mathbf{J}\mathbf{A}_k + \mathbf{1}\mathbf{a}^\top,$$

so that we have to minimize

$$\left\|\tilde{\mathbf{Z}}_2 - \mathbf{X}\mathbf{B}\mathbf{K}^\top\right\|^2,$$

under the restriction that $n^{-1}\mathbf{B}^\top\mathbf{X}^\top\mathbf{X}\mathbf{B} = \mathbf{I}$. As

$$\left\|\tilde{\mathbf{Z}}_2 - \mathbf{X}\mathbf{B}\mathbf{K}^\top\right\|^2 = \left\|\tilde{\mathbf{Z}}_2 - \mathbf{X}\mathbf{N}\right\|^2 + \left\|\mathbf{N} - \mathbf{B}\mathbf{K}^\top\right\|^2_{\mathbf{X}^\top\mathbf{X}}$$

with $\mathbf{N} = (\mathbf{X}^\top\mathbf{X})^{-1}\mathbf{X}^\top\tilde{\mathbf{Z}}_2$, the unconstrained update, an update of $\mathbf{B}$ and $\mathbf{K}$ is found by the generalized singular value decomposition of $\mathbf{N}$ [58]. These two steps can be combined, that is,

$$\mathbf{R}_x^{-1}\mathbf{X}^\top\tilde{\mathbf{Z}}_2 = \mathbf{P}\boldsymbol{\Phi}\mathbf{Q}^\top,$$

where $\mathbf{R}_x$ is the matrix square root of the matrix $\mathbf{X}^\top\mathbf{X}$, that is $\mathbf{X}^\top\mathbf{X} = \mathbf{R}_x\mathbf{R}_x^\top$. The updates for $\mathbf{B}$ and $\mathbf{K}$ can be obtained as

$$\mathbf{B}^+ = \sqrt{n}\mathbf{R}_x^{-1}\mathbf{P}_M,$$

where $\mathbf{P}_M$ denotes the M columns of $\mathbf{P}$ corresponding to the M largest singular values and

$$\mathbf{K}^+ = \frac{1}{\sqrt{n}}\mathbf{Q}_M\boldsymbol{\Phi}_M.$$

Finally, an update for **L** can be obtained from **a** and **K**. For every response variable we have

$$a_r = \sum_{m=1}^{M} k_{rm} l_{rm},$$

which is an unidentified system, that is, there are many choices of l_{rm} that provide a solution. These solutions correspond to any position on the decision line (or plane or hyperplane in higher dimensional spaces), as discussed in relation to Fig. 1. We find the position on this hyperplane that is closest to the origin of the Euclidean space, that is,

$$l_{rm}^{+} = \frac{a_r k_{rm}}{\sum_m k_{rm}^2}.$$

A summary of the algorithm can be found in Algorithm 1.

The number of parameters of the model is:

- $PM - M(M+1)/2$ for the regression weights **B**;
- RM parameters in the matrix **K**;
- R parameters in **L**;

which sum to $(P+R)M + R - M(M+1)/2$.

Data: **X**, **G**, M
Result: **B**, **K**, **L**
Compute: $\mathbf{R}_x^{-1}\mathbf{X}^\top\mathbf{G} = \mathbf{P}\mathbf{\Phi}\mathbf{Q}^\top$;
Initialize: $\mathbf{B}^{(0)} = \sqrt{n}\mathbf{R}_x^{-1}\mathbf{P}_M$;
Compute: $\mathbf{V} = \frac{1}{\sqrt{n}}\mathbf{Q}_M\mathbf{\Phi}_M$;
Initialize: $\mathbf{K}^{(0)}$ by taking the uneven rows of **JV**;
Initialize: $\mathbf{L}^{(0)}$ by taking the uneven rows of $(\mathbf{I}-\mathbf{J})\mathbf{V}$;
Compute: $\tilde{\mathbf{\Pi}}$ and $\tilde{\mathbf{\Theta}}$;
while $t = 0$ *or* $(L^t - L^{(t-1)})/L^t > 10^{-8}$ **do**
  $t = t + 1$;
  Compute: $\mathbf{Z} = \tilde{\mathbf{\Theta}} + 2(\mathbf{G} - \tilde{\mathbf{\Pi}})$;
  Compute: $\mathbf{Z}_1 = \frac{1}{2}\mathbf{Z}\mathbf{J}\mathbf{A}_k$;
  Compute: $\mathbf{a} = -\mathbf{Z}_1^\top\mathbf{1}/n$;
  Compute: $\mathbf{Z}_2 = \frac{1}{2}\mathbf{Z}\mathbf{J}\mathbf{A}_k + \mathbf{1}\mathbf{a}^\top$;
  Compute: $\mathbf{R}_x^{-1}\mathbf{X}^\top\mathbf{Z}_2 = \mathbf{P}\mathbf{\Phi}\mathbf{Q}^\top$;
  Update: $\mathbf{B}^{(t)} = \sqrt{n}\mathbf{R}_x^{-1}\mathbf{P}_M$;
  Update: $\mathbf{K}^{(t)} = \frac{1}{\sqrt{n}}\mathbf{Q}_M\mathbf{\Phi}_M$;
  Update: $l_{rm}^{(t)} = a_r k_{rm}/(\sum_m k_{rm}^2), \forall r$;
  Compute $\tilde{\mathbf{\Pi}}$, $\tilde{\mathbf{\Theta}}$, and L^t;
end

Algorithm 1: MELODIC Algorithm

3.3 Algorithm for Constrained Model

Sometimes, researchers have an idea in advance of which responses belong together in which dimensions. Let us denote the set of response variables that pertains to dimension m by $\mathcal{D}_m$. Furthermore, let us denote the set of dimensions to which response variable r pertains as $\mathcal{S}_r$. Then, for $m \notin \mathcal{S}_r$, we restrict $l_{rm} = k_{rm} = 0$.

Much of the unconstrained MM algorithm can be used, except for two aspects: (i) orthonormality restriction on **XB** needs to be relaxed and (2) the updates will be done dimension-wise. We still need a scale restriction per dimension on **XB**. The equivalent of (9) can be written

$$g_c(\mathbf{B}, \mathbf{K}, \mathbf{L}) = \left\| \frac{1}{2}\mathbf{ZJA}_k + \mathbf{1a}^\top - \sum_{m=1}^{M} \mathbf{Xb}_m\mathbf{k}_m^\top \right\|^2,$$

which shows that the last term can be decomposed in dimensional terms. To update for dimension s, we first define

$$\tilde{\mathbf{Z}}_3 = \frac{1}{2}\mathbf{ZJA}_k + \mathbf{1a}^\top - \sum_{m \neq s} \mathbf{Xb}_m\mathbf{k}_m^\top.$$

Next, we define the matrix $\tilde{\mathbf{Z}}_s$ to be the subset of the matrix $\tilde{\mathbf{Z}}_3$ consisting of the columns for which $r \in \mathcal{D}_s$.

Similar to the unconstrained model, a generalized singular value decomposition of $\tilde{\mathbf{Z}}_s$ gives updates for $\mathbf{b}_s$ and $\mathbf{k}_s$ by taking the highest singular value and the corresponding vector. The update for **a** is the same as in the unconstrained model. The update for **L** is similar to the unconstrained model, but we only give non-zero values to the dimensions to which response variable r pertains. A summary of the algorithm is given in Algorithm 2.

The total number of parameters for the constrained model depends on the specific constraints. Nevertheless, we have

- $(P - 1)M$ parameters for the regression weights **B**;
- We define an indicator matrix of size $R \times M$ indicating which response belongs to which dimension. The number of parameters in the matrix **K** equals the number of ones in that indicator matrix (see the next section for an example);
- R parameters in **L**.

definition of algorithm is difficult because of the selection of responses that belong to a given dimension.

Data: $\mathbf{X}$, $\mathbf{G}$, M
Result: $\mathbf{B}$, $\mathbf{K}$, $\mathbf{L}$
Compute: $\mathbf{R}_x^{-1}\mathbf{X}^\top\mathbf{G} = \mathbf{P}\boldsymbol{\Phi}\mathbf{Q}^\top$;
Initialize: $\mathbf{B}^{(0)} = \sqrt{n}\mathbf{R}_x^{-1}\mathbf{P}_M$;
Compute: $\mathbf{V} = \frac{1}{\sqrt{n}}\mathbf{Q}_M\boldsymbol{\Phi}_M$;
Initialize: $\mathbf{K}^{(0)}$ by taking the uneven rows of $\mathbf{JV}$;
Initialize: $\mathbf{L}^{(0)}$ by taking the uneven rows of $(\mathbf{I} - \mathbf{J})\mathbf{V}$;
Set elements in $\mathbf{K}^{(0)}$ and $\mathbf{L}^{(0)}$ to zero, following the constraints;
Compute: $\tilde{\boldsymbol{\Pi}}$ and $\tilde{\boldsymbol{\Theta}}$;
while $t = 0$ *or* $(L^t - L^{(t-1)})/L^t > 10^{-8}$ **do**
  $t = t + 1$;
  Compute: $\mathbf{Z} = \tilde{\boldsymbol{\Theta}} + 2(\mathbf{G} - \tilde{\boldsymbol{\Pi}})$;
  Compute: $\tilde{\mathbf{Z}}_1 = \frac{1}{2}\mathbf{ZJA}_k$;
  Compute: $\mathbf{a}^+ = -\tilde{\mathbf{Z}}_1^\top\mathbf{1}/n$;
  for $s = 1, \ldots, M$ **do**
    Compute: $\tilde{\mathbf{Z}}_s$;
    Compute: $\mathbf{R}_x^{-1}\mathbf{X}^\top\tilde{\mathbf{Z}}_s = \mathbf{P}\boldsymbol{\Phi}\mathbf{Q}^\top$;
    Update: $\mathbf{b}_s^{(t)} = \sqrt{n}\mathbf{R}_x^{-1}\mathbf{P}_1$;
    Update: $\mathbf{k}_s^{(t)} = \frac{1}{\sqrt{n}}\mathbf{Q}_1\boldsymbol{\Phi}_1$;
  end
  Update: $l_{rm}^{(t)} = a_r / \sum_{m=1}^{M} k_{rm}$, for $m \in \mathcal{S}_r$ and $\forall r$;
  Compute $\tilde{\boldsymbol{\Pi}}$, $\tilde{\boldsymbol{\Theta}}$, and L^t;
end

Algorithm 2: MELODIC Algorithm for Constrained Model

4 Two Empirical Applications

In this section, we discuss two empirical applications of the model. The first data set considers profiles of drug consumption; the second, profiles of mental disorders. For the first data set, we use the unconstrained model. For the second data set, we start with a set of constrained models representing different theories.

Table 1 AIC and BIC statistics for models in 1–7 dimensions for the drug consumption data

Dimensionality	Deviance	#param	AIC	BIC
1	18311	30	18371	18538
2	18117	48	18213	**18479**
3	18030	65	**18160**	18520
4	17998	81	18160	18609
5	17987	96	18179	18711
6	17980	110	18200	18810
7	17975	123	18221	18903

4.1 Drug Consumption Data

The drug consumption data [26] has records for 1885 respondents. For each respondent, nine attributes are measured. We have personality measurements based on the big five personality traits, neuroticism (N), extraversion (E), openness to experience (O), agreeableness (A), and conscientiousness (C), and two other personality characteristics, namely impulsivity (I) and sensation seeking (S). Data were also collected on age and gender.[2]

In addition, participants were questioned concerning their use of 18 legal and illegal drugs. For each drug, participants were asked whether they never used the drug, used it over a decade ago, in the last decade, in the last year, month, week, or day. In our analysis, we coded whether participants used the particular drug in the last year (yes or no). Furthermore, in our analysis, we focused on the drugs that had a minimum percentage of 10% and a maximum of 90%, which are Amphetamine, Benzodiazepine, Cannabis, Cocaine, Ecstasy, Ketamine, legal highs, LSD, Methadone, Mushrooms, and Nicotine ($R = 11$).

The first step in the analysis is to select the dimensionality. We fit models in one to seven dimensions and compute information criteria statistics for comparison. The results are given in Table 1, where we can see that either the two- or three-dimensional solution is optimal according to the AIC and BIC statistics.

We should also check the influence of the predictor variables. Each of the predictor variables is left out of the two-dimensional model. The AIC and BIC statistics are shown in Table 2, where it can be seen that only impulsivity can be considered for being left out of the model; all other predictors would lead to a substantial loss in fit. We decided to further interpret the model using all predictor variables except impulsivity.

The graphical representation of the two-dimensional model is shown in Fig. 3.[3] The first thing that catches the eye in Fig. 3 is that the categories for "yes" (labels

[2] Also level of education, ethnicity, and country of origin are available in the original database. We omitted these from the analysis.

[3] We left out the decision lines to avoid clutter.

Table 2 AIC and BIC statistics for two-dimensional models with 1 predictor left out of the model

Left out	Deviance	#param	AIC	BIC
Age	19303	46	19395	19650
Gender	18417	46	18509	18764
Neuroticism	18181	46	18273	18528
Extraversion	18134	46	18226	18481
Openess	18449	46	18541	18796
Agreeablenes	18137	46	18229	18484
Conscientiousness	18172	46	18264	18519
Impulsivity	18121	46	18213	**18468**
Sensation seeking	18409	46	18501	18756

ending with 1) are all to the right-hand side of the categories for "no" (labels ending with 0). Therefore, participants who use drugs lie on the positive side of the first dimension. It can be seen that cannabis is furthest to the left, while Ketamine, LSD, and Methadone are on the extreme right-hand side. Apparently, if participants start using drugs, they start with cannabis and only add other drugs later.

Considering the predictor side, we see that five predictor variables run from left to right: openness to experience (O), sensation seeking (S), extraversion (E), age, and gender. With respect to gender and age, boys use more drugs than girls and younger people use more drugs than the elderly. Furthermore, participants who are more open to experience (O) and less extravert (E) use more drugs. Finally, participants scoring high on sensation seeking (S) use more drugs than participants who score low on sensation seeking.

The vertical dimension is harder to interpret because the differences between the yes and no points are often small. The largest differences are for benzodiazepine and methadone (yes category has a higher coordinate) and for LSD (yes category has a lower coordinate). The predictor variable neuroticism (N) points strongly in this direction, indicating that neurotic participants tend to use benzodiazepine and methadone more frequently but LSD less frequently. The variable axis for agreeableness (A) is almost parallel to the vertical dimension, but in the opposite direction to neuroticism, indicating opposite effects.

To get a more detailed interpretation, let us look at the estimated implied logistic regression coefficients (Eq. 5) in Table 3. Since the predictor variables are standardized to have zero mean and standard deviation one, these are changes in log odds for one standard deviation increases in the predictors. The numbers in each column can be interpreted as the standardized coefficients in a single logistic regression model.

We can verify how well each response variable is represented in the low-dimensional space. To do this, we define a measure called Quality of Representation, Q_r, which is defined by

$$Q_r = (L_{(0,r)} - L_r)/(L_{(0,r)} - L_{lr}),$$

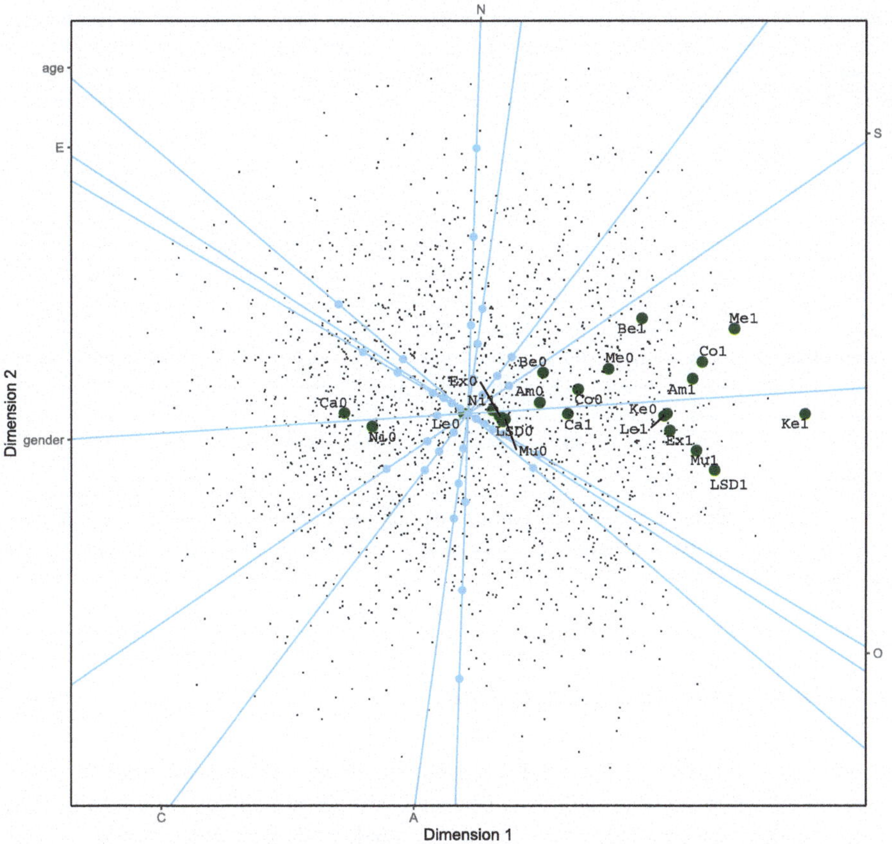

Fig. 3 Two-dimensional solution for the drug consumption data. Predictor variable labels are printed on the border of the graph where N = Neuroticism; E = Extraversion; O = Openness to experience; A = Agreeableness; C = Conscientiousness; S = Sensation seeking. We use the convention that labels are printed at the positive side of the variable. Markers on the variable axes indicate standard deviation increases/decreases from the mean. Category points are labeled by the name of the drug together with a 1 (yes) or 0 (no), that is, `Am1` denotes the class point for the use of Amphetamine and `Am0` for no use of Amphetamine. Am = Amphetamine; Be = Benzodiazepine; Ca = Cannabis; Co = Cocaine; Ex = Ecstasy; Ke = Ketamine; Le = legal highs; LSD = LSD; Me = Methadone; Mu = Mushrooms; Ni = Nicotine

where $L_{(0,r)}$ is the deviance of the intercept-only logistic regression model for response variable r, L_r is the part of our loss function for response variable r, and L_{lr} is the deviance from a logistic regression with the same predictor variables. Thus, Q_r can be interpreted as the proportion of loss in deviance imposed by the Melodic model compared to an unconstrained logistic regression for response variable r. The quality of representation for the response variables in this analysis is given in the last

Table 3 Estimated implied regression coefficients (Eq. 5) of each of the predictor variables for each of the response variables. In the columns Am = Amphetamine; Be = Benzodiazepine; Ca =Cannabis; Co = Cocaine; Ex = Ecstasy; Ke = Ketamine; LE = legal highs; LSD = LSD; Me = Methadone; Mu = Mushrooms; Ni = Nicotine. The last line shows the quality of representation (Q_r)

	Response variables										
Predictor variables	Am	Be	Ca	Co	Ex	Ke	Le	LSD	Me	Mu	Ni
Age	−0.59	−0.23	−0.99	−0.45	−0.81	−0.62	−0.90	−1.13	−0.41	−0.97	−0.47
Gender	−0.33	−0.22	−0.47	−0.27	−0.36	−0.29	−0.42	−0.44	−0.27	−0.40	−0.26
Neuroticism	0.18	0.36	0.04	0.20	−0.06	0.03	0.02	−0.26	0.28	−0.16	0.13
Extraversion	−0.08	−0.03	−0.12	−0.06	−0.10	−0.08	−0.11	−0.14	−0.06	−0.12	−0.06
Openess	0.33	0.16	0.54	0.26	0.43	0.33	0.48	0.58	0.25	0.51	0.27
Agreeableness	−0.11	−0.17	−0.07	−0.11	−0.02	−0.04	−0.06	0.05	−0.14	0.02	−0.08
Conscientiousness	−0.18	−0.17	−0.22	−0.16	−0.15	−0.14	−0.19	−0.14	−0.18	−0.15	−0.14
Sensation seeking	0.47	0.38	0.62	0.40	0.45	0.39	0.55	0.50	0.43	0.47	0.37
Quality	1.00	0.96	0.97	0.90	0.96	0.95	1.00	0.98	0.94	0.99	0.98

row of Table 3, where it can be seen that most response variables are well represented. The response variable "cocaine" (Co) is worst represented, although still with 89.8% recovered.

4.2 *Depression and Anxiety Data*

Depression and anxiety disorders are common at all ages. Approximately one out of three people in the Netherlands will be faced with them at some time during their lives. It is not clear why some people recover quickly and why others suffer for long periods of time. The Netherlands Study of Depression and Anxiety (NESDA) was therefore designed to investigate the course of depression and anxiety disorders over a period of several years. For more information about the study design, see Penninx et al. [51]. In our application, we will analyze data from the first wave, focusing on the relationship between personality and depression and anxiety disorders. The data were previously analyzed by Spinhoven et al. [55]. Data were collected from three different populations: from primary health care; from generalized mental health care; and from the general population. Our analysis will focus on the population of generalized health care.

We have data for 786 participants. The diagnoses, Dysthymia (D), Major Depressive Disorder (MDD), Generalized Anxiety Disorder (GAD), Social Phobia (SP), and Panic Disorder (PD), were established with the Composite Interview Diagnostic Instrument (CIDI) psychiatric interview. Personality was operationalized using the 60-item NEO Five-Factor Inventory (NEO-FFI). The NEO-FFI questionnaire measures the following five personality domains: Neuroticism, Extraversion, Agreeableness, Conscientiousness and Openness to Experience. In addition to these five predictors, three background variables were measured: age, gender, and education in years.

The prevalences in the data are 21.25% for dysthymia, 76.21% for major depressive disorder, 30.41% for generalized anxiety disorder, 41.6% for social phobia, and 52.8% for panic disorder. Of the 786 participants, 272 have a single disorder, and the others all have multiple disorders. There are 235 participants with two disorders, 147 with three, 96 with four, and 36 participants with five disorders.

Due to the high comorbidity among disorders, the scientific field of psychiatry developed three different theories:

1. a unidimensional structure where all the disorders are represented by a single dimension;
2. a two-dimensional structure with one dimension representing distress (D, MDD, GAD) and the other fear (SP, PD);
3. a two-dimensional structure with one dimension representing depression (D, MDD) and the other anxiety (GAD, SP, PD).

We can of course define another two-dimensional structure (Theory 4) in which dysthymia and major depressive disorder pertain to the first dimension, social phobia, and panic disorder to the second, and generalized anxiety disorder to both dimensions.

Each of the three two-dimensional theories gives rise to a different response variable by dimension indicator matrix:

$$\mathbf{D}_2 = \begin{bmatrix} 1 & 0 \\ 1 & 0 \\ 1 & 0 \\ 0 & 1 \\ 0 & 1 \end{bmatrix}, \quad \mathbf{D}_3 = \begin{bmatrix} 1 & 0 \\ 1 & 0 \\ 0 & 1 \\ 0 & 1 \\ 0 & 1 \end{bmatrix}, \quad \mathbf{D}_4 = \begin{bmatrix} 1 & 0 \\ 1 & 0 \\ 1 & 1 \\ 0 & 1 \\ 0 & 1 \end{bmatrix}.$$

We fitted the four models reflecting the four theories to the data. The fit statistics can be found in Table 4, where it can be seen that all four theories give about the same fit, but the distress-fear hypothesis (corresponding to $\mathbf{D}_2$) has a slightly lower AIC and the unidimensional model a lower BIC value than the other theories.

The graphical display for the distress-fear model (Theory 2) is given in Fig. 4. We can see that three response variables pertain to the horizontal dimension, while two pertain to the vertical dimension. The class points for dysthymia, major depressive disorder, and generalized anxiety disorder fall on the horizontal dimension (and thus have vertical decision lines), while the class points for social phobia and panic

Table 4 AIC and BIC statistics for the models reflecting the four theories

Theory/model	Deviance	#param	AIC	BIC
1	4553.34	17	4587	4667
2	4531.17	24	4579	4691
3	4533.71	24	4582	4694
4	4530.35	25	4580	4697

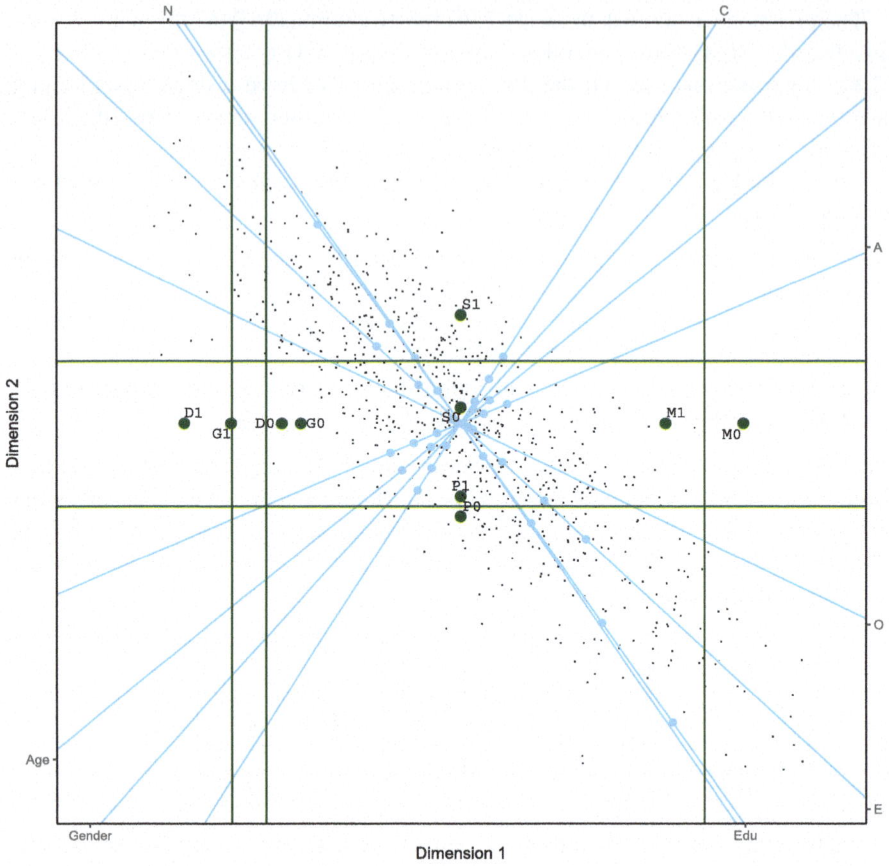

Fig. 4 Two-dimensional solution following Theory 2 for the depression and anxiety data. Predictor variable labels are printed on the border of the graph where N = Neuroticism; E = Extraversion; O = Openness to experience; A = Agreeableness; C = Conscientiousness and Edu = Education. We use the convention that labels are printed at the positive side of the variable. Markers on the variable axes indicate standard deviation increases/decreases from the mean. Category points are labeled by the name of the drug together with a 1 (yes) or 0 (no) with D = Dysthymia, M = Major Depressive Disorder, G = Generalized Anxiety Disorder, S = Social Phobia (SP), and P = Panic disorder (PD)

disorder fall on the vertical dimension (and therefore have horizontal decision lines). The decision lines partition the two-dimensional space into rectangular regions in which a certain response profile is most probable. We further see that, on the horizontal dimension, dysthymia is best discriminated as the two points lie farthest apart (distance 0.80), while the distance for major depressive disorder is 0.64 and for generalized anxiety disorder 0.56. On the vertical axis, social phobia is well discriminated (distance 0.76) while panic disorder is hardly discriminated (distance 0.16). The latter means that, using the three background variables and the five personality variables together with the imposed model structure, we have hardly any information

Table 5 Implied logistic regression coefficients for the Depression and Anxiety data. D = Dysthymia, M = Major Depressive Disorder, G = Generalized Anxiety Disorder, S = Social Phobia (SP), and P = Panic Disorder (PD). The last row represents the quality of representation for the five response variables

	Response variable				
Predictor variable	D	M	G	S	P
Gender	0.08	0.07	0.06	−0.09	−0.02
Age	0.19	0.15	0.13	−0.15	−0.03
Education	−0.15	−0.12	−0.10	−0.21	−0.04
Neuroticism	0.46	0.37	0.32	0.62	0.14
Extraversion	−0.27	−0.22	−0.19	−0.24	−0.05
Openness	−0.03	−0.02	−0.02	−0.01	−0.00
Agreeableness	−0.15	−0.12	−0.11	0.06	0.01
Conscientiousness	−0.09	−0.07	−0.07	0.14	0.03
Quality	0.98	0.90	0.85	0.99	0.17

to distinguish participants with and without panic disorder. We will come back to this issue later.

The implied logistic regression coefficients are given in Table 5. As in our previous analysis we standardized the predictor variables such that the coefficients give changes in log odds for one standard deviation changes in the predictors. The most important predictor for mental disorders is neuroticism, which concurs with the conclusion in [55]

The quality of representation for the five response variables is given in the last row of Table 5, where we see that the response variable panic disorder is poorly represented in this model. This could already be inferred from the graphical representation (the two points almost coincide) and the implied coefficients table, where most coefficients for the response variable panic disorder are very small. Apparently, when we use a standard logistic regression model for this response variable, the predictor variables discriminate the two categories much better.

Because one response variable is poorly represented in the best fitting model, we also fitted an unconstrained model in two dimensions. Such a model has 28 parameters; the value of the loss function (deviance) is 4521.81 (AIC = 4578; BIC = 4708). The AIC indicates a better fit than the previous constrained models. The quality of representation of the response variables in this model is 0.97, 0.88, 0.85, 0.86, and 0.90, no longer indicating any poorly fitting response variables anymore.

The biplot for this two-dimensional unconstrained model is given in Fig. 5, where we can see that the decision lines for major depressive disorder, dysthymia, generalized anxiety disorder, and social phobia run more or less parallel to the vertical dimension and the decision line for panic disorder is more or less horizontal. This exploratory finding suggests a new theory: that major depressive disorder, dysthymia, generalized anxiety disorder, and social phobia pertain to a single underlying dimension, but that panic disorder behaves differently.

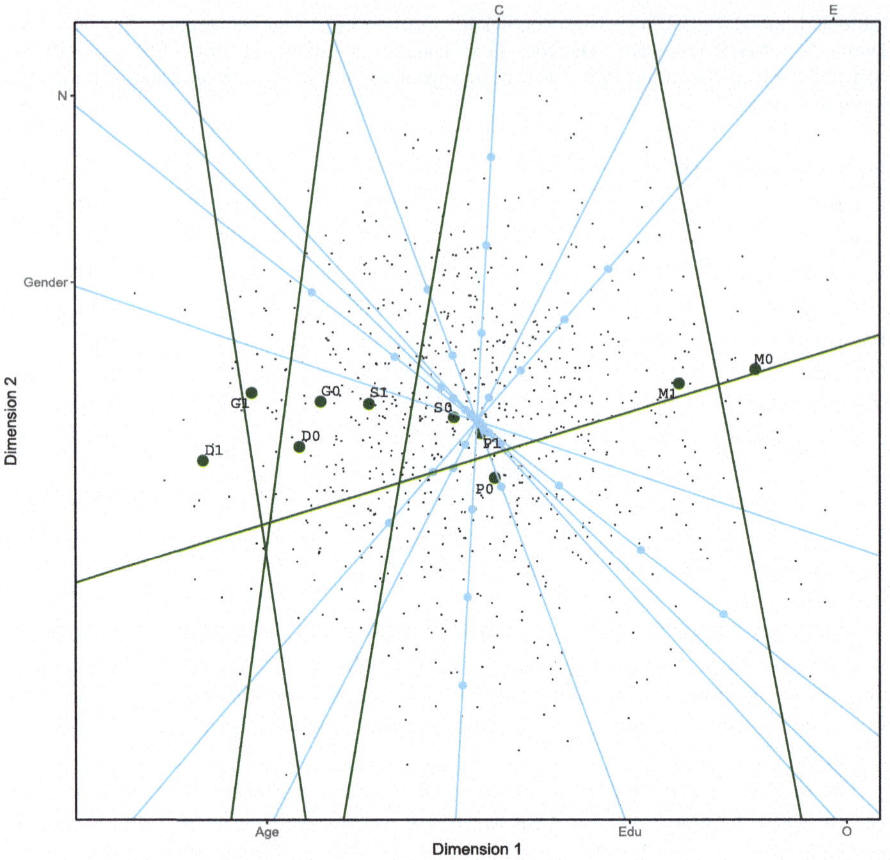

Fig. 5 Graphical representation of the two dimensional unconstrained solution for the depression and anxiety data. Predictor variable labels are printed on the border of the graph where N = Neuroticism; E = Extraversion; O = Openness to experience; A = Agreeableness; C = Conscientiousness and Edu = Education. We use the convention that labels are printed at the positive side of the variable. Markers on the variable axes indicate standard deviation increases/decreases from the mean. Category points are labeled by the name of the drug together with a 1 (yes) or 0 (no) with D = Dysthymia, M = Major Depressive Disorder, G = Generalized Anxiety Disorder, S = Social Phobia (SP), and P = Panic Disorder (PD)

5 Related and Competing Approaches

The MELODIC family is a statistical toolbox for simultaneous logistic regressions in a reduced dimensional space for the analysis of multivariate binary data. Other statistical models have been proposed for such data; in this section we show some relationships and comparisons. The related approaches can be divided in two types of models: marginal models and conditional models.

Marginal models are like standard regression models, dealing in some way with the dependency among responses. In generalized estimating equations (GEE; [47, 71]), a working correlation structure is adopted and estimation and inference are adjusted based on this structure using a sandwich estimator [68]. GEE has been mainly developed in the longitudinal context but can also be applied to multivariate responses. Maximum likelihood estimation is also possible for marginal models -see for example [5]—but this is computationally more demanding.

Our MELODIC family can be seen as a member of the GEE family, where implicitly we adopt an independent working correlation structure. Moreover, we lay a dimensional structure on the response space. A standard GEE model would be equal to our one-dimensional model, with the constraint that all responses are equally well discriminated (see [69]). Ziegler et al. [72] discuss a set-up where the predictor variables have different effects on each of the response variables. This set-up would correspond to our model in maximum dimensionality (R), where each response pertains to a single dimension. Asar and İlk [3] propose a method in which selected predictor variables have the same effect on selected response variables. This is similar to our constrained model, where predictors have a similar effect on response variables pertaining to a given dimension.

As discussed in Sect. 2.2, our general MELODIC model is equivalent to reduced rank vector logistic regression models [70]. Whereas we started our development from a distance perspective, these reduced rank models start from an inner-product perspective. De Rooij and Heiser [22] give an extensive discussion of the interpretational differences of the two perspectives. Reduced rank logistic regression models can also be visualized using biplots, as has been shown by [66]. The interpretation of these models is through projections on calibrated variable axes as in general biplots (see [30, 31]). We also proposed a constraint model, where *a priori* knowledge about the dimensional structure of the response variables can be incorporated. As far as we know, such constrains have not been proposed in logistic reduced rank models before.

In *conditional models*, latent variables are included in order to model the dependency among the responses. The main examples are generalized linear mixed models and latent class models. The family of GLMMs includes item response models and factor analysis models [54]. When these are expanded with predictor variables, explanatory item response models [16] and structural equation models are obtained. Explanatory item response models have mainly been developed for unidimensional latent variables. Some progress has been made with multidimensional item response models, but the underlying structure should be known a priori (as in our constrained models). Explanatory multidimensional item response models still need further development, partly because estimation is often quite troublesome in such models due to the intractable integral in the likelihood function [64].

Latent class models [46, 48] have been developed for multivariate binary data including predictors [65]. In latent class models, as in GLMMs, the dependency among the response variables is modeled using a latent variable, which in this case is categorical. No dimensional structure is imposed underlying the outcomes, only a choice of the number of categories of the categorical latent variable. These latent class models often require a large sample size in order to obtain stable and reliable results [36].

Hubbard et al. [42], when comparing generalized estimating equations and generalized linear mixed models, noted that "mixed models involve unverifiable assumptions on the data-generating distribution, which lead to potentially misleading estimates and biased inference". More specifically, although the distribution of the random effects cannot be identified from the data, the estimates and inference change according to different choices of the distribution of these random effects. This makes the application of conditional models problematic. Another issue for conditional models is the number of indicator or response variables. In our example on depressive and anxiety disorders, for example, there are only five response variables. Distributing these five response variables over two underlying dimensions results in a dimension with only two dichotomous indicators. It is generally acknowledged that this number is much too low for valid inference. This small number is less of an issue in our family, because we do not assume a particular distribution for the underlying dimension.

A final practical problem in these conditional models is that researchers often first try to find the dimensional structure and in a second step include the predictor variables. The measurement model (step 1), however, might change substantially when the predictor variables are included, leading to a completely different interpretation. To solve this problem, researchers have developed three-step ([7], sometimes called the BCH approach) and two-step estimators [4] within the context of latent class models which were recently adapted for other conditional models. In our family of models we have no division in measurement and structural model; the two go hand in hand.

6 Conclusion and Discussion

In this study, we presented distance models for simultaneous logistic regression analysis of multiple binary response variables based on ideas of multidimensional unfolding. Row objects (participants in our examples) are presented together with the two categories of the response variables in a low-dimensional Euclidean space, where the relative distance between a point representing a participant and the points representing the classes of a response variable determines the probability for each class. The model is estimated by minimizing a deviance function. These models take into account the dependency among the response variables by using a low-dimensional Euclidean space: with the increase in value of a predictor variable, the probabilities of all response variables change simultaneously. We christened the models the MELODIC family, that is the MultivariatE LOgistic DIstance to Categories family. We presented versions of the model both for cases in which we have an a priori theory about the dimensional structure of the response variables and for cases when we do not have such a theory. Two empirical applications are shown, one with and one without such an a priori structure.

In the case of a two-dimensional model, the result can be interpreted using a biplot. In the case of a higher-dimensional solution, similar biplots can be constructed for pairs of dimensions. A coherent interpretation of the complete model from such bi-dimensional plots, might, however, more difficult. Alternatively, the model can be interpreted by the implied logistic regression coefficients which have a change in log odds interpretation similar to ordinary logistic regression models. We illustrated both methods of interpretation in the empirical examples. The fact that the model can be interpreted using a graph and using tables is beneficial, because applicants of statistical models can be divided into two groups: those who prefer visualizations and those who prefer numbers. With the MELODIC family, an applicant can choose which mode of interpretation is most suitable.

We developed a fast iterative majorization algorithm to estimate the parameters of the model. The algorithm converges monotonically to the global optimum of the deviance function. The algorithm alternates between (1) updating an auxiliary vector $\mathbf{a}$, which is simply obtained by taking an average, and (2) updating the regression weights ($\mathbf{B}$) and item discriminations ($\mathbf{K}$), which can be obtained from a generalized singular value decomposition. All model parameters can be obtained from these updates.

A measure of quality of representation for each response variable was also proposed, ranging between 0 and 1. A higher value implies only a small loss of fit with regard to a univariate logistic regression with the specific response variable and the same set of predictor variables. This measure can be used as a diagnostic tool to assess whether response variables are well represented by the model. In our second application, we saw that for one response variable the quality of representation was low. Further exploratory analysis suggested a different substantial theory.

In applications, we need to select the predictor variables as well as the dimensionality of the model. We used information criteria such as the AIC and BIC in the empirical applications. Alternatively, cross-validation or other model selection criteria can be used. We did not discuss uncertainty estimation for our model. Assuming the model is true, we can compute the Hessian matrix and derive standard errors for the parameter estimates from this matrix. Following the GEE setup, we could develop a sandwich estimator for the covariance matrix of the parameters. Alternatively, the bootstrap can be used [23]. The two latter approaches acknowledge the fact that the model is an approximation [9, 10] and probably not a completely accurate representation of a population model. In that sense, the sandwich estimator and the bootstrap can estimate uncertainty with respect to a target model in the population. Focusing on predictive accuracy instead of explanatory value (cf. [53]), the performance of our model in comparison to that of independently fitted logistic regression models is expected to be higher [8].

In this manuscript, we only focused on linear effects of the predictor variables. Non-linear effects of predictor variables on the responses can easily be incorporated as long as they can be translated into a design matrix $\mathbf{X}$, such as with the use of quadratic and cubic effects or with splines defined in terms of a truncated power basis [27]. Non-linear variable axes can be presented in the graphical representation as smooth curves, where effects are still additive. Interactions can also be included in the model. In the graphical representation, *conditional variable axes* need to be represented. In the case of an interaction between variables X_1 and X_2, the graphical representation has a variable axis for X_1 for *each* value of X_2 (or the other way around). For an example of biplots with such interactions among predictors, see [21]. Note that both the nonlinearity and the interactions are effects with respect to all response variables.

The past two decades have seen a rise in penalized estimation methods, which are methods that impose penalties on the parameters of the model. For the MELODIC family, these could be penalties on the regression weights to generalize the model to the case where $P >> n$, such as L_1 (Lasso penalty; [61]) or L_2 penalties (Ridge penalty; [40]). To implement these in the MELODIC family, we would need to alter the identification constraints. In the current algorithm, we used $n^{-1}\mathbf{B}^\top\mathbf{X}^\top\mathbf{X}\mathbf{B} = \mathbf{I}$ to identify the model, but this scaling does not seem to be in line with a penalty on **B**. Therefore, the fixed scaling normalization should be placed on the discrimination values (**K**). Another potential type of penalty is a nuclear norm penalty [25]. In the outlined algorithm (see Algorithm 1), we use a singular value decomposition. If we apply an L_1 penalty to the singular values, the discrimination (**K**) between the categories of response variables slowly diminishes. Because the singular values are ordered, the discrimination in the higher dimensions becomes zero first, and later the discrimination of the first dimensions also becomes zero. If we choose the optimal value of the penalty parameter by cross-validation, such a penalty could be used for dimension selection

We are currently building an R-package that enables empirical researchers to apply the models to their own data. For the moment, the R-code of the examples presented can be found on GitHub (https://github.com/mjderooij/melodic).

References

1. Agresti, A. (2003). *Categorical data analysis*. Wiley.
2. Anderson, J. A. (1984). Regression and ordered categorical variables. *Journal of the Royal Statistical Society: Series B (Methodological), 46*(1), 1–22.
3. Asar, Ö., & İlk, Ö. (2014). Flexible multivariate marginal models for analyzing multivariate longitudinal data, with applications in R. *Computer methods and programs in biomedicine, 115*(3), 135–146.
4. Bakk, Z., & Kuha, J. (2018). Two-step estimation of models between latent classes and external variables. *Psychometrika, 83*(4), 871–892.
5. Bergsma, W., Croon, M., & Hagenaars, J. (2009). *Marginal models: For dependent, clustered, and longitudinal categorical data*. Springer.
6. Berkson, J. (1944). Application of the logistic function to bio-assay. *Journal of the American statistical association, 39*(227), 357–365.
7. Bolck, A., Croon, M., & Hagenaars, J. (2004). Estimating latent structure models with categorical variables: One-step versus three-step estimators. *Political Analysis, 12*(1), 3–27.
8. Breiman, L., & Friedman, J. H. (1997). Predicting multivariate responses in multiple linear regression. *Journal of the Royal Statistical Society: Series B (Statistical Methodology), 59*(1), 3–54.

9. Buja, A., Brown, L., Berk, R., George, E., Pitkin, E., Traskin, M., Zhang, K., Zhao, L., et al. (2019). Models as approximations I: Consequences illustrated with linear regression. *Statistical Science, 34*(4), 523–544.
10. Buja, A., Brown, L., Kuchibhotla, A. K., Berk, R., George, E., Zhao, L., et al. (2019). Models as approximations II: A model-free theory of parametric regression. *Statistical Science, 34*(4), 545–565.
11. Busing, F. M. T. A. (2010). *Advances in multidimensional unfolding*. Doctoral thesis, Leiden University.
12. Coombs, C. H. (1950). Psychological scaling without a unit of measurement. *Psychological Review, 57*(3), 145.
13. Coombs, C. H., & Kao, R. (1955). Nonmetric factor analysis. University of Michigan. Department of Engineering Research. Bulletin.
14. Cox, D. R. (1958). The regression analysis of binary sequences. *Journal of the Royal Statistical Society: Series B (Methodological), 20*(2), 215–232.
15. Cramer, J. S. (2002). The origins of logistic regression. Tinbergen Institute Discussion Paper, 02–119(4).
16. De Boeck, P., & Wilson, M. (2004). *Explanatory item response models: A generalized linear and nonlinear approach*. Springer Science & Business Media.
17. De Leeuw, J. (2005). *Gifi goes logistic: Scasa keynote*.
18. De Leeuw, J. (2006). Principal component analysis of binary data by iterated singular value decomposition. *Computational Statistics and Data Analysis, 50*(1), 21–39.
19. De Leeuw, J., & Heiser, W. J. (1977). Convergence of correction matrix algorithms for multidimensional scaling. In J. Lingoes, E. Roskam, & I. Borg (Eds.), *Geometric Representations of Relational Data* (pp. 735–752). Mathesis Press.
20. De Rooij, M. (2009). Ideal point discriminant analysis revisited with a special emphasis on visualization. *Psychometrika, 74*(2), 317.
21. De Rooij, M. (2011). Transitional ideal point models for longitudinal multinomial outcomes. *Statistical Modelling, 11*(2), 115–135.
22. De Rooij, M., & Heiser, W. J. (2005). Graphical representations and odds ratios in a distance-association model for the analysis of cross-classified data. *Psychometrika, 70*(1), 99–122.
23. Efron, B., & Tibshirani, R. (1986). Bootstrap methods for standard errors, confidence intervals, and other measures of statistical accuracy. *Statistical Science*, 54–75.
24. Evans, G. W. (2014). *Logistic Gifi: A Logistic Distance Association Model for Exploratory Analysis of Categorical Data*. Ph.D. thesis, UCLA.
25. Fazel, M. (2002). *Matrix rank minimization with applications*. Doctoral thesis, Stanford University.
26. Fehrman, E., Muhammad, A. K., Mirkes, E. M., Egan, V., & Gorban, A. N. (2017). The five factor model of personality and evaluation of drug consumption risk. In: F. Palumbo, A. Montanari, M. Vichi (Eds.), *Data science. Studies in classification, data analysis, and knowledge organization* (pp. 231–242). Cham: Springer.
27. Friedman, J., Hastie, T., & Tibshirani, R. (2001). *The elements of statistical learning*. Springer series in statistics.
28. Friendly, M., & Kwan, E. (2011). Comment-why tables are really much better than graphs. *Journal of Computational and Graphical Statistics, 20*(1), 18.
29. Gelman, A. (2011). Why tables are really much better than graphs. *Journal of Computational and Graphical Statistics, 20*(1), 3–7.
30. Gower, J., & Hand, D. (1996). *Biplots*. Taylor & Francis.
31. Gower, J., Lubbe, S., & Roux, N. (2011). *Understanding biplots*. Wiley.
32. Gower, J. C. (1966). Some distance properties of latent root and vector methods used in multivariate analysis. *Biometrika, 53*(3–4), 325–338.
33. Groenen, P. J. F. (1993). *The majorization approach to multidimensional scaling*. DSWO Press Leiden.
34. Groenen, P. J. F., Giaquinto, P., & Kiers, H. A. L. (2003). Weighted majorization algorithms for weighted least squares decomposition models. Econometric Institute Research Papers EI 2003–2009, Erasmus University Rotterdam.

35. Groenen, P. J. F., & Josse, J. (2016). Multinomial multiple correspondence analysis. arXiv:1603.03174.
36. Gudicha, D. W., Tekle, F. B., & Vermunt, J. K. (2016). Power and sample size computation for wald tests in latent class models. *Journal of Classification, 33*(1), 30–51.
37. Guttman, L. (1968). A general nonmetric technique for finding the smallest coordinate space for a configuration of points. *Psychometrika, 33*(4), 469–506.
38. Heiser, W. J. (1981). *Unfolding analysis of proximity data*. Doctoral dissertation, Leiden University.
39. Heiser, W. J. (1995). Convergent computation by iterative majorization: Theory and applications in multidimensional data analysis. In W. J. Krzanowski (Ed.), *Recent advances in descriptive multivariate analysis* (pp. 157–189). Clarendon Press.
40. Hoerl, A. E., & Kennard, R. W. (1970). Ridge regression: Biased estimation for nonorthogonal problems. *Technometrics, 12*(1), 55–67.
41. Hotelling, H. (1936). Simplified calculation of principal components. *Psychometrika, 1*(1), 27–35.
42. Hubbard, A. E., Ahern, J., Fleischer, N. L., Van der Laan, M., Satariano, S. A., Jewell, N., Bruckner, T., & Satariano, W. A. (2010). To gee or not to gee: comparing population average and mixed models for estimating the associations between neighborhood risk factors and health. *Epidemiology*, 467–474.
43. Hunter, D. R., & Lange, K. (2004). A tutorial on MM algorithms. *The American Statistician, 58*(1), 30–37.
44. Izenman, A. J. (1975). Reduced-rank regression for the multivariate linear model. *Journal of multivariate analysis, 5*(2), 248–264.
45. Jolliffe, I. T. (2002). *Principal Component Analysis*. Springer.
46. Lazarsfeld, P. F., & Henry, N. W. (1968). *Latent structure analysis*. Houghton Mifflin Co.
47. Liang, K.-Y., & Zeger, S. L. (1986). Longitudinal data analysis using generalized linear models. *Biometrika, 73*(1), 13–22.
48. McCutcheon, A. L. (1987). *Latent class analysis*. Sage.
49. Molenberghs, G., & Verbeke, G. (2006). *Models for discrete longitudinal data*. Springer Science & Business Media.
50. Pearson, K. (1901). Principal components analysis. *The London, Edinburgh, and Dublin Philosophical Magazine and Journal of Science, 6*(2), 559.
51. Penninx, B. W., Beekman, A. T., Smit, J. H., Zitman, F. G., Nolen, W. A., Spinhoven, P., et al. (2008). The Netherlands study of depression and anxiety (NESDA): rationale, objectives and methods. *International Journal of Methods in Psychiatric Research, 17*(3), 121–140.
52. Roskam, E. E. (1968). *Metric analysis or ordinal data in psychology*. Vam.
53. Shmueli, G. (2010). To explain or to predict. *Statistical Science, 25*, 289–310.
54. Skrondal, A., & Rabe-Hesketh, S. (2004). *Generalized latent variable modeling: Multilevel, longitudinal, and structural equation models*. Crc Press.
55. Spinhoven, P., De Rooij, M., Heiser, W., Smit, J. H., & Penninx, B. W. (2009). The role of personality in comorbidity among anxiety and depressive disorders in primary care and specialty care: A cross-sectional analysis. *General Hospital Psychiatry, 31*(5), 470–477.
56. Stein, C. et al. (1956). Inadmissibility of the usual estimator for the mean of a multivariate normal distribution. In *Proceedings of the Third Berkeley Symposium on Mathematical Statistics and Probability, Volume 1: Contributions to the Theory of Statistics*. The Regents of the University of California.
57. Takane, Y. (1987). Analysis of contingency tables by ideal point discriminant analysis. *Psychometrika, 52*(4), 493–513.
58. Takane, Y. (2013). *Constrained principal component analysis and related techniques*. CRC Press.
59. Takane, Y., Bozdogan, H., & Shibayama, T. (1987). Ideal point discriminant analysis. *Psychometrika, 52*(3), 371–392.
60. Ter Braak, C. J., & Looman, C. W. (1994). Biplots in reduced-rank regression. *Biometrical Journal, 36*(8), 983–1003.

61. Tibshirani, R. (1996). Regression shrinkage and selection via the lasso. *Journal of the Royal Statistical Society: Series B (Methodological), 58*(1), 267–288.
62. Torgerson, W. S. (1952). Multidimensional scaling: I. theory and method. *Psychometrika, 17*(4), 401–419.
63. Torgerson, W. S. (1958). *Theory and methods of scaling*. Wiley.
64. Tuerlinckx, F., Rijmen, F., Verbeke, G., & De Boeck, P. (2006). Statistical inference in generalized linear mixed models: A review. *British Journal of Mathematical and Statistical Psychology, 59*(2), 225–255.
65. Vermunt, J. K. (2010). Latent class modeling with covariates: Two improved three-step approaches. *Political Analysis,* 450–469.
66. Vicente-Villardón, J. L., & Hernández-Sánchez, J. C. (2020). External logistic biplots for mixed types of data. In *Advanced studies in classification and data science* (pp. 169–183). Springer.
67. Vugteveen, J., De Bildt, A., Hartman, C., & Timmerman, M. (2018). Using the dutch multi-informant strengths and difficulties questionnaire (SDQ) to predict adolescent psychiatric diagnoses. *European Child and Adolescent Psychiatry, 27*(10), 1347–1359.
68. White, H. (1980). A heteroskedasticity-consistent covariance matrix estimator and a direct test for heteroskedasticity. *Econometrica: Journal of the Econometric Society,* 817–838.
69. Worku, H. M., & De Rooij, M. (2018). A multivariate logistic distance model for the analysis of multiple binary responses. *Journal of Classification, 35*(1), 124–146.
70. Yee, T. W., & Hastie, T. J. (2003). Reduced-rank vector generalized linear models. *Statistical Modelling, 3*(1), 15–41.
71. Zeger, S. L., & Liang, K.-Y. (1986). Longitudinal data analysis for discrete and continuous outcomes. *Biometrics,* 121–130.
72. Ziegler, A., Kastner, C., & Blettner, M. (1998). The generalised estimating equations: an annotated bibliography. *Biometrical Journal: Journal of Mathematical Methods in Biosciences, 40*(2), 115–139.

Examining Cross-Cultural Value Questionnaires with Quantitative Methods

Reliability and Scalar Equivalence of the Measurement Instruments of Human Values in Germany and Japan

Wolfgang Jagodzinski, Kazufumi Manabe, Hermann Dülmer, Carola Hommerich, and Eldad Davidov

Abstract Good measurement instruments have to meet certain necessary conditions, especially if they are to be used in cross-cultural research. In this study, we examine whether the German and Japanese translations of Shalom Schwartz's PVQ-57 fulfill two of these conditions. The survey measures 19 human values by 57 indicators. First, we examine whether these human values are reliably measured in a student survey which was carried out in autumn/winter 2018 at German and Japanese universities. Second, we investigate whether the instruments allow the comparison of latent value means across the German and Japanese student samples. Such a comparison is meaningful only if scalar equivalence, a high form of measurement equivalence, is reached. Results suggest that at least partial scalar invariance can be established across the samples. We discuss the findings in the broader perspective of value research and outline topics for further investigation.

The authors would like to thank Lisa Trierweiler for the English proof of the manuscript. Eldad Davidov would like to thank the University of Zurich Research Priority Program "Social Networks" for their support during work on this study.

Supplementary Information The online version contains supplementary material available at https://doi.org/10.1007/978-981-99-2240-6_5.

W. Jagodzinski (✉) · H. Dülmer · E. Davidov
University of Cologne, Cologne, Germany
e-mail: jagodzinski@uni-koeln.de

K. Manabe
The Institute of Statistical Mathematics, Tachikawa, Japan

C. Hommerich
Sophia University, Tokyo, Japan

E. Davidov
URPP Social Networks and University of Zurich, Zurich, Switzerland

© The Author(s), under exclusive license to Springer Nature Singapore Pte Ltd. 2023
A. Okada et al. (eds.), *Facets of Behaviormetrics*, Behaviormetrics: Quantitative Approaches to Human Behavior 4, https://doi.org/10.1007/978-981-99-2240-6_5

1 Introduction

Philosophers, historians, and scholars of law often consider values as eternal objects in a metaphysical sphere. The empirical social sciences, by contrast, conceptualize values as properties of human beings or groups. This paper focuses on values, which can be loosely described as stable orientations, goals, or standards of persons. We infer these human values from various forms of behavior, for example, from nonverbal behavior in everyday life or in experiments, from verbal self-descriptions, or from answers to questions. In survey research, mostly responses to standardized items are used for the measurement of values. In cultural research, the database often consists of responses and questionnaires in different languages. In the latter type of research, comparisons between countries and cultures are not always admissible. This paper investigates two necessary conditions for meaningful comparisons of group means: the items in the surveys must be reliable, and at least some of these items must be scalar equivalent measures of the underlying values.

Reliability and scalar equivalence tests are quantitative approaches to research human cognition and behavior. They impose mathematical restrictions on the measurement instrument and can also be applied to the measurement of cognitions and human behavior. Values are not cognitions in the narrow sense, but they are often described as elements of the cognitive system. Apart from that, both methods impose mathematical restrictions on the interpretation of response behavior because they tell us under which conditions and to what extent responses can be interpreted as expressions of values.

Specifically, we will examine whether a German and a Japanese translation of Shalom Schwartz's Portrait Values Questionnaire consist of scalar equivalent, reliable indicators. The Japanese questionnaire has been translated by Kazufumi Manabe in cooperation with Carola Hommerich. Our analyses provide initial findings on the suitability of Schwartz's Portrait Values Questionnaire for measuring values in Japan.

We will briefly sketch the value theory of Shalom Schwartz and his conceptualization of 19 values (Sect. 2), describe our data set and operationalizations (Sect. 3), and present the results of the reliability analysis (Sect. 4). The following Sect. 5 elaborates implicit assumptions of simple mean difference tests, briefly outlines the basic features of our scalar equivalent models, and summarizes the core results of parameter estimation. The German and Japanese means of 17 human values, which have been estimated with these scalar equivalent models, are compared in Sect. 6. Finally, after discussing the results (Sect. 7), we present our summary and conclusions in Sect. 8.

2 The Value Theory of Shalom Schwartz

Survey research largely agrees on four properties of values. P1: *They cannot be directly observed but must be inferred from the responses of people.* For this reason, we often speak of latent or underlying values. P2: *They are stable.* It is assumed that values are internalized in the early stages of life and that they only exceptionally or slowly change in adulthood. P3: *Values are overarching orientations, goals, or standards, which influence a broad spectrum of attitudes and behaviors.* People who place a high emphasis on personal security, for instance, will not only strive for a secure job but also try to insure themselves against all kinds of risks or spend holidays only in secure countries. P4: *The relationship between values and behavior is complex.* We expect people to be guided by their values and to show value-consistent behavior. We observe, however, that convinced pacifists suddenly join the army to protect their country against an external aggressor. Affluent and highly educated people may present themselves as more open, tolerant, and egalitarian in an interview than they really are because they know the socially desired responses, are less affected by the negative effects of social change, and can better protect themselves. There are many other reasons, why the behavior of people is sometimes inconsistent with their internalized values or their responses in surveys.

Schwartz [9] initially elaborated 10 values that he later expanded into 19 values [12] which share these four characteristics. By referring to earlier work, Schwartz and colleagues define values as "trans-situational goals, varying in importance, that serve as a guiding principle in the life of a person or social entity" [12]: 664. There are several important aspects in his theory [12].

(a) *Values form a continuum and a semi-circular structure.* Values that are located close to each other share similar motivations and may be pursued simultaneously. Opposing values have conflicting motivations, and individuals may find it difficult to pursue them at the same time.

(b) *The values may be divided into more or less specific dimensions depending on the research purpose.* In 1992, Schwartz differentiated between two dimensions: Self-transcendence included the values universalism and benevolence vs. self-enhancement which included the values power and achievement. Conservation included the values conformity, tradition, and security and it opposed openness to change which included the values, stimulation and self-direction. The value hedonism was located between openness to change and self-enhancement. In 2012, Schwartz and colleagues proposed another differentiation: They distinguished between social values (self-transcendence and conservation) versus personal values (openness to change and self-enhancement) and between self-protective/anxiety-avoidance values (self-enhancement and conservation) and growth/anxiety-free values (self-transcendence and openness to change). Table 1 lists the different values and their corresponding dimensions, and Fig. 1 presents their position in the circle.

From the value properties P1–P4 we can infer how the ideal measurement instrument of values should look like: Each value should be measured by a broad spectrum

Table 1 Three levels of values: four higher-order values, the 10 basic values, and 19 more narrowly defined values in the refined theory of values [12]

4 higher values[a]	10 basic values and underlying motivations[b]	19 narrowly defined values and underlying motivation[c]
Self-transcendence	Benevolence—Preservation and enhancement of the welfare of people with whom one is in frequent personal contact	Benevolence-Dependability (BED)—Being a reliable and trustworthy member of the ingroup
		Benevolence-Caring (BEC)—Devotion to the welfare of ingroup members
	Universalism—Understanding, appreciation, tolerance, and protection for the welfare of *all*people and of nature	Universalism-Tolerance (UNT)—Acceptance and understanding of those who are different from oneself
		Universalism-Concern (UNC)—Commitment to equality, justice, and protection for all people
		Universalism-Nature (UNN)—Preservation of the natural environment
Conservation		Humility (HUM)[d]—Recognizing one's insignificance in the larger scheme of things
	Conformity—The restraint of actions, inclinations, and impulses that are likely to upset or harm others and violate social expectations or norms	Conformity-Interpersonal (COI)—Avoidance of upsetting or harming other people
		Conformity-Rules (COR)—Compliance with rules, laws, and formal obligations)
	Tradition—Respect, commitment, and acceptance of the customs and ideas that traditional culture or religion provides	Tradition (TR)—Maintaining and preserving cultural, family, or religious traditions
	Security—Safety, harmony, and stability of society, relationships, and self	Security-Societal (SES)—Safety and stability in the wider society
		Security-Personal (SEP)—Safety in one's immediate environment

(continued)

Table 1 (continued)

4 higher values[a]	10 basic values and underlying motivations[b]	19 narrowly defined values and underlying motivation[c]
Self-enhancement		Face (FAC)[d]—Security and power through maintaining one's public image and avoiding humiliation
	Power—Social status and prestige, control, or dominance over people and resources	Power-Resources (POR)—Power through control of material and social resources
		Power-Dominance (POD)—Power through exercising control over people
	Achievement—Personal success through demonstrating competence according to social standards	Achievement (AC)—Definition unchanged
Openness to change	Hedonism—Pleasure and sensuous gratification for oneself	Hedonism (HE)[d]—Definition unchanged
	Stimulation—Excitement, novelty, and challenge in life	Stimulation (ST)—Definition unchanged
	Self-Direction—Independent thought and action, choosing, creating, and exploring	Self-Direction-Action (SDA)—The freedom to determine one's own actions
		Self-Direction-Thought (SDT)—The freedom to cultivate one's own ideas and abilities

[a][9, 10, 12]
[b][9]
[c][12]
Adapted from [2, 3, 12]
[d]Face and humility are newly introduced values. Face is located between the higher-order values, conservation and self-enhancement. Hedonism is located between the higher-order values, self-enhancement and openness to change. Humility is located between the higher-order values, self-transcendence and conservation. The overlaps are accurately visualized in Fig. 1

of items so that the answers to these items allow us to reliably infer the distribution of the respondents on the latent value continuum (P1, P3). This value measurement should be regularly repeated, best in panel studies, so that the stability of the value can be assessed (P2). The measurement should take place under conditions where differences in behavior are most likely a consequence of value differences (P4) and not of differences in beliefs and perceptions. Because such an instrument would impose tremendous costs on nonexperimental value research, the standards for appropriate measurement are considerably lower in practice.

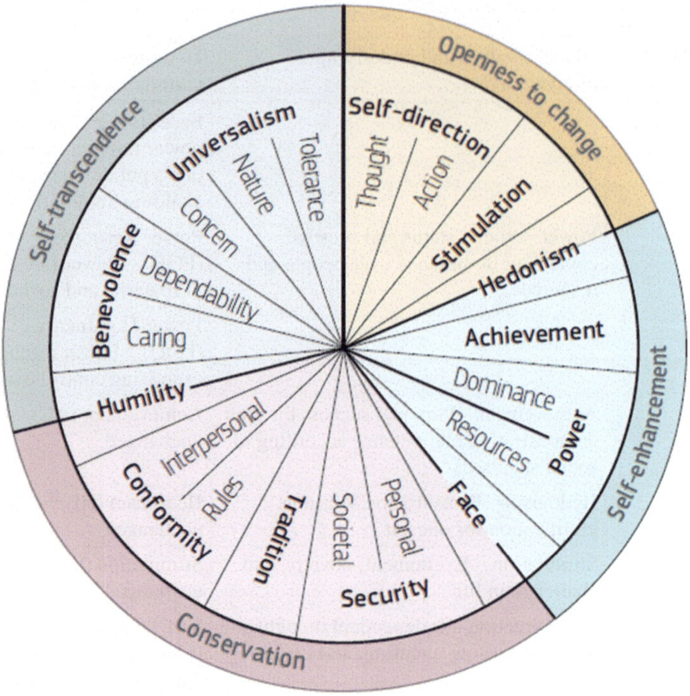

Fig. 1 The circular motivational continuum of 19 values in the refined value theory (adapted from [8], with permission of the authors)

3 Data, Operationalizations, and Descriptive Analysis

Among the various value questionnaires elaborated by Schwartz, the Portrait Values Questionnaires (PVQs) are probably those most frequently used [11]. A condensed version of the PVQ has been included in the core questionnaire of the European Social Survey but Schwartz and colleagues [12] recommend the use of a longer version with 57 items for a comprehensive measurement of values. Each of the 19 values of the expanded value theory is assessed by three questions. Across several countries and languages, this new questionnaire has demonstrated good reliability and validity and its ability to differentiate between the 19 values. It has displayed high levels of comparability (see, e.g., [2, 3]).

The PVQ-57 does not consist of questions asking about broadly described desirable states or goals; rather each of the 57 items offers a brief, vivid portrayal of the stable goals or behaviors of imaginary persons, such as: "She"—or "he" in the male version—"always tries to help others" and "She is always concerned about the environment". Respondents are asked how similar they are to these imaginary focal persons. They can answer on a six-point scale ranging from "not like me at all" (=1) to "very much like me" (=6). The more similar the respondent feels to the imaginary person, the more importance is attached to the underlying value. Each value is operationalized by three items or indicators of this type. To avoid halo effects, the three indicators of a value are not placed directly one after the other but are distributed over the entire battery of 57 items. The three indicators of self-direction-thought, for example, are questions number 1, 23, and 39 in the PVQ-57.

Altogether, the PVQ is, so to speak, an ongoing project where indicators, question wordings, and translations are sporadically changed and improved. While we were able to adopt an earlier German translation with some minor changes, a completely new Japanese translation had to be prepared.[1] The Japanese translation is based on the PVQ-RR (10/2013). In contrast to the German version, later changes of two items in the PVQ 57-RR (6/2014) were not considered. We therefore cannot expect these two indicators to have the same loadings and to be scalar equivalent in Germany and Japan, but they can still be indicators of the same value. Our analysis allows to examine this question more closely.

In autumn/winter 2018, the male and female versions of the German questionnaire were presented to 313 students at the University of Cologne, and the gender-neutral version of the Japanese questionnaire to 193 students at Hokkaido University and 102 students at Aoyama Gakuin University in the Metropolitan Area of Tokyo. In our analyses, we collapsed the surveys of the two Japanese universities. The reader can download the English master questionnaire, the two German and the gender-neutral Japanese translations, the pooled data set, and a methodological appendix from Electronic Supplementary Material (see Appendix).

Before we turn to the empirical analyses, some terminological clarifications are necessary. Although our samples consist exclusively of university students, we will speak of Germans and Japanese for the sake of brevity. We distinguish between the questions in the surveys, the written responses, and their numerical representation in the data sets, which are called variables. Questions are identified by their position in the questionnaire. Question 14 is the 14th question in the questionnaire and V14 is the corresponding numerical representation of the responses. The terms "item" and "indicator" can refer to the question, the variable, or both. Numbers under the heading "Items" in Tables 2 or 3 refer to variables.

While it is typically easy to determine the specific meaning of the term "item", it may be more difficult to disambiguate the term "value". Value can be a human value like security or freedom, the numerical representation of a human value, i.e., the value scale, or the specific numerical value on the value scale. Accordingly, "the value of a value" can be the numerical value on a human value scale, but also the higher order value of a human value. In the latter sense, we may call hedonism of lesser value and benevolence of greater value. To reduce these ambiguities, with "value $\boldsymbol{V}$" we refer to the numerical representation of an unspecified human value.[2] We use capital letters when we refer to one of the values of Shalom Schwartz.

---

[1] See [4].

[2] Bold letters in italics indicate that values are, technically speaking, vectors. We never refer to a set of several values so that it is unnecessary to distinguish between matrices and vectors.

Table 2 Means and standard deviations of the value indicators

Item no.		Germany			Japan		
	Item label: It is important to her/him	N	$\widehat{\mu}$	$\widehat{\sigma}$	N	$\widehat{\mu}$	$\widehat{\sigma}$
	SDT: Self-Direction Thought						
1	… to form her/his views independently	312	5.03	0.860	295	4.31	1.008
23	… to develop her/his own opinions	313	5.29	0.803	294	4.42	1.015
39	… to figure things out her-/himself	313	4.87	0.963	295	4.63	0.973
	SDA: Self-Direction Action						
16	… to make her/his own decisions about her/his life	313	5.26	0.802	295	4.98	0.956
30	… to plan her/his activities independently	313	4.77	0.956	295	4.43	1.040
56	… to be free to choose what she/he does by her/himself	313	5.61	0.626	295	4.93	0.893
	ST: Stimulation						
10	… always to look for different things to do	311	4.32	1.054	295	4.12	1.257
28	… to take risks that make life exciting	310	4.19	1.144	294	4.46	1.238
43	… to have all sorts of new experiences	311	4.99	0.943	295	4.74	1.029
	HE: Hedonism						
3	… to have a good time	311	5.42	0.753	295	5.33	0.867
36	… to enjoy life's pleasures	313	5.46	0.767	295	5.26	0.871
46	… to take advantage of every opportunity to have fun	313	4.83	1.016	294	4.87	1.019
	AC: Achievement						
17	… to show her/his abilities[(a)]	312	4.66	1.002	294	4.49	1.199
32	… to be very successful	310	4.06	1.234	294	4.06	1.133

(continued)

Means and standard deviations of all 57 items in the German and Japanese sample are displayed in Table 2. The indicators are grouped into 19 blocks. Each block starts with the name and label of the value and is completed by the three associated indicators.

Together with the covariances among the three indicators (I_1–I_3) of each value, the means and standard deviations in Table 2 form the observed information on which our ML estimations with the programs AMOS and MPLUS are based.

Table 2 (continued)

Item no.		Germany			Japan		
48	… that people recognize what she/he achieves	312	4.50	1.014	294	4.30	1.156
	POD: Power Dominance						
6	… that people do what he/she says they should	311	3.32	1.225	294	3.00	1.170
29	… to have the power to make people do what she/he wants	311	2.73	1.341	295	3.42	1.189
41	… to be the one who tells others what to do	311	2.57	1.200	294	3.10	1.092
	POR: Power Resources						
12	… to have the power that money can bring	312	3.16	1.302	295	4.19	1.009
20	… to be wealthy	312	2.94	1.312	295	4.23	1.081
44	… to own expensive things that show her/his wealth	312	2.30	1.216	294	2.77	1.244
	FA: Face						
9	… that no one should ever shame her/him	312	4.17	1.271	295	3.73	1.181
24	… to protect her/his public image	313	3.97	1.143	295	3.87	1.168
49	… never to be humiliated	309	4.22	1.245	293	3.83	1.212
	SEP: Security Personal						
13	… [very] … to avoid disease and protect her/his health	312	4.90	1.041	295	4.78	1.061
26	… to be personally safe and secure	313	4.96	0.928	295	4.75	1.055
53	… not to feel threatened[(b)]	313	4.99	1.000	295	4.16	1.047
	SES: Security Society						
2	… that her/his country is secure and stable	313	5.40	0.807	295	4.79	1.126

(continued)

4 Reliabilities of Sum Scores and Indicators

It follows from P1 to P4 that changes or differences in ***V*** should have similar effects on all three indicators. All three should increase if ***V*** becomes more important and should decrease in the opposite case. The better the indicators, the higher their intercorrelations and the higher the reliability of a value score.

a. Most reliability estimations are explicitly or implicitly based on the equation

$$\boldsymbol{I}_j = \lambda_j \boldsymbol{V} + \boldsymbol{\epsilon}_j \text{ with } j = 1, \ldots, J \tag{1}$$

where $\boldsymbol{I}_j$ is the j-th out of J indicators of a latent variable $\boldsymbol{V}$, $\boldsymbol{\epsilon}_j$ is the measurement error, and λ_j is the strength of the effect of $\boldsymbol{V}$ on $\boldsymbol{I}_j$ or the loading of the indicator on the latent variable. $\boldsymbol{V}$ and $\boldsymbol{\epsilon}_j$ have linear and additive effects on the indicator. The measurement errors of different indicators have zero means and

Table 2 (continued)

Item no.		Germany			Japan		
35	… to have a strong state that can defend its citizens	310	4.42	1.274	295	4.21	1.121
50	… that her/his country protects itself against all threats	312	4.37	1.249	295	4.06	1.145
	TR: Tradition						
18	… to maintain traditional values and ways of thinking	312	3.13	1.446	295	3.24	1.239
33	… to follow her/his family's customs or the customs of a religion	313	2.65	1.520	295	2.79	1.288
40	… to honor the traditional practices of her/his culture	312	3.16	1.435	295	3.44	1.173
	COR: Rule Conformism						
15	… never to violate rules or regulations	313	3.42	1.266	295	4.13	1.116
31	… to follow rules even when noone is watching	311	3.23	1.345	292	3.92	1.224
42	… to obey all the laws	310	3.30	1.366	294	3.70	1.220
	COI: Interpersonal Conformism						
4	… to avoid upsetting other people	312	4.01	1.233	295	4.09	1.224
22	… never to annoy anyone	311	4.25	1.207	295	3.98	1.226
51	… never to make other people angry	313	3.73	1.323	295	3.92	1.256
	HU: Humility						
7	… never to think she/he deserves more than other people	310	3.82	1.262	295	3.53	1.115
38	… to be humble	312	4.33	1.147	295	4.53	1.175
54	… to be satisfied with what she/he has and not ask for more	313	4.04	1.254	295	3.13	1.204

(continued)

are uncorrelated with the latent variable $\boldsymbol{V}$ and among each other. Let

$$\boldsymbol{S} = \sum_{j=1}^{J} \boldsymbol{I}_j \qquad (2)$$

be the unweighted sum of all indicators. Then the reliability of the sum score can be defined as the proportion of variance explained by the underlying value $\boldsymbol{V}$ divided by the total variance of $\boldsymbol{S}$.[3] Table 3 displays for each of the 19 values

[3] If the latent variable has unit variance ($\sigma_v^2 = 1$), then under the model assumptions the explained variance is simply the squared sum of the factor loadings. Cronbach's α is calculated under the

Table 2 (continued)

Item no.		Germany			Japan		
	UNN: Universalism Nature						
8	… to care for nature	312	4.48	1.205	295	3.32	1.172
21	… to take part in activities to defend nature	312	3.63	1.338	295	3.05	1.150
45	… to protect the natural environment from destruction or pollution	313	4.64	1.225	294	3.87	1.076
	UNC: Universalism Concern						
5	… that the weak and vulnerable in society be protected	313	5.06	0.896	295	4.30	1.072
37	… that every person in the world have equal opportunities in life	313	5.35	0.868	294	4.52	1.222
52	… that everyone be treated justly, even people she/he doesn't know	313	5.12	0.954	295	4.34	1.141
	UNT: Universalism Tolerance						
14	… to be tolerant toward all kinds of people and groups	313	5.54	0.724	293	4.66	1.050
34	… to listen to and understand people who are different from her/him	313	4.96	0.893	294	4.72	1.038
57	… to accept people even when she/he disagrees with them	312	5.02	0.923	295	4.80	0.976
	BEC: Benevolence Caring						
11	… to take care of people she/he is close to	312	5.59	0.702	295	4.20	1.178
25	… [very] … to help the people dear to her/him	313	5.69	0.591	295	4.84	0.961

(continued)

Cronbach's α, the reliability of sum scores which are calculated from the estimates of 17 scalar equivalent models, and finally the standardized factor loadings of the indicators of these models. For easier identification, the column numbers are displayed in brackets in the table header. Numbers and names of the values appear in col. (1), followed by the numbers of the indicators in col. (2), and the reliabilities of sum scores in cols. (3)–(6). In addition, we report the standardized loadings of the indicators in our scalar equivalent models in cols. (7)–(12) because these loadings reflect the measurement quality of each indicator. Apart from Cronbach's α, all estimates refer to the scalar equivalent models in the methodological appendix.

b. The estimates of Cronbach's α are displayed in col. (3) for Germany (DE) and in col. (4) for Japan (JP). In experimental research, $\alpha \geq 0.7$ is usually regarded as a good reliability. To us, the softer criterion of 0.6 seems more appropriate for nonexperimental research, and in this case even more so because we had to

simplifying assumption that all loadings are equal ($\lambda_1 + \cdots + \lambda_J = \lambda$) so that the explained variance becomes $(J\lambda)^2$.

Table 2 (continued)

Item no.		Germany			Japan		
47	… to concern her-/himself with every need of her/his dear ones	312	4.99	1.041	294	4.70	0.987
	BED: Benevolence Dependability						
19	… that people she/he knows have full confidence in her/him	313	5.39	0.809	294	3.90	1.177
27	… to be a dependable and trustworthy friend	312	5.62	0.620	295	4.72	1.024
55	… that all her/his friends and family can rely on her/him completely	313	5.36	0.847	293	3.57	1.244

Note
(1)Abbreviations in the table header:
No => Number of the item in the questionnaire
N => Number of respondents
$\widehat{\mu}$=>Estimated mean
$\widehat{\sigma}$ = Estimated standard deviation
(2)The English master questionnaires on which the German and the Japanese translations were based differed in Items 17 and 53. The wording of the German translation basis is displayed in the Table.
(a)Item 17—(English master questionnaire of the Japanese translation): … to have ambitions in life.
(b)Item 53—(English master questionnaire of the Japanese translation): … to avoid anything dangerous.

confine ourselves with three indicators per value. If further indicators with similar inter-correlations were added, the reliabilities would increase.

The threshold of 0.6 is reached by all but two human values in Japan. In Germany, self-direction action is a borderline case ($\alpha = 0.58$) which surpasses the threshold in an appropriately specified model (0.618 in col. (5)). In 7 out of 19 cases the alpha reliability is slightly higher in Japan than in Germany, in the remaining 12 cases the opposite holds.

We exclude humility ($\alpha = 0.387$) and achievement ($\alpha = 0.396$) from further analyses because their reliabilities are disappointingly low in Japan. We can hardly imagine that the translation of the older version of V17 alone has caused the low reliability of the Japanese achievement scale. A general problem of the measurement of achievement is that the items mostly address extrinsic goals.[4] Whether this is the reason for the low reliability in Japan has to be further examined.

The major problem of humility is V54, which shows an extremely low correlation with the other two indicators in Japan. Presently, we do not know the reason for this failure. In any case, the three indicators seem to describe humility in Japan less well than in Germany.

[4] The German version of V17 [show her/his abilities] and V48 [achievements are recognized]—clearly indicate an extrinsically motivated orientation. The third indicator, V32 [be successful], is ambiguous in this respect: we may strive for success because it gives us an inner feeling of satisfaction or because our success is socially recognized.

Table 3 Reliability Analysis

			Sum scores				Full or partial scalar equivalent models					
			Alpha[a]		ScEqM[b]		Standardized Loadings of the indicators					
(1)		(2)	(3)	(4)	(5)	(6)	(7)	(8)	(9)	(10)	(11)	(12)
No	Value	Items[c]	DE	JP	DE	JP	DE	JP	DE	JP	DE	JP
1	SDT	**1**, 23, 39	0.611	0.670	0.617	0.673	0.60	0.57	**0.77**	**0.68**	*0.43*	0.65
2	SDA	**16**, 30, 56	0.580	0.645	0.618	0.659	**0.72**	0.56	0.61	0.50	*0.37*	**0.84**
3	ST	**10**,43,28	0.613	0.806	0.644	0.796	0.55	0.64	**0.75**	**0.86**	0.57	0.77
4	HE	**3**, 36, 46	0.764	0.713	0.778	0.714	0.68	0.63	**0.79**	0.69	0.74	**0.70**
5	AC	(17,32,48)	0.630	***0.396***	–							
6	POD	**29**, 41, 6	0.785	0.678	0.793	0.683	0.73	0.63	**0.88**	**0.74**	0.63	0.58
7	POR	**12**,20,44	0.854	0.666	0.855	0.686	0.78	0.67	**0.92**	**0.80**	0.73	0.50
8	FA	**9**,24, 49	0.731	0.739	0.765	0.756	0.71	0.70	0.50	0.49	**0.91**	**0.90**
9	SEP	**26**, 13, 53	0.686	0.635	0.696	0.661	**0.88**	**0.81**	*0.45*	*0.46*	0.64	0.59
10	SES	**35**, 50, 2	0.782	0.713	0.809	0.741	0.79	**0.82**	**0.80**	0.81	0.66	*0.42*
11	TR	**18**,33, 40	0.868	0.828	0.868	0.830	0.82	0.79	0.82	0.75	**0.84**	**0.84**
12	COR	**15**, 31,42	0.867	0.771	0.870	0.774	**0.91**	**0.85**	0.76	0.68	0.83	0.67
13	COI	**4**, 51, 22	0.658	0.823	0.692	0.828	0.57	0.63	**0.83**	**0.91**	0.53	0.78
14	HU	(7,38,54)	0.641	***0.387***	–							
15	UNN	**8**, 45, 21	0.885	0.822	0.877	0.831	**0.90**	0.75	0.80	0.76	0.82	**0.85**
16	UNC	**52**,37, 5	0.803	0.771	0.803	0.778	0.77	**0.79**	**0.81**	0.72	0.69	0.68
17	UNT	**34**, 57, 14	0.731	0.767	0.735	0.771	**0.74**	**0.81**	0.64	0.77	0.70	0.61
18	BEC[d]	**11**, 25, 47	0.681	0.718	0.610	0.672	**1.0**	**1.0**	0.63	0.80	0.36	0.38

(continued)

c. Cronbach's α is based on the restrictive assumption that ***V*** has the same effect on all indicators. If different effects of ***V*** on the indicators are admitted and all other constraints are kept, a congeneric model with usually slightly higher reliabilities is obtained. Cronbach's α is a lower bound reliability estimate of these models. In cols. (5) and (6), we report the reliability of sum scores, which are calculated from the parameters of the scalar equivalent models (ScEqM) that we will discuss later. These models differ in a variety of constraints. They all impose equality constraints across countries but only a few of them set the loadings of different indicators equal. Some models allow for correlated errors. Therefore, the ScEqM–sum score reliabilities often are higher than Cronbach's α but sometimes also below (cf. value 18 (BEC) in Table 3). Even in the latter case, they remain above the threshold of 0.6. Overall, the sum score reliabilities of the 17 values are satisfactory.

Table 3 (continued)

			Sum scores				Full or partial scalar equivalent models					
			Alpha[a]		ScEqM[b]		Standardized Loadings of the indicators					
(1)		(2)	(3)	(4)	(5)	(6)	(7)	(8)	(9)	(10)	(11)	(12)
19	BED[d]	**19**,27, 55	0.729	0.679	0.764	0.635	0.76	**0.79**	0.62	0.45	**0.92**	0.75

Notes
[a]Cronbach's α
[b]ScEqM: Sum scores calculated from the estimates of full or partial scalar equivalent models
[c]The item numbers in col. (2) correspond to the numbers of the items in the questionnaire and the number of the variable. The bold number indicates the reference indicator. The order of the item numbers in each row of col. (2) corresponds to the order of the indicators on the right-hand side of the table (cols. (7)–(12)). Thus, we get the following information about SDT (*Self-Direction—Thought)*in the first row of the table: The value is measured by the three variables V1, V23, and V39. V1 is the reference indicator because **1** is a bold letter. V1 is also the first variable on the right side of the Table (cols. (7), (8)), V23 the second (cols. (9), (10)), and V39 the third (cols. (11), (12)). Therefore, 0.60 in cell (1, 7) and 0.57 in cell (1, 8) are the standardized loading loadings of V1 in Germany and Japan
[d]In these models, (positively) correlated measurement errors have been admitted. As a consequence, the generalized congeneric reliability may even be lower than Cronbach's α
[e]Cols. (7)–(12): The largest standardized factor loading in each country is always depicted in bold numbers. The best indicator (with the highest loadings) of SDT is V23 whose standardized loading is 0.77 in Germany (cell (1, 9)) and 0.68 in Japan (cell (1, 10)). The best indicator of SDA is V16 in Germany (standardized loading = 0.72—cell (2, 7)) and V56 in Japan (standardized loading = 0.84—cell (2, 12))

The reliability of each indicator is equal to its squared standardized loadings. We follow the widespread practice and report the unsquared standardized loadings in cols. (7) to (12). They vary between 0.36 (indicator V47 of BEC in Germany) and 1.0 (indicator V11 of BEC in Germany and Japan). According to the relative weak criterion of [1], they are all above 0.3 and therefore acceptable. However, a loading of 0.4 would mean that only 16% of the total variance of an indicator is explained by the latent value. There is still room for improvement particularly regarding benevolence-caring (BEC) where the standardized loading of the third indicator (V47) is below 0.4 in both countries.

5 Scalar Equivalence: The Problem, Model Specification, and Estimation

The comparison of group means is presumably one of the most common test procedures in empirical research. It is often overlooked, however, that such comparisons are only admissible if the conditions of scalar measurement invariance are met. Multiple indicators allow the test of these conditions, single indicators do not.

5.1 *Why Mean Difference Tests Can Lead to False Conclusions*

The descriptive analysis of indicators, which always should precede the advanced testing, can already show why simple mean difference tests can lead to erroneous conclusions. Line graphs and histograms are useful tools[5] for a check. Figure 2 presents the line graphs of the three indicators of *Stimulation* (***ST***) with the six answer categories on the horizontal axis and the percentages of answers in each answer category on the vertical axis.[6] The German and Japanese percentages appear as numerical values in small boxes above each category. The Japanese percentages are connected by a dotted line, and the German percentages are connected by a solid line. Means and standard deviations of the Japanese and the German sample are displayed beside the graphs. For ease of comparison, we have also displayed the means as vertical lines in the diagrams, with the Japanese mean again as a dotted line and the German mean as a solid line.

The charts show that German students place slightly more emphasis on diversity (V10) and new experiences (V43) on average and Japanese students place more emphasis on excitement (V28). If Eq. (1) still holds and each indicator has the same positive, unstandardized loading on ***ST*** in Germany and Japan, then the mean difference must have the same sign for all three indicators. In Fig. 2, however, we see that the German mean is higher than that of Japan for two variables (V10, V43) and lower for one variable (V28). This finding led us to conclude that these three variables cannot be indicators of ***ST*** or that our assumptions must be wrong.

Let us investigate the latter possibility. Even if the signs of the three mean differences differ, the three variables can still be reliable indicators of ***ST***. Here are three reasons why we might get the reversed sign for the mean difference of V28: (a) Measured variables, so far not considered, may differently affect the German and Japanese responses to V28. (b) Unmeasured variables may have the same effect. (c) The unstandardized factor loadings of V28 may differ in Germany and Japan. If V28 has a large unstandardized loading on ***ST*** in Japan and a small one in Germany, the Japanese mean of V28 can be larger than the German mean even though Germans score higher on Stimulation on average. Let us take a closer look at these three possibilities.

(a) Risk-taking can be related to another human value included in our survey: Persons who care deeply about their Personal Security (***SEP***) may also be risk aversive. Anxious people try to avoid larger risks. Accordingly, V28 might be affected by two values, positively by ***ST*** and negatively by ***SEP***. Germans may score higher on ***SEP***, and only for this reason, score lower on V28 than Japanese on average.
(b) In our survey, the value of Personal Security is measured but many other factors are not. Responses to V28 may be dependent on whether we live in a secure

[5] See the line graphs in [4], further descriptive analyses of the PVQ in [5, 6].

[6] The charts are produced with SPSS, V24.

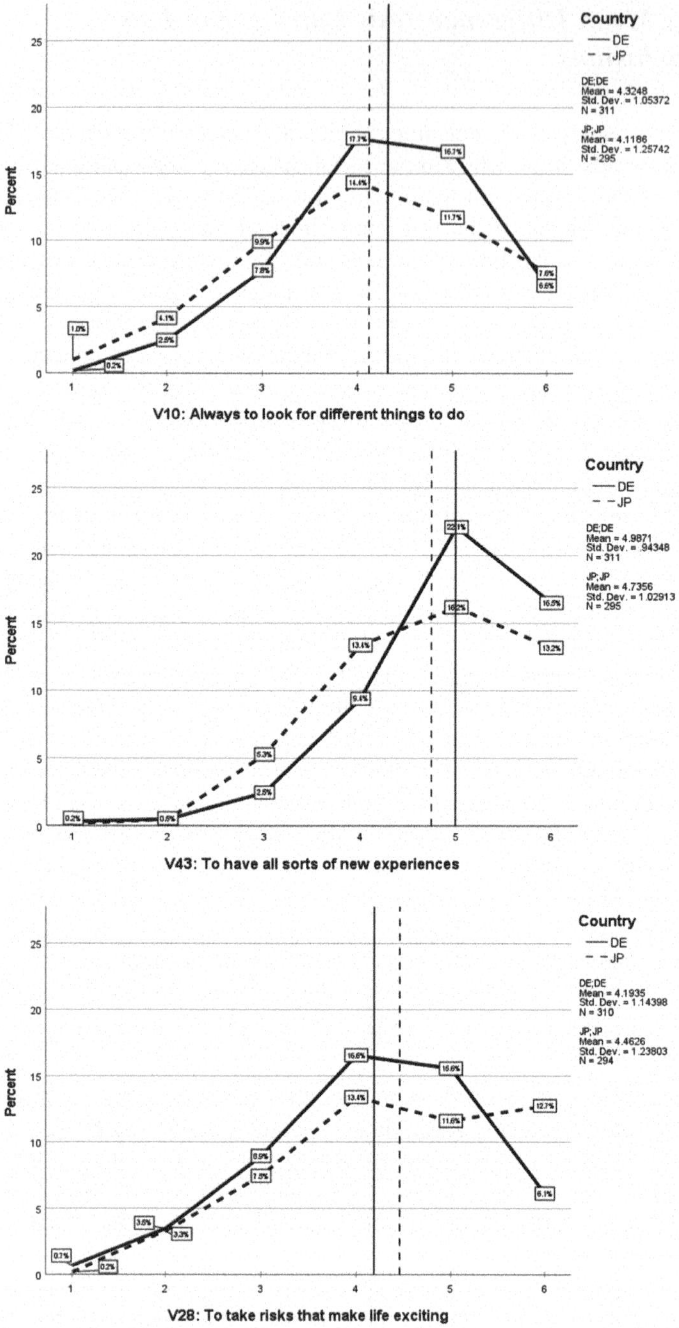

Fig. 2 Frequency distributions of the three indicators of *stimulation* (***ST***)

or insecure environment, for instance. Similarly, customs and regulations as well as social pressure or social desirability may differently affect Germans and Japanese. Suppose that risk-takers are more admired in Japan than in Germany, at least among younger people. Taking risks would be socially desirable. Young Japanese, therefore, would portray themselves in the interview as more risk-taking than they actually are. Accordingly, we would overestimate the Japanese willingness to take risks and obtain an upward-biased Japanese mean. As this hypothetical example shows, simple mean difference tests can lead to false conclusions if the goals and behaviors under study are differently valued in the compared cultures. In fact, our data do not indicate that risk-taking is differently valued in Germany and Japan.

(c) Finally, the indicator loadings can differ between groups. In the PVQ, the likelihood is especially large because the behaviors and goals of the imaginary focal person are often only broadly and vaguely described. Suppose we want to measure risk-taking during high diving and portray the imaginary focal person C as follows: "He likes to take risks in high diving." Arthur, who has only made headers from the one, imagines C as person who jumps from the five. Bill has already made headers from the five and imagines C as a swimmer who jumps from the ten. In the interview, Arthur and Bill choose response category 3 (a little like me). Apparently, their answers are based on different scales because they have compared themselves with different imaginary persons. If cautious Arthur had the numerical value of 2 and the slightly more courageous Bill the value of 5 on a latent risk-taking scale, we would have to multiply Arthur's value by 1.5 and Bill's value by 0.6 to obtain the value of 3 on the observed response scale. In this hypothetical example, we know the scores of the latent value scale. In practice, we have to estimate the mean of $\boldsymbol{V}$, and this can only be done if at least some indicators have the same[7] loadings in all groups and are not differently affected by the factors mentioned above under (a) and (b).

5.2 *Model Equation and Estimation*

Accordingly, we must solve two problems: (a) State a theoretically plausible but testable model equation and (b) make sure that the latent value scale or at least its moments can be estimated.

(a) Our simplified model assumes that J indicators are available for measuring the value $\boldsymbol{V}$ in G groups and that the following equation holds for the relationship between indicators and value $\boldsymbol{V}$:

$$\boldsymbol{I}_j^g = c_j^g + \lambda_j^g \boldsymbol{V}^g + \boldsymbol{\varepsilon}_j^g \text{ with } j = 1, \ldots, J \text{ and } g = 1, \ldots, G \tag{3}$$

The superscript g indicates the relationships in multiple groups as a single equation. In our case, g represents the two groups, Germany and Japan ($g \in \{DE, JP\}$).

[7] Or nearly the same.

Just as in Eq. (1), we further assume that the observed indicator score is the sum of the weighted value ***V*** and measurement error. If not stated otherwise, we also keep the former assumptions about measurement errors. As the PVQ-57 conceptualizes the items as pure indicators of a single value, we assume that correlations between a second value, say ***SEP***, and the indicators of a reference value, say ***ST***, are due to the correlation between the two values. Direct effects of several values on an indicator are not admitted.

If unmeasured variables affect the response behavior of groups differently, the intercepts of the respective indicators can vary. As already pointed out, social desirability and many other socio-economic and cultural factors can produce such a variation. If risk avoidance were a more desirable virtue in Germany than in Japan, the German intercept of V28 may be larger ($c_{28}^{DE} > c_{28}^{JP}$).

(b) It follows from the previous considerations that the mean of a value ***V*** can only be estimated if the parameters in Eq. (3) are identified and the intercepts and the loadings of the indicators are the same or minimally different from each other in the groups under investigation. Identification can be reached by fixing the values or by imposing equality constraints on the parameters. Models are fully scalar invariant if each indicator has the same loadings and the same intercepts in all groups. Under these conditions, the models of all groups are full scalar equivalent (FSE). If only a subset of at least two indicators meets these requirements, the models are partial scalar equivalent (PSE) (cf. [13]). We always begin with a test for full scalar equivalence. If the models do not fit the data, we specify an appropriate PSE model by selecting the indicator that most likely does not meet the criteria of scalar equivalence Then, to obtain fitting models, we relax the intercept or the factor loading of this indicator or both. As an exception, we allow also for correlated measurement errors. The results of our estimation and testing are reported in the appendix.[8]

We established FSE for four values (FA, TR, UNC, and BED) and PSE for the remaining human values. In all models, the comparative fit index (CFI), the normalized fit index (NFI), and the relative fit index (RFI) were above 0.95, which indicates a good fit. The root mean square error of approximation (RMSEA) remained below 0.06 in 15 models and below 0.07 in all 17 models. Overall, the fit is good or at least acceptable.

We had anticipated that V53 might not be a scalar invariant measure of Personal Security because different items were translated in Germany and Japan. Similarly, Fig. 2 led us to expect that Risk Taking (V28) would be differently associated with Stimulation in Germany and Japan. Both expectations were confirmed. According to our estimation (see Appendix, Table A2), the intercepts of V28 are equal in both countries but the unstandardized loadings of V28 differ. If the mean of ***ST*** has been appropriately estimated from V10 and V42, we can conclude that the value of Stimulation has a stronger impact on risk-taking in Japan than in Germany.

[8] If the equality constraint is relaxed in the PSE models, the estimates of the freed parameters of the groups may largely differ. Other approaches—the Bayesian approximation approach or alignment optimization approach—may produce smaller differences. For a comparison of these approaches in a Monto Carlo simulation, see [7]. We cannot say, however, that one of the approaches is superior to the others. Ultimately, the theoretically most plausible model should be selected.

6 Mean Difference Tests of Human Values

After the full or partial scalar equivalence has been established in 17 models, a further necessary condition for the comparison of latent means is fulfilled. Technically, the test can be performed by imposing an equality constraint on both means. Thereby, we gain one degree of freedom and observe an increase of the χ^2 value of the model. A χ^2 increase > 2 with df = 1 is interpreted as a significant increase. We would like to note that these comparisons are not based on representative samples. However, they do suggest some directions and illustrate how such comparisons may be done using population data.

On the left side of Table 4, the relevant information is displayed. For each model, the estimated German and Japanese mean is reported in cols. (1) and (2), the difference between these means in column (3), and the increase in χ^2, induced by the equality constraint on the means, in column (4).

We find 12 instances of a positive and 5 instances of a negative sign for the differences in the structural latent means between Germany and Japan reported in col. (3). Japanese seem to be slightly more traditional (TR) and more conformist (COI,

Table 4 Mean difference tests: structural means versus mean scores

	Structural (Latent) means				Mean scores		
	DE	JP	Diff	χ^2increase	DE	JP	Diff
	(1)	(2)	(3)	(4)	(5)	(6)	(7)
M1: FA	4.11	3.79	0.32	16.954	4.12	3.81	0.31
M2:TR	3.09	3.27	−0.18	3.466	2.98	3.15	−0.17[a]
M3: UNC	5.15	4.31	0.84	112.292	5.18	4.39	0.79
M4: BED	5.38	3.91	1.47	339.211	5.46	4.07	1.39
M5: SDA	5.27	4.97	0.30	23.431	5.21	4.78	0.43
M6: BEC	5.59	4.2	1.39	230.78	5.43	4.58	0.85
M7 SDT	5.03	4.3	0.73	142.457	5.06	4.45	0.61
M8: SEP	4.96	4.75	0.21	7.274	4.95	4.56	0.39
M9: ST	4.33	4.11	0.22	10.287	4.50	4.44	0.06[a]
M10: HE	5.44	5.3	0.14	7.231	5.24	5.15	0.09[a]
M11: POD	2.79	3.35	−0.56	42.949	2.87	3.17	−0.3
M12: POR	3.15	4.2	−1.05	160.151	2.80	3.73	−0.93
M13: SES	4.45	4.19	0.26	8.383	4.73	4.36	0.37
M14: COI	3.99	4.11	−0.12	3.072	3.99	4.00	−0.01[a]
M15: COR	3.41	4.14	−0.73	57.205	3.31	3.92	−0.61
M16: UNN	4.45	3.39	1.06	118.291	4.25	3.41	0.84
M17: UNT	4.97	4.72	0.25	11.738	5.18	4.72	0.46

Note
[a]Not significant at the 5% level in a 2-tailed t-test

COR), and they consider both forms of power (POD, POR) as more important. The fulfillment of two necessary conditions does not prove, of course, that German and Japanese really differ with regard to the underlying values. But the test results can stimulate further research if we transform them into research-leading hypotheses for broader, systematic studies. With regard to traditionalism, it is probably true that Japanese celebrate historical events by parades and public rites more often than Germans, but Germans love their medieval cities and castles. A closer examination may uncover the finding that traditionalism is not an overarching value but rather a domain-specific attitude: A person may be very traditional in one domain of life but very modern in others. And if we look only at a society as a whole, we usually ignore composition effects: Traditionalism of a person may be more widespread in some geographical areas, some professions, some age groups, etc. than in others.

The same may apply to conformism. Mask wearing in the pandemic is not the only example where rule conformism seems to be more widespread in Japan than in Germany. But there may be counterexamples, too. As long as we have not systematically examined the whole domain of conformist behavior, we should be cautious with conclusions about the value of conformism. The power hypothesis (POD, POR) may inspire an investigation of the question whether Germans—after World War II—have developed a broken relationship with power, which also results in an ambivalent response behavior.

All forms of universalism (UNC, UNN, UNT), benevolence (BED, BEC), self-direction (SDT, SDA), and security (SEP, SES) are more important to Germans than to Japanese. But is it really so? It would be necessary to investigate daily behavior of people before such hypotheses can be convincingly confirmed. In general, but in particular, with regard to universalism and self-direction, the question arises whether German students pay lip service to what they have learned in their postwar education. And if benevolence in everyday behavior should be lower in Japan than in Germany, the causes of the difference are of interest: Is it possible that Japanese are less willing to help because they do not want to make mistakes and take actions that are not desired by the person being helped? Even though the comparison of working hours and leisure time clearly seems to support the hypothesis that Germans are more hedonistic (HE) than Japanese, the question remains how hedonism becomes manifest in everyday life. Is the number of working hours correlated with hedonism, for instance?

These few remarks may show that difference in means tests with the improved measurement instrument may only support common stereotypes as long as they are not supplemented by comprehensive studies including factors, which cause or produce differences in human values. In this way, we may also find out why, contrary to our expectation, saving face (FA) is also more important to Germans than to Japanese.

How much do our results differ from mean difference tests of sum scores or unweighted indices? To answer this question, we display the results of these tests on the right side of Table 4. For ease of comparison, we did not present sum scores, but the averages of the three indicator means. Col. (5) displays the German mean, col. (6) the Japanese mean, and col. (7) the mean differences. Four tests, indicated by the footnote [a], are not significant on the 5% level. The remaining 13 tests are not only significant but also display similar mean differences as indicated by the latent means in col. (3). Apparently, our FSE and PSE models are more sensitive to minor differences in the observed means. This is advantageous if the models are correctly specified; if not, it may turn into a disadvantage. A major difference between the structural (latent) means reported on the left side of the table and the mean differences reported on the right is that our PSE models detect significant differences even in those four instances where a test of the unweighted indices on the right side was not significant.

7 Discussion

Two scores, Achievement (AC) and Humility (HU), showed such a low reliability in Japan that we excluded them from further analysis. It is not yet possible to say whether we are dealing with a translation problem or with a general problem of conceptualization. In any case, some of us believe that the operationalization of both values is improvable also in the master questionnaire. The model of Benevolence-Care (***BEC***) suffers from two shortcomings: a low standardized factor loading of V47 in both countries and a high measurement error correlation. Here, a different operationalization should also be considered. The remaining 16 values show an acceptable reliability and could be operationalized in full or partial scalar equivalent models. The mean difference tests of the latent values provided theoretically interesting results. Given the tenet that values should influence a wide range of behaviors, it is somewhat disappointing that only four of 17 values can be measured by three scalar equivalent indicators. The remaining 13 rest on two scalar equivalent indicators.

Our analyses do not prove scalar equivalence of course. As always, in statistical modeling, we cannot exclude that equally or even better fitting competing models come to the opposite conclusions. For instance, we have estimated the models for universalism under the untested assumption that social desirability is uncorrelated with the three values of universalism. This may be wrong. If social desirability could be appropriately measured, it may turn out that it is correlated with the values of universalism and that the differences between Germany and Japan are partly caused by socially desirable response behavior.

8 Summary and Conclusions

One of the many advantages of the application of mathematical models to human behaviors is that these models force us to fully disclose our untested model assumptions—assumptions about the effects of measurement error, latent and unmeasured variables, scale invariances, etc. They also provide us with transparent criteria of confirmation and refutation. We have utilized structural equation models for examining two issues that are widely neglected in comparative nonexperimental research:

the reliability and the scalar equivalence of measurement instruments. Our focus was the Japanese and German translations of the PVQ-57 questionnaire of Shalom Schwartz. Translations meeting high standards would allow researchers the opportunity to study the value theory of Shalom Schwartz and his colleagues more systematically and accurately in Germany and Japan. We tested the reliability and validity of the PVQ-57 value measures, and we further examined whether the two translations display high levels of equivalence.

Our results suggested that the two translations of the scale exhibit satisfactory levels of reliability and validity as well as satisfactory levels of comparability across samples thus allowing the comparison of the value means across the samples. Further analyses demonstrated similarities and differences in the values across samples. We illustrated how such comparisons may be carried out.

We would like to note that we examined two necessary conditions as a kind of initial quality check; we have not conducted systematic tests of a value theory. Neither the stability of values nor their influence on a broad spectrum of behaviors was assessed. We have completely ignored the factors that influence the internalization and change of human values and thereby produce value differences between societies and cultures. The analyses were limited to the relation between 19 human values that are measured by 57 items. We made several simplifying assumptions. Some of them could have been subjected to a test. For instance, we excluded direct effects of several human values on an indicator. We also did not examine possible differences between smaller groups—between students of different academic subjects, male and female students, or students at different universities. Our samples are not random but arbitrarily selected classes in Hokkaido, the Tokyo area, and Cologne. Random samples of the German and Japanese populations were not available. Indeed, while the comparisons we conducted are not based on representative samples, they do raise some directions for future research and illustrate how such comparisons may be performed using population data.

Notwithstanding these limitations, this is—to the best of our knowledge—the first study to examine the reliability, validity, and comparability of the Japanese translation of the PVQ with the German version. Our findings show that the Japanese translation of the PVQ, with some refinements, is suitable for comparing Japanese value priorities with those of German respondents. We hope that this finding will stimulate the use of the PVQ questionnaire and research on human values and their effect on behavior in Japan, and the further examination of their cross-cultural comparability with values measured using other languages.

Appendix

Zipped file of the English master questionnaire, the two German and the gender-neutral Japanese translations, the pooled data set, and a methodological appendix (named BM50th_JetAl.zip) can be downloaded from Electronic Supplementary

Material [Springer will write the URL at Electronic Supplementary Material]. If difficulties should occur the reader may contact the co-author Herman Dülmer, University of Cologne [www.hduelmer@uni-koeln.de].

References

1. Brown, T. A. (2015). *Confirmatory factor analysis for applied research* (2nd ed.). The Guilford Press.
2. Cieciuch, J., Davidov, E., Vecchione, M. & Schwartz, S. H. (2014a). A hierarchical structure of basic human values in a third-order confirmatory factor analysis. *Swiss Journal of Psychology, 73*(3), 177–182.
3. Cieciuch, J., Davidov, E., Vecchione, M., Beierlein, C. & Schwartz, S. H. (2014b). The cross-national invariance properties of a new scale to measure 19 basic human values: A test across eight countries. *Journal of Cross-Cultural Psychology, 45*(5), 764–776.
4. Manabe, K. (2018). Methodological examination of Schwartz values survey (in Japanese with English abstract). *Kwansei Gakuin University School of Sociology Journal, 129*, 75–94.
5. Manabe, K., Jagodzinski, W., Davidov, E., Dülmer, H. & Hommerich, C. (2020a). Cross-national comparisons of human values based on the conceptual framework of S. Schwartz: The student surveys in Germany and Japan (I) (in Japanese with English abstract). *Kwansei Gakuin University School of Sociology Journal, 133*, 87–107.
6. Manabe, K., Jagodzinski, W., Davidov, E., Dülmer, H. & Hommerich, C. (2020b). Cross-national comparisons of human values based on the conceptual framework of S. Schwartz: The student surveys in Germany and Japan (II) (in Japanese with English abstract). *Kwansei Gakuin University School of Sociology Journal, 135*, 1–20.
7. Pokropek, A., Davidov, E. & Schmidt, P. (2019). A Monte Carlo simulation study to assess the appropriateness of traditional and newer approaches to test for measurement invariance. *Structural Equation Modeling: A Multidisciplinary Journal.*
8. Scharfbillig, M., Smillie, L., Mair, D., Sienkiewicz, M., Keimer, J., Pinho Dos Santos, R., Vinagreiro Alves, H., Vecchione, E. & Scheunemann, L. (2021). Values and identities—a policymaker's guide, EUR 30800 EN, Publications Office of the European Union, Luxembourg, ISBN 978-92-76-40966-3. https://doi.org/10.2760/022780, JRC126150.
9. Schwartz, S. H. (1992). Universals in the content and structure of values: Theoretical advances and empirical tests in 20 countries. In Zanna, M. P. (Ed.), *Advances in experimental social psychology* (Vol. 25, pp. 1–65). Academic.
10. Schwartz, S. H. (1994). Are there universal aspects in the structure and contents of human values? *Journal of Social Issues, 50*(4), 19–45.
11. Schwartz, S. H. (2003). A Proposal for Measuring Value Orientations across Nations (ESS, Ed.; pp. 259–319). https://www.europeansocialsurvey.org/docs/methodology/core_ess_questionnaire/ESS_core_questionnaire_human_values.pdf
12. Schwartz, S. H., Cieciuch, J., Vecchione, M., Davidov, E., Fischer, R., Beierlein, C., Ramos, A., Verkasalo, M., Lönnqvist, J.-E., Demirutku, K., Dirilen-Gumus, O., & Konty, M. (2012). Refining the theory of basic individual values. *Journal of Personality and Social Psychology, 103*(4), 663–688.
13. Steenkamp, J.-B. E. M., & Baumgartner, H. (1998). Assessing measurement invariance in cross-national consumer research. *Journal of Consumer Research, 25*, 78–90.

Handbooks as a Means of Mapping a Discipline. An Exploration of the World of Attachment

P. M. Kroonenberg and M. Stoltenborgh

Abstract Handbooks generally consist of a qualitative view of the major research areas of a discipline and an authoritative collection of state-of-the-art reviews of a discipline. This paper presents a quantitative overview of the discipline based on such reviews. It is created by using frequency of citations, co-authorships, and numbers of papers referenced in the handbook's chapters; here in particular of the Handbook of Attachment. The results show the existence of a detailed network of attachment researchers and outlines the subdisciplines they work in.

Keywords Handbooks · Quantitative review · Multidimensional scaling · Correspondence analysis · Attachment research

1 Introduction

This paper presents a quantative view of attachment research, its researchers, and their interrelationships. It is a vigorous research area with a long tradition which started with the pioneering work of Bowlby and Ainsworth [1, 2, 6–8]. A sign of the maturity of the research field was the appearance of the *Handbook of Attachment* in 1999 [10]. Nine years later the second edition of the *Handbook of Attachment* appeared [11], and again eight years later the third edition [12]. Our paper deals only with the first two editions as our database was finalised before the third edition appeared. Google Scholar (accessed April 2002) lists around 5100 citations to the three editions jointly, and a further number citations refer to the separate chapters of the three editions. In their Preface the editors state that.

The authors thank Prof. Judi Mesman for her insightful and constructive comments on the manuscript.

P. M. Kroonenberg (✉) · M. Stoltenborgh
Leiden University, Leiden, The Netherlands
e-mail: p.m.kroonenberg@fsw.leidenuniv.nl

© The Author(s), under exclusive license to Springer Nature Singapore Pte Ltd. 2023

A. Okada et al. (eds.), *Facets of Behaviormetrics*, Behaviormetrics: Quantitative Approaches to Human Behavior 4, https://doi.org/10.1007/978-981-99-2240-6_6

> It seems unlikely that either John Bowlby [...], or Mary Ainsworth [...] dreamed for a moment that their theoretical efforts would spawn one of the broadest, most profound, and most creative lines of research in 20th-century psychology. (p. x).

The number of chapters increased from 37 to 40 and to 43 in the third edition, and the number of pages from 888 to 974 and to 1068 in the third edition. As the reason for this increase the editors noted in their introduction to the second edition, that "[..] psychology has moved in the direction of neuroscience and behavioral genetics" (p. xii), which necessitated the additional chapters. Translational research, i.e. putting theory into clinical practice via intervention studies was a reason for an increase as well. The aim of the editors for the editions of the *Handbook* was to cover the most important areas in attachment research and to have overviews presented by leading researchers in the various subdomains. As formulated by Mary Ainsworth, the godmother of the attachment theory:

> This [first edition of the] *Handbook* is certain to be a rich source of information and ideas for years to come, providing experts with solid reviews of topical areas, and helping newcomers understand what has been accomplished and what remains to be done. Attachment researchers have a great deal to be proud of – and many exciting directions to pursue.[1]

The authoritativeness of the *Handbooks* derives from the breadth of their coverage, the standing and expertise of the chapter authors, and the sources relied upon in these volumes. Because of their standing the *Handbooks* can be used a basis for exploring how the field of attachment and the researchers working in it are organised and linked. In other words they give a view on the *World of Attachment*, who lives in it, how they are related and what their main interests are. The aim of this paper is to provide such a numerical and graphical view of its inhabitants and their major areas of research as an example of similar endeavours in other disciplines.

The basic information for this exercise are the bibliographic references at the end of the chapters and the author indices. In particular, one may use the number of citations to a specific author in a chapter as an indication of the relevance of this author's work to the topics discussed in the chapter. If an author is cited in all chapters, the work of that author is relevant to all areas of the discipline. If an author is only cited in a single chapter, her or his work is primarily relevant for the topics discussed in the chapter in question. Furthermore, the references per chapter also allow the investigation of the relatedness of research domains, as researchers working in related domains will tend to be cited in chapters dealing with these domains.

Another view of a discipline can be acquired by examining co-authorships, as it can provide insight into the productive relationships between authors. Joint publications are indicative of common interests and cross-fertilisation, but they are only a sufficient indication. Mutual inspiration and influence do not necessarily have to result in joint publications. Bowlby, the founding father of attachment research, is a case in point.

---

[1] This quote was found on http://www.amazon.co.uk/Handbook-Attachment-Research-Clinical-Applications/dp/1572308265 (Retrieved 22/4/2009), whether it is also published elsewhere is unknown to us.

The two ways of investigating publications, i.e. examining citations in chapters and examining co-authorships, are providing different perspectives on the discipline.

The investigation in this paper is structured around a number of questions: The first question is directed towards the examination of the web of linkages between researchers (*Who has published with whom?*). This allows the identification of so-called *spiders*, i.e. the researchers who are central in this web. The next and most important question is the determination of the characteristic domains, and which researchers and *spiders* are mostly strongly associated with those domains (*Who published what?*). As we are basing ourselves on two editions of the *Handbook of Attachment* about a decade apart, we can pose the question in which way the world of attachment has changed in that decade both with respect to joint research and the fields that are being investigated (*How has the world of the discipline changed?*).

2 Method

2.1 Data Collection and Research Questions

Who is included?

The basic data sources considered in this paper are the first and second editions of the *Handbook of Attachment*; as indicated the third edition came too late to be included in the database. As is evident from the author indices at the end of the volumes a large number of researchers are cited in the *Handbooks* so that it was impossible to include all of them in this study. Moreover, we were after the more influential investigators to give an adequate picture of the *World of Attachment*. Therefore, we decided to limit ourselves to those researchers mentioned most often in the author indices. There are certain limitations to this approach as we will discuss in detail below. In order to make a selection we counted the number of times a person was mentioned in the author index and decided somewhat arbitrarily to only include those authors who were cited on 30 pages or more in either *Handbook*.

Who published with whom?

To answer this question we counted how often papers of the selected authors were cited basing ourselves on the reference lists at the end of each chapter. In particular we counted the numbers of times articles of two joint authors were listed in each of these reference lists, and then added these numbers over all chapters of each *Handbook*. The data form a square matrix with authors as rows and the same authors as columns. A cell of the matrix for author A and author B contains the number of times that A and B's joint papers appear in the chapter reference lists (see Fig. 1, left-hand panel). Using joint authorship it is possible to establish which researchers are central in the *Web of Attachment*. Such authors are considered the *spiders of the web*, where *spiders* are defined as authors who have at least six co-authors who also

Research question	Who published with whom?	Who published what?
Data format	Similarity matrix of authors by authors 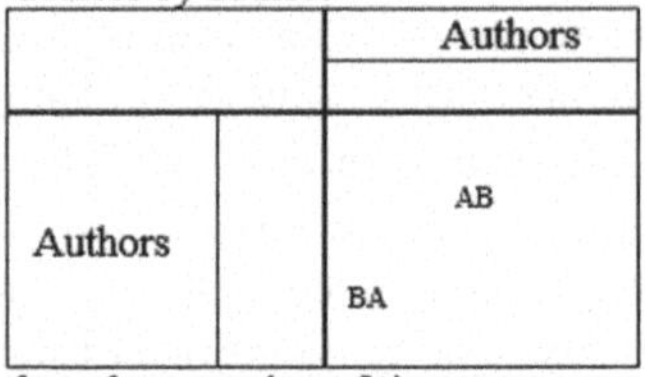 $f_{BA} = f_{AB}$ = number of times papers written by (co-)authors A and B have been cited in an entire *Handbook*.	Cross-tabulation of authors by chapters 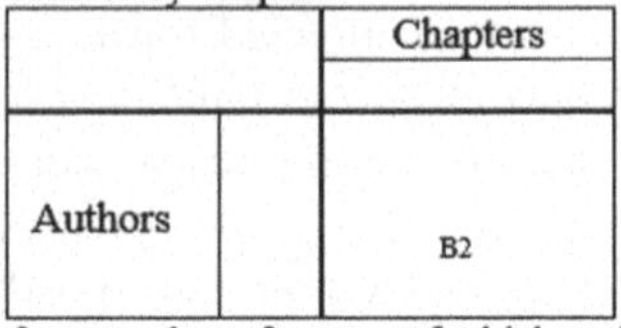 f_{B2} = number of papers of which author B was a co-author cited in Chapter 2. (f_{A2}, f_{B2}, f_{C2}) is the profile of Chapter 2; (f_{B1}, f_{B2}, f_{B3}) is the profile of Author B
Analysis technique	Multidimensional scaling	Correspondence analysis

Fig. 1 Overview of data structures, research questions and analysis methods

belonged to our selected authors. They are almost surely the most influential in the field.

Who published what?

This question could be answered by checking in the reference list of each chapter how often articles of a particular author were cited irrespective the order in the author list. The data form a cross-tabulation of authors by chapters with in the cells authors' citation frequencies per chapter (see Fig. 1, right-hand panel). The basic idea behind these data is that the chapters in the *Handbooks* each represent a particular substantive domain, so that many citations in a chapter indicate that the work of the author is relevant for the subject areas treated in that chapter. Especially the *spiders*' research domains are of particular interest.

How has the world of attachment changed?

Because we have information from both *Handbooks* which are about a decade apart it will be possible to comment on emerging areas in attachment research and their researchers, as well as on differential contributions of different authors to the field by comparing the results from the analysis across editions.

Drawbacks of the method of data collection

The way of constructing the database, i.e. using the reference lists in combination with the author index, is not entirely without problems. In particular a number of discrepancies between the two should be mentioned.

First. Citations to chapters in the same edition of a *Handbook* are not listed in the chapter reference lists. This resulted in lower citation counts from the reference lists compared to those based on the author index for those researcher(s) who were author of a chapter in the same edition.

Second. Citations to different articles or book chapters of the same author on the same page resulted in a higher count from the reference lists than from the author index as the author index only indicates that an author is mentioned on a particular page and not how often this was done. In theory these discrepancies can be resolved by manually scanning all 1863 pages in the *Handbooks* but given that these pages were not available at the time in a digital format, it was considered a too onerous task given the expected rewards. This may, however, be different at the present time for studies of other domains.

Third. Often the first time a reference is cited in a chapter all authors are mentioned but later citations in the same chapter are only indicated as *First* Author et al. if there are three or more authors of the reference. This reduces the counts in the author index for all second and later authors. Apart from going through the entire texts by hand, there is little one can do about this. Again, if such information is available in a digital format, resolving this issue becomes available.

Fourth. A final point is the relationship between content and citations. One may divide papers roughly in theoretical investigations, empirical research papers, narrative or qualitative reviews, meta-analyses or quantitative reviews, and methodological papers presenting methods of collecting and/or, analysing data and validation studies of such methods, such as the present one. The expected citation rates of papers in the various categories are most likely not equal. It is expected that especially papers presenting important advances in methodology, and narrative and quantitative reviews will tend to attract relatively larger numbers of citations. Such citation tendencies will undoubtedly have had an influence on the results presented here, but what these influences are and how they colour the analyses presented here requires separate studies; see for an analysis of differential citation rates, [23].

3 Analysis Methods

The two basic questions (*Who published with whom?* and *Who published what?*) were tackled with multidimensional scaling and correspondence analysis respectively using the appropriate modules in SPSS (see [21]), description of the latest version of the software: https://www.ibm.com/docs/en/SSLVMB_26.0.0/pdf/en/IBM_SPSS_Categories.pdf). Both techniques aim to provide a graphical representation of the matrix or table to be analysed. The four matrices with the information on which this paper is based can be requested from the first author.

Multidimensional Scaling (MDS)

MDS is a technique which operates on a similarity matrix or on a dissimilarity matrix (e.g. [5]). Our co-author matrix is a similarity matrix in which a high number indicates that papers written jointly by the row and column authors were often referenced in the *Handbook* chapters. The main purpose of the technique is to find a low-dimensional spatial representation, mostly a two-dimensional graph, such that the co-authorship patterns embodied in the similarity matrix are represented as well as possible. In the graph the similarities are inversely represented by distances: the greater the similarities the smaller the distances between two authors. Groups of authors who publish nearly exclusively within their own research group will lie closely together in the space with few links to the outside world, but authors who do not publish together (generally) lie at larger distances from each other. Because in this paper we do not use the full-dimensional space but only the two most important dimensions, there is a certain amount of distortion in that some authors seem closer together than they really are, especially in the middle of the graph. To counteract this we have drawn connecting lines between some of the authors (for details see below).

Researchers were designated as *spiders* if they have published papers with six or more of the other selected authors. However, note that there is a category of authors who are influential but have only published on their own so that they cannot be represented in the Web because their co-author count is zero.

Correspondence Analysis (CA)

CA is a technique for (large) rectangular contingency tables with positive numbers, mostly frequencies (e.g. [4, 17, 18]). In large tables a test of independence between rows and columns is nearly always significant and thus not very informative. The interesting part of (large) contingency tables is the interaction between rows and columns, and it is this interaction which is displayed in correspondence-analysis graphs. Our contingency table consists of authors (rows) and chapters (columns), so that the graphs help interpreting the interaction between authors and chapters. As chapters in the *Handbooks* almost exclusively deal with a single topic or attachment subdomain we may proceed with equating chapters with subject areas.

Contingency tables may be examined both row-wise (here, by authors) and column-wise (here, by chapters). In the latter case we refer to *chapter profiles*, which consist of the number of times (or proportion of times) that individuals have been an author of papers mentioned in the chapter references. Similarly an author profile consists of the number of times (or proportion of times) an author appears in the references of the chapters (see Fig. 1, right-hand panel). When chapters have similar profiles largely the same authors are referred to in these chapters so that we may assume these chapters treat similar topics. If authors have similar profiles across chapters we will assume that they work in the same subdomains of attachment.

The results of a correspondence analysis can be portrayed in three maps or graphs which show either (1) the relationships between the authors via the author profiles, or (2) between the chapters showing the chapter profiles or (3) between authors and between chapters showing both type of profiles in a single graph. Here we will present the first two graphs explicitly and the third one only implicitly (see below).

In these correspondence graphs[2] the origin represents the model of independence or lack of interaction so that all authors and chapters would lie at the origin if there was no relationship between authors and chapters. The further away chapters and

---

[2] *Technical note*: The correspondence analysis graphs in this paper are shown in their symmetrical representations,. This makes for more easy comparisons between column points and row points (for details see [18], Chaps. 6 and 13).

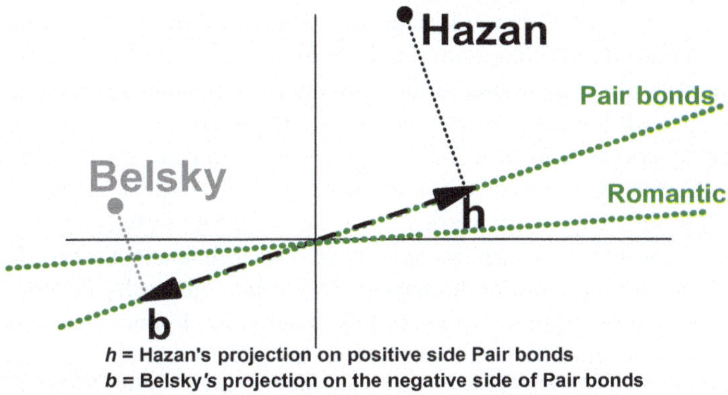

Fig. 2 Interpretation of correspondence analysis biplot of chapters and authors

authors lie from the origin the greater their deviations from independence. Positive deviations indicate that authors are referenced more often than expected on the basis of independence between chapters and authors, and negative deviations indicate that authors are referenced less than expected. Chapters are generally represented as lines passing both through the origin and the chapter points. However, to avoid clutter in the graphs we will only show part of these lines, in particular the part from the origin to the chapter point (see Fig. 2). The part shown is the positive side of the line and refers to positive deviations from independence, while the part on the other side of the origin is the side of the negative deviations. When chapter lines point in the same direction and have small angles, the chapter profiles are similar, when they are at right angles the chapter profiles are independent of each other. Authors profiles are shown as points (see Fig. 2). The closer the points lie together the more similar their profiles are.

To evaluate the relationships between authors and chapters one should superimpose the author graph and the chapter graph. However, this makes the graph unreadable due to the large number of objects and labels. Therefore, we have decided to present these graphs in close proximity requiring the reader to mentally superimpose them. To assist in this we have drawn lines for the major research domains in the author graphs (see Figs. 8 and 10). To evaluate which authors have a comparatively higher number of references in a particular chapter one should project each author on the line of that chapter (see Fig. 2). The order of the projections of the authors on the chapter lines indicates the relative size of the authors' numbers of references in a chapter bearing in mind the distortions due to the two-dimensional approximation. When an author point projects highly and positively on a chapter line the individual has a relatively large number of references in this chapter. When an author point projects highly on the negative part of the line (not shown in our graphs) very few or none of that author's publications are present in the reference list of that chapter.

A tiny subset of the data portrayed in Fig. 2 may serve as an example. The angle between the chapter on Pair bonds and the chapter on Romantic relationships is small so that authors who are more frequently referenced in the Pair Bonds chapter tend to be referenced more frequently in the Romantic relationships chapter as well. Hazan is an example of an author with relatively large number of references (6 and 7, respectively), while Belsky is an example of a relatively less referenced author in both these chapters (once in each chapter). The origin indicates the point of independence of authors and chapters. Author points projecting into the origin have a profile proportional to the marginal or average author profile. Similarly chapters located at the origin have profiles proportional to average or marginal chapter profile.

One technical point about both analyses is that they generally do not very well represent rows and/or columns with very low or zero total frequencies [19]. In order to get appropriate outcomes, authors without co-authors cannot be included and authors with very low co-authors scores are not always optimally placed in the graph of the multidimensional scaling analyses due to the lack of information about their relationships with other authors. The prime example is Bowlby, one of the founding fathers of attachment theory, whose paper with Ainsworth (mentioned once in 1999 and twice in 2008) is the gossamer thread which connects him to the web of attachment determined by co-authorship (Figs. 5 and 6). Without these three references he could not have been included in the graph. In the correspondence analysis, authors who did not feature in the chapter reference lists could not be included in the correspondence analysis solution. In particular, this was the case for Roisman and Munholland in the first edition of the *Handbook*.

4 Results

4.1 Authors: Page and Reference Counts

Page counts

Fifty-two authors appear on at least 30 pages of either *Handbook.* As was mentioned in the method section an author's page count is an underestimate of the times the author has been cited, because the number of times an author appears on the same page is not included, nor are second and later authors included more than once for publications with three or more authors in the same chapter.

The dots in Fig. 3 represent these 52 authors. The axes show the percentages of numbers of pages on which these authors were referenced for each of the two editions. The percentages were taken with respect to the total number of pages mentioned in the index for the selected authors i.e. 2670 in the first edition and 3650 in the second edition.

Figure 3 shows that there is a great stability in the page counts with virtually all authors differing less than one percent between the two editions. Visually there seem to be three groups of authors: the *First-Generation* or *Pioneers* of attachment research (Bowlby, Ainsworth, and Main), the *Second-Generation* researchers (circled in Fig. 3) and other frequently cited authors.

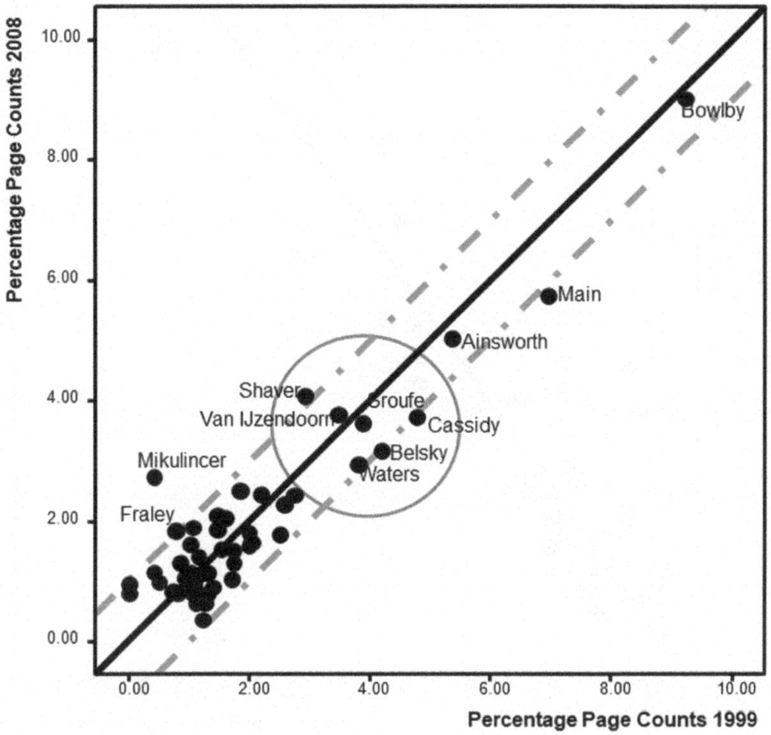

Fig. 3 Page counts. Percentages of *Handbook* pages on which the selected authors are mentioned. The central solid line indicates equal percentages in both *Handbooks*; lines parallel indicate ±1%. Authors above the central line have increased percentages in 2008, those below have decreased percentages

Reference counts

Figure 4 represents another way to evaluate the influence of authors' work. It shows for both editions the percentage of the times that authors occurred in the reference lists of the chapters—*reference counts*. Interestingly, all three *First*-Ge*neration* researchers were referenced comparatively less in 2008 than in 1999.

The correlations between the two sources of information about the influence of the selected authors (page counts and reference counts) are 0.91 for the first edition and 0.82 for the second edition, indicating that the two measures express both similar but also different aspects of the citation process (see Table 1). The page counts are an estimate of the overall citation frequency and are sensitive to multiple references to the same paper, while reference counts are more influenced by the number of citations to different papers. Many references in a chapter to the same paper have no influence on the reference count.

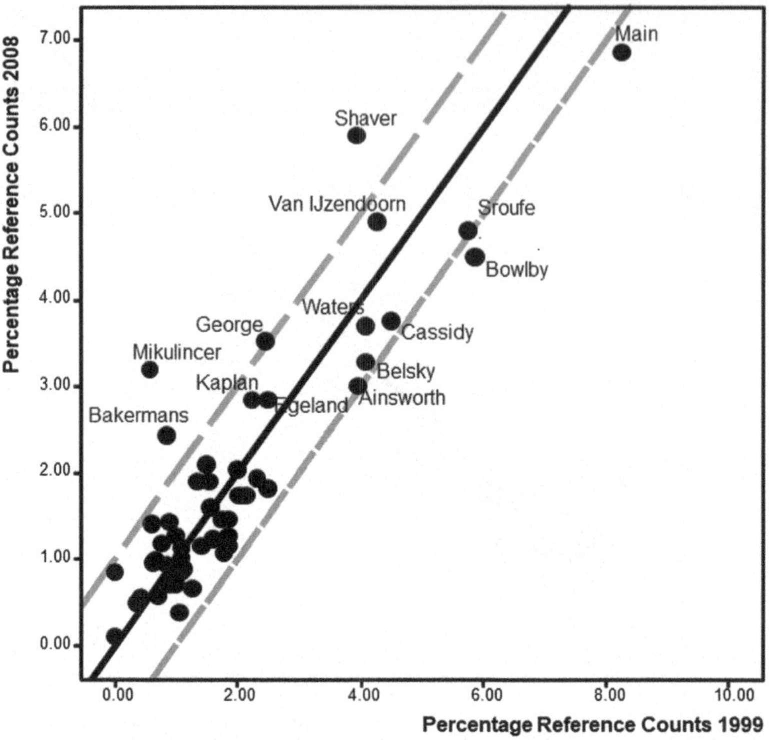

Fig. 4 Reference counts. Count of selected authors' chapter references. The central solid line indicates equal percentages in both *Handbooks*; lines parallel indicate ±1%. Authors above the central line have increased percentages in 2008, those below have decreased percentages

4.2 *The Web of Attachment and Its Spiders: Who Publishes with Whom?*

The co-authorship matrices of both *Handbooks* were subjected to a multidimensional scaling analysis using the SPSS program PROXSCAL [9, 13]. The matrix itself is largely empty as most authors have only published with few of the other selected authors. The results of the multidimensional scaling are portrayed in Figs. 5 and 6, for the 1999 and 2008 editions, respectively. Short distances indicate more co-authored papers by a pair of authors.

Figures 5 and 6 give a representation of the Web of Attachment based on co-authorship. It can be observed that this web has become more dense over time. The number of *spiders* increased from 8 to 11, the total number of co-authors increased from 156 (mean 3.0) to 192 (mean = 3.7), the number of connections between *spiders* increased from 9 to 17, the number of authors not connected to any *spider* decreased from 8 to 5, and the number of authors without co-authors decreased from 4 to 2.

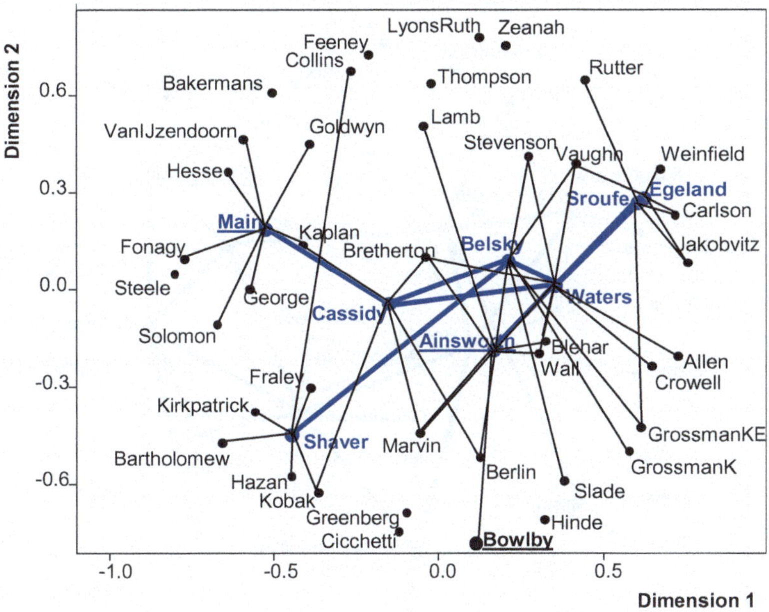

Fig. 5 Co-authorship 1999. *Spiders*: solid blue dots; *First-Generation* researchers: underlined. Heavy blue lines connect *spiders*. Other lines connect *spiders* with their co-authors. Mikulincer, Munholland, Roisman, and Simpson are not included due to zero co-author counts[3]

In the figures we have underlined the *First-Generation* researchers and marked the *spiders* (in solid blue dots). The fatter solid (blue) lines connect *spiders*. In addition, we have connected the *spiders* with all their co-authors. This shows that virtually all selected authors are connected with at least one spider. In other words, relatively few papers were written by isolated authors. In addition, it is evident that there is also a subweb of the *spiders* themselves.

Partnerships with highest number of citations (Based on the 2nd Edition)

Figures 5 and 6 indicate the links between authors but do not indicate how successful these co-operations were in attracting a high number of references in the chapter bibliographies. This is indicated in Table 2. Note that these values should not be interpreted as an indication of overall citation rate of a research couple in the literature. After all a single publication can achieve at most as many references as there are chapters. Thus the number of references for *Patterns of Attachment* [3] is at most 37 for the first edition (it is referenced in 32 chapters) and 40 for the second edition (31 chapters), while the book itself was cited an astounding 33,400 by April 2022 (Google Scholar).

---

[3] Due to space considerations Ms. Bakermans-Kranenburg and Ms. Stevenson-Hinde are indicated in the graphs as Bakermans and Stevenson, respectively.

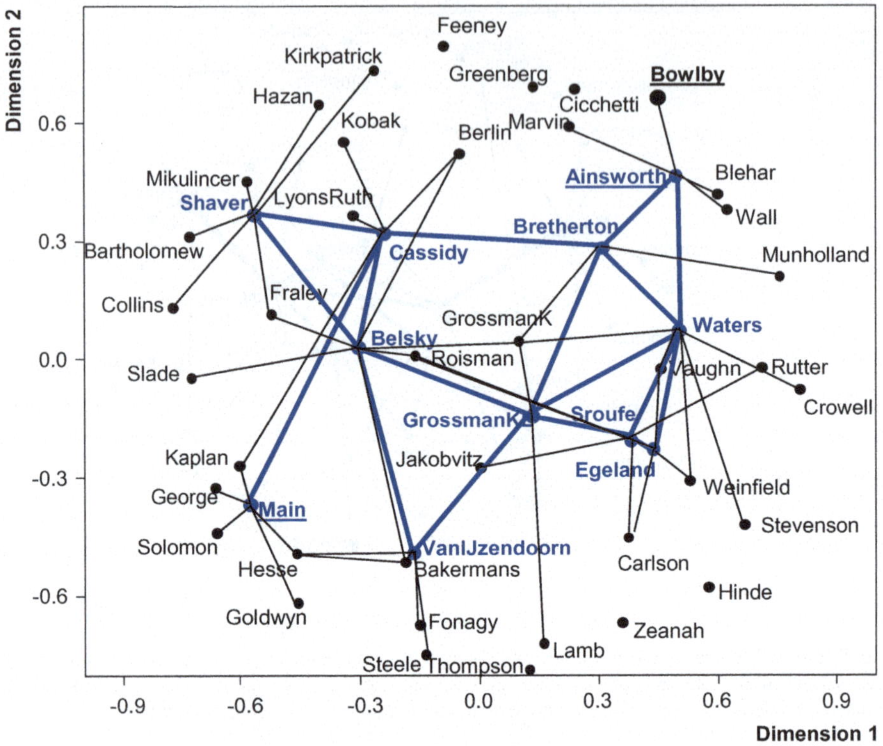

Fig. 6 Co-authorship 2008. *Spiders*: solid blue dots; *First-Generation* researchers: underlined. Heavy lines connect *spiders*. Other lines connect *spiders* to their co-authors

Table 1 Correlations between *Page counts* and *Reference counts*

		Page counts		Reference counts	
		1999	2008	1999	2008
Page counts	1999	1.00			
	2008	0.93	1.00		
Reference counts	1999	0.91	0.83	1.00	
	2008	0.76	0.82	0.89	1.00

High values in Table 2 occur if authors have co-authored several papers which are referenced in many chapters, so that Table 2 provides an impression of how successful author duos have been in attracting references to their joint papers across the *Handbooks*.

The most successful in terms of partnerships is clearly Main who wrote five papers with various co-authors [14–16, 22]. The major topics of these papers were the theoretical considerations, detailing and development of the Adult Attachment

Table 2 Number of references in chapters bibliographies to pairs of co-authors

Co-authors	1st Ed	2nd Ed	Rel. Increase
Main and Kaplan	58	99	+
Van IJzendoorn and Bakermans-Kranenburg	24	87	+++
Main and George	34	70	++
Sroufe and Egeland	45	77	+
Shaver and Mikulincer	0	68	+++
George and Kaplan	29	67	++
Main and Hesse	28	51	+
Grossmann, K. E. and Grossmann, K	32	45	+
Egeland and Carlson	13	44	+++
Shaver and Hazan	51	43	–
Sroufe and Carlson	22	42	+
George and Solomon	32	42	
Main and Goldwyn	29	32	–
Ainsworth and (Blehar, **Waters**, and Wall)	31	31	–
Main and Cassidy	37	31	–
Fonagy and Steele	42	29	–

Note Selection criterion: 1% of occurrences, i.e. 27 in 1999 and 37 in 2000. **Bold authors** are *spiders*; **Bold numbers** meet the 1%-selection criterion; +++ = three-fold increase or more; ++ = two-fold increase; + = increase larger than increase on the basis of increased citations (=1.36); (–) = decrease with respect to 1st Edition

Interview which has become an extremely powerful instrument in the field of adult attachment and in clinical studies and practice. *Patterns of Attachment* by [3] fulfilled a similar role for the Strange Situation Procedure. Powerful theoretically based and well-constructed research instruments are fundamental to any science and therefore attract many citations.

Ainsworth's citations for her book *Patterns of Attachment* in Table 2 may not seem impressive at first site, but the feat of being referenced in practically every chapter of both *Handbooks* underlines the book's importance to the field.

4.3 Who Publishes What?

The basis for answering this question is the contingency table of authors (rows) and chapters (columns) in which the cells represent how many publications of a particular author are referenced in that chapter. To investigate the patterns we will look at both chapter profiles (Figs. 7 and 9), at the author profiles (Figs. 8 and 10), and at the figures together to assess which authors are especially referred to in which chapters of each edition.

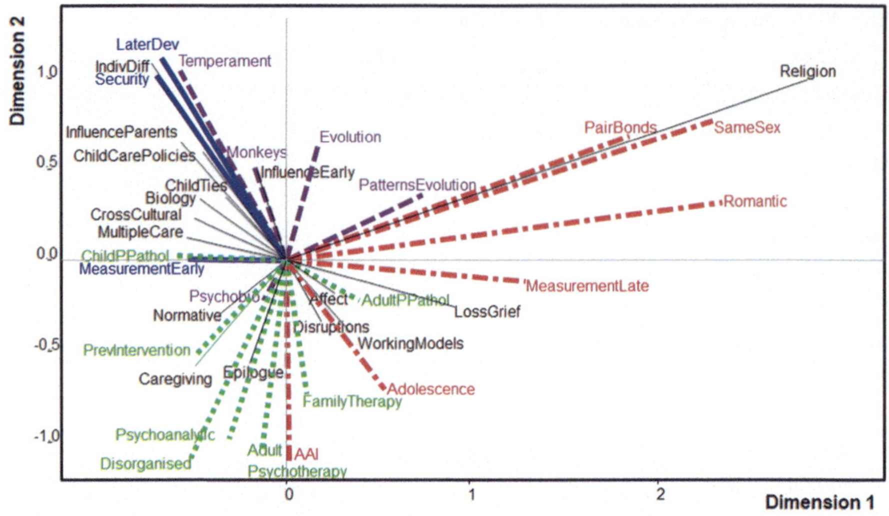

Fig. 7 Chapter space 1999. Based on the authors-by-chapters contingency table constructed from the references in the chapters of the 1999 edition of the *Handbook*. The axes from the correspondence analysis explain 20% and 11% of the variability (inertia), respectively. Red dash-dot lines = attachment style (adults); thick blue solid lines = developmental attachment; green dotted lines = clinical applications; purple dashed lines = biological aspects; thin black solid lines = other

1999 Chapter profiles

In both *Handbooks* the chapters have been divided into a number of parts. The first part in the 1999 edition consists of introductions and overviews (Part I) and the last part of emerging topics (Part VI). The middle chapters are organised in substantive domains within attachment research, in particular Biological (Part II), Infancy and Childhood (Part III), Adolescence and Adulthood (Part IV), and Clinical Applications (Part V).

The first two major areas of attachment research that emerge from Fig. 7 reflect the distinction between (1) *developmental attachment research* (children) and (2) *attachment style research* (adults). On a theoretical level, research regarding attachment patterns in children is rooted in the traditions of developmental psychology, whereas research examining the development of adult attachment styles originates from a social psychology framework. The third major area of attachment research represents (3) *clinical applications* with a focus on attachment representations and internal working models of attachment that are generally targeted in therapies and interventions.

What can also be derived from the graph is where one should position the chapters of the first and the last Parts of the *Handbook*. For instance Weinfeld, Sroufe, Egeland and Carlson's chapter (Chap. 4; label in the graph: *IndivDiff*) on the nature of

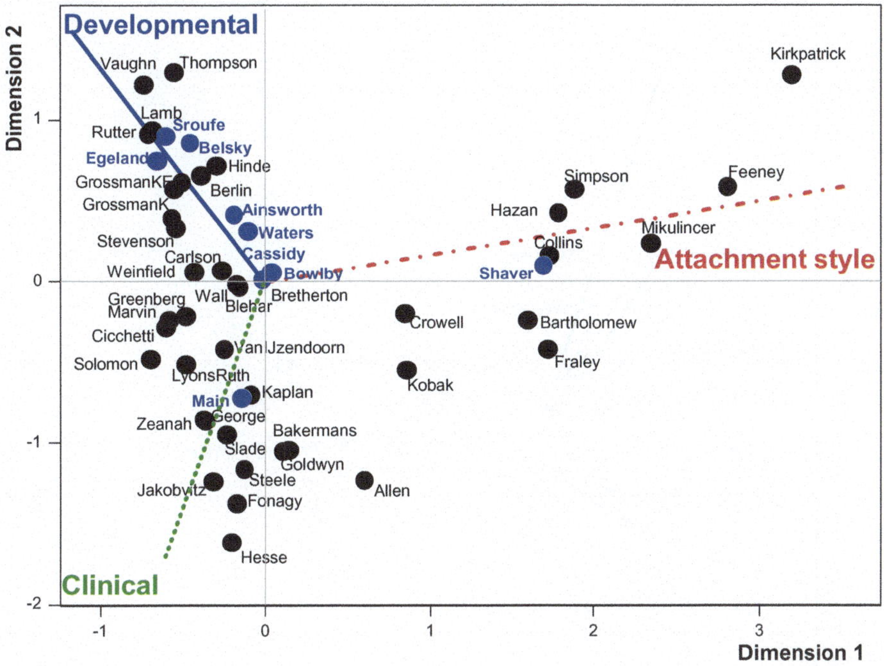

Fig. 8. Author space 1999. Based on the authors-by-chapters contingency table constructed from the references of the chapter of the 1999 edition of the *Handbook*. The lines represent chapter groups (see text)

individual differences in infant-caregiver attachment is firmly located in the developmental area, while Kirkpatrick's chapter (Chap. 35; *Religion*) on attachment and religious representations and behaviour is a clear representative of the adult attachment-style domain. On the other hand the figure also shows that based on their references some chapters do not necessarily fall within their allotted Part. For instance temperament research (Chap. 10; *Temperament*) which is positioned in the 1999 *Handbook* in the Biological perspectives part can be found within the developmental attachment domain. Similarly, Chap. 19 (*AAI*) dealing with the Adult Attachment Interview and positioned as belonging to the part on attachment in adolescence and adulthood can be found in Fig. 6 within the attachment representation and clinical applications domain.

The Bretherton and Munholland's chapter (Chap. 5; *WorkingModels*) on internal working models has references in common with both the attachment style research with adults and clinical applications.

The chapters grouped under the heading Biological Perspectives are not located in a specific well-defined area in the graph, which suggest that the research on biological perspectives does not have a clearly defined group of references in 1999, or at least not to our selected authors. Only the citation profiles of the evolutionary perspective chapters are fairly similar to each other.

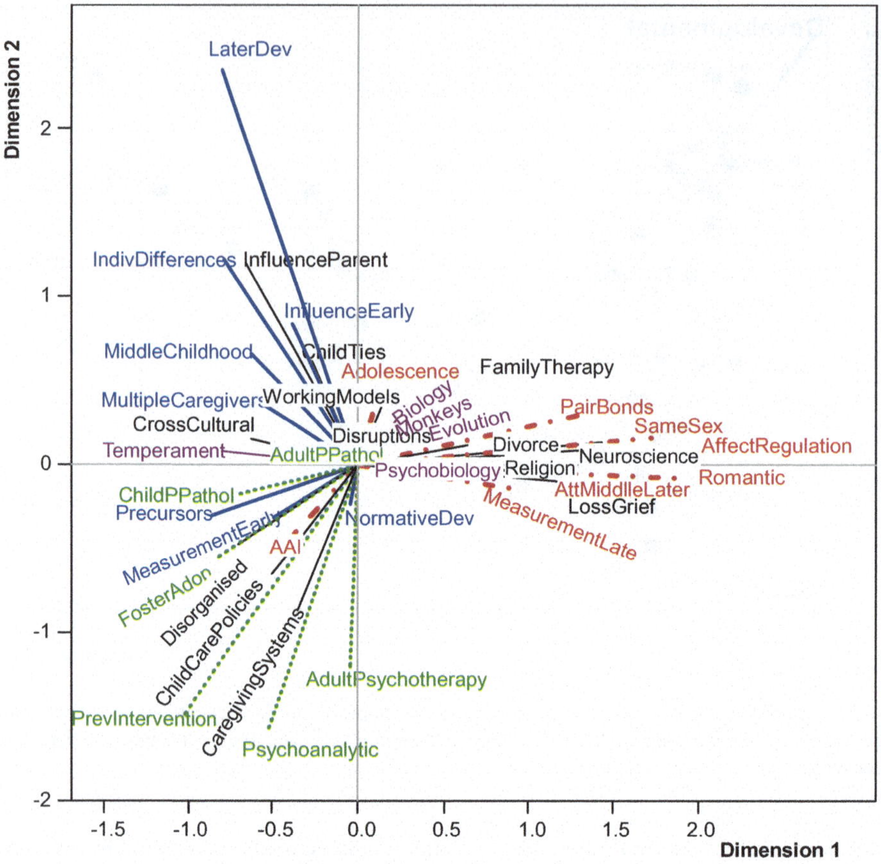

Fig. 9 Chapter space 2008. Based on the authors by chapters contingency table constructed from the references of the chapter of the 2008 edition of the *Handbook*. The axes from the correspondence analysis explain 20% and 10% of the variability (inertia), respectively. Red dash-dot lines = attachment style (adults); thick blue solid lines = developmental attachment; green dotted lines = clinical applications; purple dashed lines = biological aspects; thin black solid lines = other

Chapter 8 (Polan & Hofer; *PsychoBio*) and Chap. 9 (Suomi; *Monkeys*) were written from a perspective outside the standard attachment domains as they only mention 12 and 10 papers by our core authors. This shows that they contribute new perspectives which at the time were not really embedded in the standard literature on attachment. Their placings in the graph is therefore not a very firm one.

A final point is that measurement in attachment research is not a topic in itself. The two chapters on measurement, Chaps. 19 (Adult Attachment Interview; *AAI*), and 20 (Individual differences in adult attachment; *MeasureLate*) are not located close together but they are positioned within their substantive subdomains.

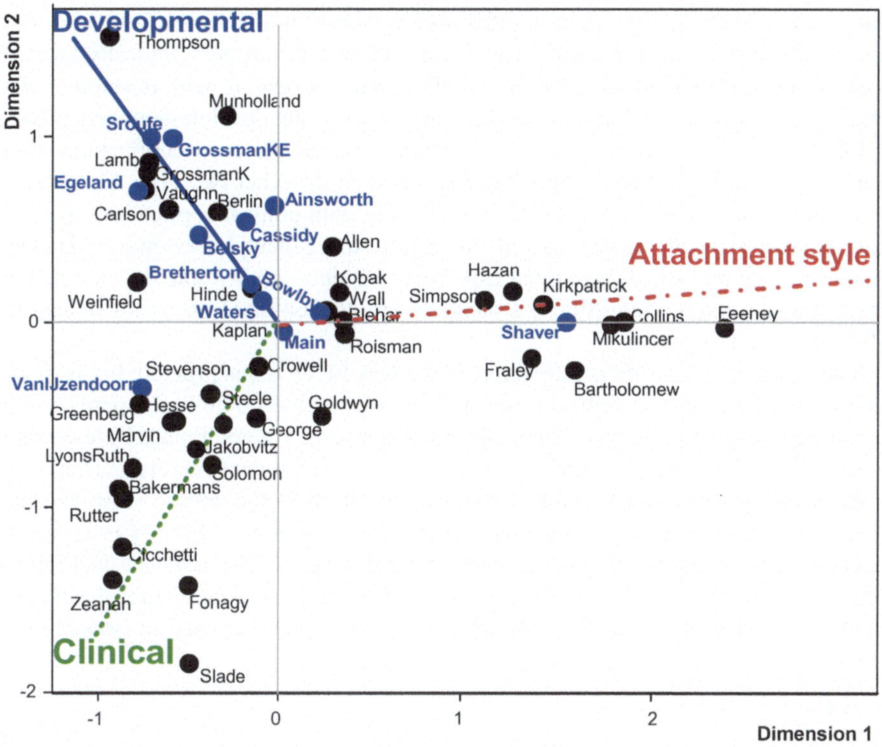

Fig. 10 Author space 2008. Based on the authors by chapters contingency table constructed from the references of the chapter of the 2008 edition of the *Handbook*

1999 Author profiles

Author profiles can best be discussed in direct relation to the subdomains for which they are mainly cited. In Fig. 8 we have drawn lines for the three major attachment domains rather than lines for all chapters. The reader should mentally add the line for a relevant chapter when reading the ensuing discussion. The figure shows the similarity of the authors' profiles over the chapters. Thus authors who have a similar profile across chapters are located close together in the plot. For instance, we observe that Feeney, Simpson, Mikulincer, Hazan, Collins, and Shaver all deal with issues from attachment style research on adults as they all have long positive projections on the lines of chapters concerned with this topic (see also the explanation in Fig. 2). Authors who are cited in (nearly) all chapters are located near the origin. Thus, not surprisingly, Bowlby takes up a central position in this graph as his work is fundamental to all attachment domains. Furthermore, authors like Bretherton, Greenberg, Cassidy and Van IJzendoorn have written papers relevant to various domains of attachment research which is confirmed by their central location in the graph. The wide-ranging book *Patterns of Attachment* (1978) is centrally located in the graph, and as such illustrates the book's influential position in attachment research. The two *spider* authors of this book (Ainsworth and Waters) are less centrally located because their other work is primarily cited in chapters dealing with developmental attachment research, so that their profiles are closer to the developmental chapters.

Of the *First-Generation* researchers, Main is rather more off centre indicating that she is more often cited in specific chapters than the other two *First-Generation* researchers, particularly in those chapters dealing with clinical applications as clinical research typically makes use of the Adult Attachment Interview (AAI). The *spiders* Sroufe and Belsky are cited especially for their work in the developmental attachment research while Shaver's citations are concentrated on issues related to adult attachment style.

The position of Cassidy is illustrative of the way Fig. 8 is constructed. In Fig. 5 we see that she takes up the central position, because she has co-authored papers with five other *spiders*, i.e. Shaver, Main, Belsky, Sroufe and Waters. Figure 8 shows that these authors together span the entire spectrum of attachment research. Therefore a reasonable position for Cassidy is somewhere close to the centre. However, the exact location will depend on the chapters in which she is most heavily referenced. Inspection shows her work to be referenced in 29 of the 37 chapters, but the highest frequencies occur in the developmental and clinical chapters rather than in the attachment style chapters, so that she is further away from the latter and in between the former two.

2008 Chapter and author profiles

In the 2008 *Handbook* the chapters have again been divided into six parts. The first part consists of overviews of attachment theory (Part I). The middle chapters are organised in substantive domains within attachment research, in particular Biological Perspectives (Part II), Infancy and Childhood (Part III), Adolescence and Adulthood (Part IV), and Clinical Applications (Part V). The last part contains chapters on special topics dealing with Systems, Culture and Context (Part IV).

Rather than describing the 2008 chapter space completely, we will primarily discuss differences with the 1999 chapter space. Clearly the reference patterns as a whole are extremely similar, so that we may conclude that in the last decade the world of attachment has not dramatically shifted. Again the space is dominated by the three domains: (1) developmental attachment research, (2) attachment style research with adults and (3) attachment representations or clinical applications.

4.4 Changes in Chapter Profiles

Notwithstanding the overall consistency over the years, a limited number of changes have occurred in the chapter space. Especially notable is the shift of the chapter on Adult psychopathology which was located with the attachment style research with adults but has moved completely to the centre of the graph. This is remarkable because the adult psychopathology chapter was written by the same authors in both editions. When one compares the references of the 1999 and 2008 chapters, there is a marked increase in references to authors in the developmental and clinical area which is the cause for the observed shift.

The new chapter on the neuroscience of attachment (Chap. 11; *Neuroscience*) predominantly refers to the adult attachment style literature which explains its location in the graph.

Again the chapters on psychobiology and monkeys have very few references to papers of our core researchers, only 9 and 16 references, respectively. Therefore, the authors of these chapters have not integrated more literature from the attachment domains than a decade ago.

4.5 Changes in Author Profiles

Not surprisingly the graphs of the author profiles for the two editions are as similar as are the chapter graphs. Looking at the *First-Generation* researchers and the *spiders* it is clear that a number of them publish primarily in the developmental field (Ainsworth, Belsky, Cassidy, Egeland, K. E. Grossman, Sroufe, Waters) and form a close-knit group who publish together and cite each other frequently. Shaver remains the only *spider* in the field of adult attachment style research. It is interesting to note that whereas Main through her work on the Adult Attachment Interview was a *spider* located well into the clinical application or attachment representation area in 1999, her work now seems somewhat less central to this area.

5 Conclusion

On the basis of the two editions of the *Handbook of Attachment* [10, 11] this paper has provided an overview of the relationships between the various domains in attachment research, has attempted to answer questions about the coherence of the group of attachment researchers, and has singled out those authors who are central to the field. An exercise such this is of course only as good as the data used. In the method section, it was already pointed out that there are some inherent inaccuracies due to way the database has been constructed. Moreover, several somewhat arbitrary decisions had to be made about key issues, such as who is considered a core author and who can be designated as a spider. Notwithstanding, the results present an insightful view of the world of attachment. The results point to interesting patterns about how this world is constituted and who does what with whom in this world. The dynamics of the relationships among authors could be examined via the two 'webs of attachments' on the basis of the co-authorships.

Among the more frequently cited authors one could distinguish three groups: the *First-Generation* researchers (Bowlby, Ainsworth and Main), the *Second-Generation* authors (Belsky, Cassidy, Shaver, Sroufe, Van IJzendoorn and Waters) and a group of about 40 frequently cited authors. The *Second-Generation* authors also function (together with Bretherton, Egeland, and K.E. Grossman) as *spiders* in the web of attachment by having at least six co-authors among our core set of authors. Nearly all of the authors were connected with at least one of the *spiders*, even more so in 2008 than in 1999, signalling the increasing cohesion among the attachment researchers.

In the decade between the two editions, the *spiders* have generally consolidated their positions, be it that there is a weak tendency for some of the original publications by the *First-Generation* researchers to become less referenced.

This attachment case study will hopefully lead to further similar studies in other fields of science. The first author has already made such a provisional attempt for the field of bibliometry [20].

References

1. Ainsworth, M. D. S. (1964). Patterns of attachment behaviour shown by the infant in interaction with his mother. *Merrill-Palmer Quarterly, 10*, 51–58.
2. Ainsworth, M. D. S. (1967). *Infancy in uganda: Infant care and the growth of attachment*. John Hopkins University Press.
3. Ainsworth, M. D. S., Blehar, M., Waters, E., & Wall, S. (1978). *Patterns of attachment: A psychological study of the strange situation.* Erlbaum.
4. Beh, E., & Lombardo, R. (2014). *Correspondence analysis: Theory, practice and new strategies.* Wiley. https://doi.org/10.1002/9781118762875
5. Borg, I., & Groenen, P. J. F. (2005). *Modern multidimensional scaling: Theory and applications* (2nd ed.). Springer.
6. Bowlby, J. (1969). *Attachment and loss: Vol. 1. Attachment.* Basic Books.
7. Bowlby, J. (1973). *Attachment and loss: Vol. 2. Separation.* Basic Books.
8. Bowlby, J. (1980). *Attachment and loss: Vol. 3. Loss.* Basic Books.
9. Busing, F. M. T. A., Commandeur, J. J. F., & Heiser, W. J. (1997). PROXSCAL: A multidimensional scaling program for individual differences scaling with constraints. In W. Bandilla & F. Faulbaum (Eds.), *Advances in statistical software 6* (pp. 67–73). Lucius & Lucius.
10. Cassidy, J., & Shaver, P. R. (Eds.). (1999). *Handbook of attachment. Theory, research, and clinical applications.* The Guilford Press.
11. Cassidy, J., & Shaver, P. R. (Eds.). (2008). *Handbook of attachment. Theory, research, and clinical applications* (2nd ed.). The Guilford Press.
12. Cassidy, J., & Shaver, P. R. (Eds.). (2016). *Handbook of attachment. Theory, research, and clinical applications* (3rd ed.). The Guilford Press.
13. Commandeur, J. J. F. and Heiser, W. J. (1993). *Mathematical derivations in the proximity scaling (PROXSCAL) of symmetric data matrices* (Technical Report. No. RR-93–03). Leiden, The Netherlands: Department of Data Theory, Leiden University.
14. George, C., Kaplan, N, & Main, M. (1984). *Adult Attachment Interview Protocol.* Unpublished manuscript. University of California at Berkeley.
15. George, C., Kaplan, N, & Main, M. (1985). *Adult Attachment Interview Protocol* (2nd ed.). Unpublished manuscript. University of California at Berkeley.
16. George, C., Kaplan, N, & Main, M. (1996). *Adult Attachment Interview Protocol* (3rd ed.). Unpublished manuscript. University of California at Berkeley.
17. Greenacre, M. J. (1984). *Theory and applications of correspondence analysis.* Academic.
18. Greenacre, M. J. (2008). *Correspondence analysis in practice* (2nd ed.). Chapman & Hall/ CRC Press.

19. Greenacre, M. J. (2013). Contribution biplots. *Journal of Computational and Graphical Statistics, 22*, 107–122.
20. Kroonenberg, P. M. (2018). Bibliographic analysis of handbooks: A proposal. Handbooks as a source of data for mapping of scientific disciplines. *Proceedings of the 23rd International Conference on Science and Technology Indicators (STI2018)*, 12–14 September 2018. Leiden, The Netherlands, pp. 1609–1618. https://scholarlypublications.universiteitleiden.nl/handle/1887/65225.
21. Meulman, J. J., & Heiser, W. J. (2004). *SPSS Categories 14*. SPSS.
22. Main, M., Kaplan, N. & Cassidy, J. (1985). Security in infancy, childhood, and adulthood: a move to the level of representation, (pp. 66–104). *Monographs of the Society for Research in Child Development*, *50*, No. 1/2.
23. Miranda, R., & Garcia-Carpintero, E. (2018). Overcitation and overrepresentation of review papers in the most cited papers. *Journal of Informatics, 12,* 1015–1030.

Behaviormetrics in Blockchains: A Novel Protocol for Recording, Processing, and Rewarding Valuable Behavior

Francisco Javier Ortín Cervera, Honorio Ros Múgica, and Innar Liiv

Abstract Feedback is always needed but not always received, especially when it is not required. But often it is very valuable, most notably when it is spontaneous and active feedback. This can help any entity and the society as a whole grow, and that is why it makes sense to reward it. The work here proposes a new blockchain protocol called the Proof of Feedback, which customers can use to provide feedback that is evaluated by a set of jurors and rewarded. In this paper, a prototype has been presented with the help of smart contracts in an EVM blockchain and a DApp has been developed for the users to interact with the protocol. The proposed protocol is technologically feasible, secure and economically viable. Such an approach and technology encourages novel aspects of transparency and reproducibility for the wider behaviormetrics community as well.

1 Introduction

Feedback is something that is always needed in any organization that wants to advance and serve well the society by fulfilling its mission. Feedback can be very valuable, depending on whether it is well-founded and whether it helps the feedback recipient get better. There are several solutions in the business world to manage specific feedback such as order process evaluation, product reviews, etc. In addition, there exists another kind of feedback that is not so easy to measure, but which can be even more valuable than the product feedback, especially when one considers the company's reputation or business model, and that is the one that comes spontaneously as well as proactively from the user/consumer of a given service or product.

F. J. Ortín Cervera · I. Liiv (✉)
Tallinn University of Technology, Tallinn, Estonia
e-mail: innar.liiv@taltech.ee

F. J. Ortín Cervera
e-mail: fortin@taltech.ee

H. Ros Múgica
Easy Feedback, Universidad de Navarra, Pamplona, Spain
e-mail: hros@easyfeedback.com

© The Author(s), under exclusive license to Springer Nature Singapore Pte Ltd. 2023
A. Okada et al. (eds.), *Facets of Behaviormetrics*, Behaviormetrics: Quantitative Approaches to Human Behavior 4, https://doi.org/10.1007/978-981-99-2240-6_7

Feedback is a necessary metric of an entity performance. There are standards, like the NPS (Net Promoter Score) [16] that simply measures the behaviour of the customers in their relation with a product or service. This paper aims to present a more thoughtful way to measure the behaviour of entities and how customers react to it. Since feedback can create great value by helping entities grow and improve what they have to offer to the society, it makes sense to reflect it in the form of a tokenized reward.

It is very complex to achieve it with traditional economic tools and without an interest of conflicts. For this purpose, a decentralized protocol called "Proof of Feedback" is designed to work in a decentralized manner with the help of blockchain technologies. This original concept was first presented by a startup Easy Feedback Token, and sketched in its whitepaper [6], but additional research is necessary to demonstrate that the implementation is feasible and reasonable with the help of blockchain technologies. One of the authors of this paper is a board member of the company.

This work explores and investigates the technical possibilities of such protocol, and how the previously presented idea could be further developed and implemented.

This paper aims to discuss the following questions:

- Is it possible to develop a protocol for rewarding valuable spontaneous feedback using blockchain technologies? Is it worth operating such a system?
- Is it possible to create a decentralized application (DApp) that allows users getting rewarded?
- Could this protocol be ruled and maintained by a decentralized autonomous organization (DAO)? What kind of functions could it assume and how could it be governed?

The Proof of Feedback is a protocol that allows customers send feedback to any entity in a private and non-anonymous manner and get rewarded for it when there is value created. For that a court of jurors evaluating the feedbacks needs to be designated and operate in a transparent and decentralized way to judge what is valuable feedback that needs to be rewarded.

Decentralization of a rewarding system can help to prevent conflict of interests and promote transparency. Also, deleting the middleman (entities managing all that process of rewarding economically) can help make it cheaper and faster to process. In order to achieve this goal, of course, the system needs to be feasible and the idea of the protocol needs further research in multiple aspects, such as how the feedback is transmitted from clients (customers) to oracles (jurors), whether that transfer is done in or out of the blockchain; there is also a need for research on how to use blockchain oracles that supply smart contracts with data from the outside world.

This paper is meant to explore the possibilities of using these concepts involved in Web 3 for tokenizating processes that previously would have been impossible to be monetized. The presented research explores further uses of blockchain technology other than decentralized finance (DeFi) that has become so popular in recent years. We view this as one of the pioneering works in the field of behaviormetrics in blockchains, combining cutting-edge technology with feedback research.

2 Novel Blockchain Protocol: Proof of Feedback

This work proposes a new blockchain protocol called the Proof of Feedback (PoF). The main idea of this protocol is to reward valuable private feedback that is given from customers to different entities. That feedback needs to be evaluated in an open and transparent manner, which is one of the concerns on the viability of this protocol.

The first idea and draft of this protocol was proposed by the Estonian startup Easy Feedback Token in their whitepaper [6] for the ICO of the EASYF, an ERC-20 token deployed in the Polygon blockchain to be the rewarding token of the PoF. The idea in the whitepaper [6] is conceptually drafted and the technicalities on how to implement that protocol, as well as, what things are possible need to be investigated. The proposal is also open to be changed and adapted to the results of this research.

The PoF is a protocol that requires expertise in each industry within the context of a given jurisdiction so, it needs an implementation in each sovereign country. In the Sect. 4.2 about the architecture of the protocol will be discussed what parts of the protocol are implemented locally. This will be further discussed in Sect. 4.4.

The main purpose of this work is to design a prototype, studying the different technical difficulties and obstacles that might arise during its implementation.

3 Existing Related Works

To the best knowledge of the authors, no previous research exists on specifically rewarding feedback with the help of blockchain technology. However, several systems have implemented non-native consensus mechanism for connecting the blockchain with end users (mainly in the healthcare field, e.g [4, 22]).

There is an interesting project interacting with the blockchain with a consensus mechanism called Proof of Disease [22], whichs aims to solve different challenges that have not been solve by previous electronic health records and health exchange information systems. Several properties of the project are similar to the protocol presented in this paper, since the Proof of Disease protocol needs to store data using a public ledger with different external parties (oracles) taking part in that process.

In addition, a research group at the University of Glasgow has implemented a prototype system in the form of a DAO to grade students with the help of blockchain [20] and there are public rewards depending on those grades.

4 Architecture of the Protocol

4.1 The Original Idea

The minimal version of the Proof of Feedback protocol can be observed in Fig. 1 from the whitepaper [6], including all the main components and their interaction.

A user gives feedback through some user interface (UI) and then that feedback is passed to the correspondent validator (Company validator in the case of company registered in a paid service where they evaluate the feedback they have received; Internal validator for a non-registered company; and the lawyer validator when the feedback is a complaint needing legal intervention for solving the case). Those validators are composed of different juries that evaluate the feedback and pass that score to the regulator node that rewards the user and necessary participants of the internal validator.

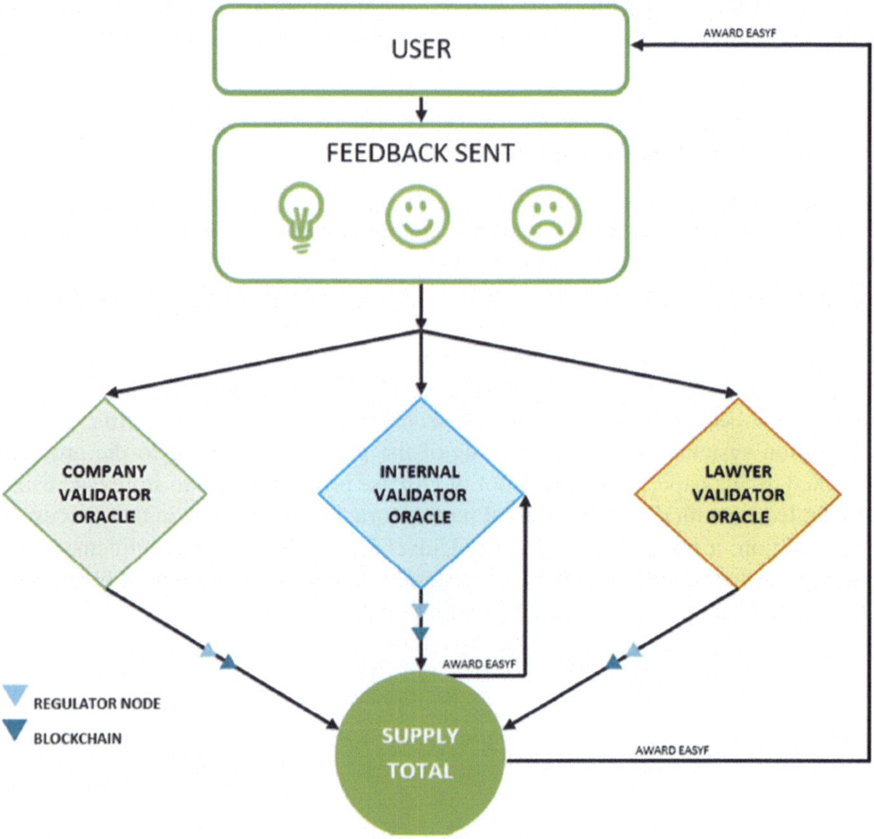

Fig. 1 PoF original concept from the Easy Feedback Token whitepaper [6]

In the original proposal, there is no interaction with the blockchain until the last step. Making it a totally centralized process until the owner of the ERC-20 EASYF token minting contract interacts with the blockchain. This is done by a regulator node that would mint EASYF only in the last step. This work aims to study possible ways to implement such system with the help of blockchain technology. A more decentralized and transparent alternative would be to have all the validators as smart contracts in the blockchain.

In the next Sect. 4.2 different protocol architecture possibilities will be described and analysed.

4.2 Possible Architectures

There are mainly three layers with a possible fourth in the Proof of Feedback that will be later discussed in more detail but will be briefly introduced here:

1. **Feedback origin layer**: here is where starts the interaction from the end-user with the PoF by writing feedback to an specific company through a DApp for providing feedback (in the original concept shown in Fig. 1 the first 2 blocks: user and feedback). That feedback is then passed to the next layer.
2. **Evaluation layer**: once the feedback from the user is received and registered it needs to be evaluated by a court of jurors that will decide if the feedback is worth a reward, and they will give an evaluation to it.
3. **Rewarding layer**: in this last layer, primarily the customer receives the reward in form of EASYF tokens according to the evaluation of the feedback. At the same time, jurors receive a small fee for their work.
4. **Minting layer**: In 2 of the proposed models, this one is part of the rewarding layer. This level is composed by the contract of the protocol in charge of minting the ERC-20 tokens, either to directly reward a customer and jurors, or to provide liquidity to a contract.

In the aim of decentralizing this process as much as possible and making everything transparent as well as publicly traceable, the evaluation layer would be already in direct interaction with the blockchain by implementing the different validators in smart contracts.

Also giving companies the possibilities to directly mint EASYF tokens would lead to the possibility of unbalance or totally biased rewarding, since it could be in their own interest to reward more their customers. This could motivate other companies might start doing the same up to a point where either the total supply is quickly minted making the protocol useless and the token losing all its value in the market. That is why it is proposed that the companies that want to reward using the PoF, can deploy their own company node and provide it with some EASYF token liquidity that they might acquire in the market.

4.2.1 PoF Architecture 1

In the architecture shown in Fig. 2 the oracle functions that are specific for each country are concentrated in one regulator node which is unique for each country.

Since the minting of the EASYF can be done only by the owner of the contract and that is only one address, there would be the need to have an intermediary "minter" contract that would own the ERC-20 token contract. Then the different regulator nodes would have access to call this minter contract which, in turn, would mint the necessary tokens.

4.2.2 PoF Architecture 2

In this second proposal, the regulator and minter node would be fused in one, decreasing the number of internal participants in the protocol and calls between each other. This has the downside that each internal and lawyer validators need to implement its own oracle calls to get external knowledge about what is the regional purchasing power. The oracle for the price of the EASYF could remain in the regulator node, since it should be the same across all the global market.

As it can be seen in Fig. 3 the rewarding and minting layer are the same in this architecture.

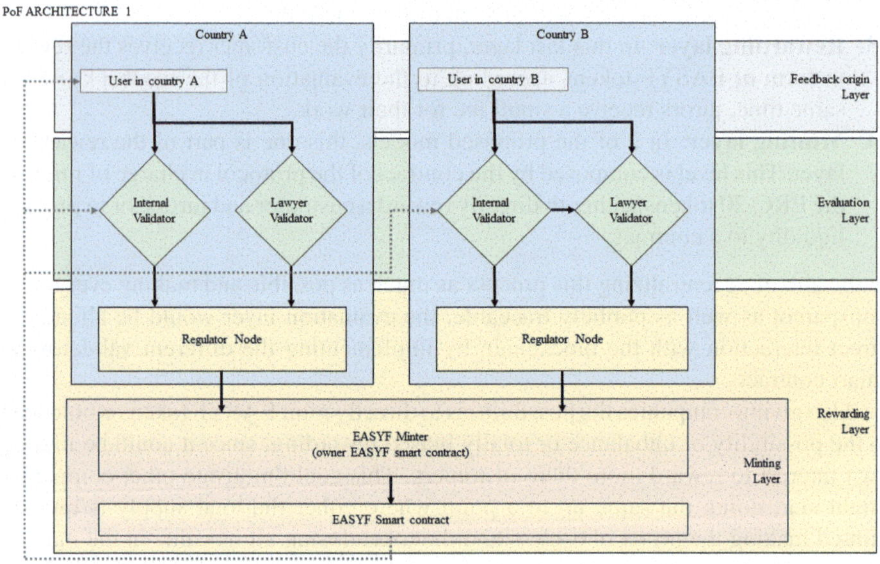

Fig. 2 PoF architecture model 1

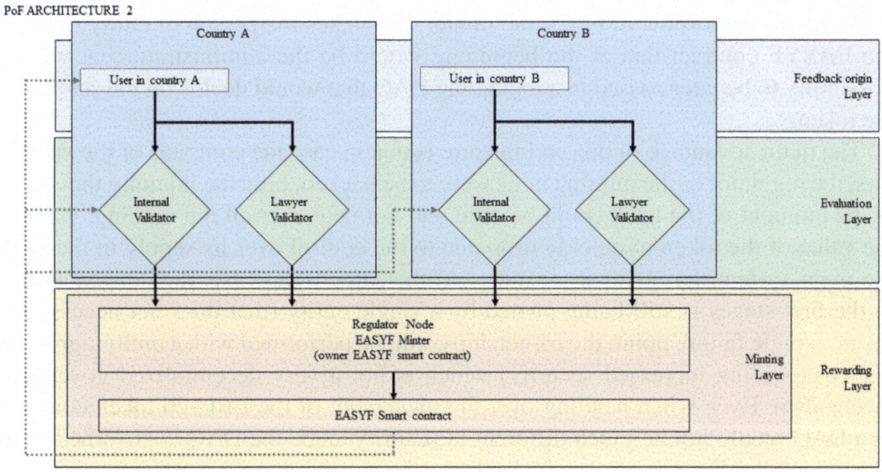

Fig. 3 PoF architecture model 2

4.2.3 PoF Architecture 3

The third variant proposed in Fig. 4 is very similar the first one in Fig. 2 but in this one the regulator nodes are provided some liquidity of EASYF and they can directly reward the customers and the jurors with their contract balance.

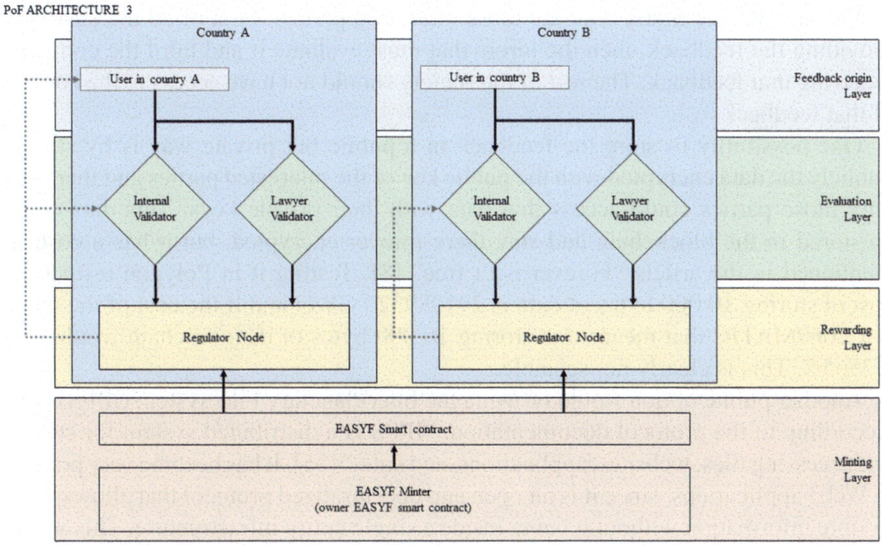

Fig. 4 PoF architecture model 3

In the diagram show in Fig. 4 the minter designates the person or entity owning the EASYF contract that at the beginning would be the administrator but with a possibility to be later passed to a regulating DAO that would decide on the supply of the token.

The main advantage of this architecture is that in case the contracts of the validators, the regulator or the minting node were to be hacked, only the liquidity they have at the moment of the hack could be used and not take over all the minting, leaving the value of the token to quickly drop due to the control over its supply in the PoF, the main system that gives it meaning. At the same time it has the drawback that in the first stages is still being owned by a single entity (that does not necessarily mean a single failure point, the ownership could be performed with a multi-signature wallet providing increased security) which is not a very decentralized governing mechanism. Even when handing over the ownership of the ERC-20 token contract to a DAO would not be much different in security since the DAO is governed by a smart contract that, in turn, can be hacked.

4.3 Feedback Storage

The matter on how to store the feedback provided by the customer is very delicate, since it needs to be private but at the same time interact with a system that is public and transparent. How can data be stored in a public ledger but at the same time be private, meaning that only the interested parties can have access to it, is the key question here.

We can analyze firstly who are those interested parties. First of all the customer providing the feedback, then the jurors that must evaluate it and third the company receiving that feedback. The rest of the society should not have access to the content of that feedback.

One possibility to store the feedback in a public but private way is by storing publicly the data encrypted with the public key of the interested parties and therefore only those parties could access that data with their private keys. This data could be stored in the blockchain and stay there forever encrypted, but it has a cost, as mentioned in the article "Forever isn't free" [8]. Testing it in Polygon testnet the cost of storing 10 000 bytes of data is 297089520 GWei and if the cost of the token is 1.20€/MATIC that means that storing 10 000 bytes of data on-chain would cost 0.3565€. This is clearly not scalable.

Another public option would be using the InterPlanetary File System (IPFS) [13]. According to the protocol documentation, "IPFS is a distributed system for storing and accessing files, websites, applications, and data." [25]. It has become very popular in Web3 applications, since it is an open and decentralized protocol that allows users to store information without it being inside a single entity infrastructure. This would be a more optimal solution. But still would require being encrypted with the public keys of the parties allowed to access that data (this requirement also applies to the on-chain storage solution).

The third option is keeping records of the feedback in a traditional and centralized manner using a database. This option is easier to handle, especially when taking on account data privacy laws. It is also true that feedback does not have to be stored forever. Its storage is essentially needed to be forwarded to the interested parties. This has been the option implemented in the prototype developed during this paper.

4.4 Jurors

The jurors are external parties, persons interacting with the protocol through externally owned accounts (EOAs), as oracles in the sense that they provide real world data, that is external to the blockchain. They are assigned feedback to evaluate according to a pre-established criteria.

There are some proposed requirements [6] that the jurors should fulfill to be deemed trustworthy and competent in the evaluation of feedback:

1. Having set up a company and/or to have been self-employed.
2. Have working experience in at least three sectors of activity. We might also have the possibility of having jurors specialized in a single sector of activity, who will only have the option of evaluating feedback in that sector.
3. Have more than 15 years of work experience.

These requirements are a proposal not yet tested. The initial proposal for choosing the jurors is that the documentation will be sent to a notary for controlling the compliance with these requirements and is the entity administrating the Proof of Feedback that will elect the jurors. The number of jurors can grow overtime as long as they meet the requirements. The Court of Jurors is elected for every country where the PoF is implemented so that they can understand the local environment (language, market, legislation, etc).

This selection part is the one that can be delegated to a DAO and will be discussed in Sect. 4.9

4.5 Regulator Node

The rewarding of the customer and the jurors happen through this node. The validators provide the evaluation that the jurors have given and is the regulator that calculates what should be the right amount rewarded to the customer.

4.6 *Validators*

The role of the validators is to be the connection with the blockchain. Firstly, the user interacts with them in a direct or indirect way when registering feedback, and then are the jurors that provide their evaluations inserting them through the validators. There are different validators depending on two things: the relationship of the company with the administrator of the protocol; and the kind of feedback given.

4.6.1 Internal Validator

The internal validator receives the feedback (customer address, feedback id) and needs to randomly select 3 jurors to assign them the evaluation task. This is an extra challenge since the EVM is a deterministic computer and performing a random selection on-chain is an easily hackable task in the sense, that it could be predicted who are going to be the jurors, and therefore, the hacker could input information in a way that the agreed juror would get to evaluate the feedback previously set.

To avoid such previously fixed performance, there would be needed an oracle to provide off-chain randomness that cannot be predicted. This kind of oracle already exits [11], nevertheless it needs an extra call to an external contract and pay the fee for such call. Solving this problem will be discussed in the implementation Sect. 5.2.1 solving this issue.

4.6.2 Lawyer Validator

When the feedback is a complaint/claim that requires legal advice or remedies during the process, the lawyer validator will be the only one in charge of managing the rewards to users. In each jurisdiction, a renowned legal firm specializing in consumer protection will form the lawyer validator. Users that offer input that meets the following conditions would receive tokens from this legal validator oracle:

1. The feedback sent has a chance of being dealt with as a judicial or extrajudicial claim.
2. When a customer engages a law company to handle the claim, the claim is processed, and then after the judge has ruled the case, the lawyer validator would compensate the user with EASYF tokens.
3. Once the judge has handed down his sentence, the lawyer validator oracle would award EASYF tokens equal to 10% of the compensation received.

4.6.3 Company Validator

In the original whitepaper [6] there was another kind of validator called company validator that was used for companies that are customers of Easy Feedback. This kind of validator led to have a company with minting capabilities and with no control by the rest of the network. This capability has been considered unsuitable for the system, removed and the possibility for a company to reward EASYF has been left in a way that they can have a company node, but they have to provide it themselves with the liquidity by means of purchasing EASYF in the market.

4.7 Feedback Evaluation

The final result of the evaluation will be the average of the 3 jurors' scores [6].

For each of the three categories of the feedback that are evaluated: usefulness, originality and execution, the Court of Jurors will use the following grading scale (Table 1).

And, therefore, the maximum total sum of the 3 jurors would be 36 points, giving a final result of an average of 12 points. This average result needs to be translated to a value in EASYF tokens. That amount needs to depend on what is the current price of EASYF in the market and the Big Mac index [27] of the country where the company receiving the feedback is located.This will be further discussed int the subsection about the different oracles Sect. 4.8

All scores will be transformed into tokens by the equivalence of 12 points means the dollar equivalent of the US price of 2 Big Macs which according to the current Big Mac index for Estonia (2022) published by The Economist magazine $4.43, would mean a reward of $8.86.

This value of around 8 dollars with the equivalence 1 EASYF = 0.05 USD, would correspond initially to 160 EASYF, at the moment of listing in trading exchanges. The 160 tokens will vary depending on the value at which they are traded on the market.

4.8 Oracles

Once the feedback is in the blockchain there are still two important data needed from the real world, and therefore two external oracles are needed to provide them to the smart contracts.

Table 1 Table showing the corresponding evaluation grades

Very low	Low	High	Medium	Very high
0	1	2	3	4

4.8.1 Token Market Price Feeding

The ideal technical architecture would be to have an external decentralized oracle hosted by a service like Chainlink [1], but since the token is still not traded in the market, such an implementation is not feasible in the scope of this work. As of now, the functionality is implemented using a smart contract.

4.8.2 Big Mac Index Feeding

This index is well-known and there is at least one public oracle implementation provided by Chainlink. The goal is to have such public oracle feeding information to the PoF, but such implementation has been left out of the scope of this work and it has been implemented manually with a smart contract. In addition, it is necessary to consider that since this index is updated once per year, it might not be worth implementing an automated call to the oracle system only for this.

4.9 Possibility of a Ruling DAO

There are many aspects of this system that would still be managed and ruled by a single entity, the company implementing the system. This is not what is sought with the development of DApps in a blockchain (what is called Web3). This is a bottleneck for decentralization. Many of the functions executed by the initial administrator of the protocol could be given to a DAO.

4.9.1 Governing Scope of the DAO

The following list presents the different functions the DAO could perform:

- Approve new jurors for the different internal validators. That would require the development of an additional system to proposed new jurors and a mechanism for voters to revise their application and make a voting decision accordingly. This presents an extra difficulty due to the regional implementation of the internal validators.
- Revise and update the requirements for new jurors applications.
- Revise and update the evaluation criteria and methods.
- Audit the different systems of the Proof of Feedback (e.g.: draw system, the maximum rewards, the evaluation layer).
- Audit the amount of feedback each customer and each juror has received. This audits can be held in a decentralized way with the help of the Graph, a decentralized protocol for indexing data of the blockchain [12].

- Provide liquidity to the different regulator nodes by minting new EASYF-s. This would mean giving the DAO the ownership of the EASYF smart contract.

It must be noted that these functions are just a draft and they require further research.

4.9.2 Possible Voting Powers

There are several voting powers that can be granted by:

- EASYF token holdings.
- Some certification that the participant has given feedback using the PoF (one voting power). This could be recorded by giving a feedback NFT (Non-Fungible Token) to each customer that has given feedback.
- Amount of times interacted with the PoF. This could also be implemented with the help of an NFT.
- Amount of times with successful feedback in the PoF.
- Delegated staking of EASYF.
- Using LP (Liquidity Provider) tokens that have been accumulated by staking.

5 Implementation Prototype

For testing this protocol, a proof of concept version has been implemented. The proof of concept version includes only one internal validator from the regional layer that covers one sector of the economic system; and one regional regulator node. In the prototype, the lawyer validator has not been implemented since it involves more parties and legal matters out of the scope in this paper. The smart contracts implementation will be discussed in Sect. 5.2.

The system has a traditional client-server architecture with an added blockchain interaction for tokenization the process of rewarding feedback.

5.1 Polygon Blockchain

The prototype has been implemented using the EVM compatible Polygon blockchain. There are mainly two reasons to use this blockchain: the community behind the blockchain (we could say its popularity) and the cost per transaction. The native token of this blockchain is called MATIC.

Ethereum is a very popular blockchain in terms of market capital and daily users, but at the time of writing this article, it has excessively high gas fees (over \$63 on average during November 2021) [5]. The effect of this has been scarcely studied, but it has been researched how it has a slight negative impact in the activity of DAOs

[7]. The design of the proposed system in this paper requires small fees for it to be feasible. This will be part of the validation process as discussed in Sect. 7.1.

Nonetheless, there are several alternative blockchains that are easily portable from one to another, due to the fact of being EVM compatible like Polygon, Binance Smart Chain (BSC), Avalanche, Fantom Opera etc. Polygon often comes as a popular alternative that is cheaper and faster than Ethereum as per 2022.

5.2 PoF Contracts

During the course of this paper several smart contracts have been implemented to test the different architectures. Using the proposed architecture shown in Fig. 4 there are mainly 3 different contracts: InternalValidator, RegionalRegulator and EASYFPrice-Feed. The last being the oracle code written for this paper, since the implementation of a decentralized oracle has become very expensive [14] and the requirements of the price feed oracle for this protocol are simple.

The process for registering feedback in the blockchain shown in Fig. 5 starts by a user, represented by a blockchain address, asking for evaluation of the feedback and for that needs to provide the ID of the feedback that has been previously registered in a storage system. It starts by checking that the contract has enough jurors registered to start working and that the feedback ID is new. Then the address of the feedback provider is stored in a structure of the state data in the internal validator contract. That structure is stored in a map where the index is the feedback ID. Using the provided feedback ID the jurors are assigned randomly and this data is also stored in

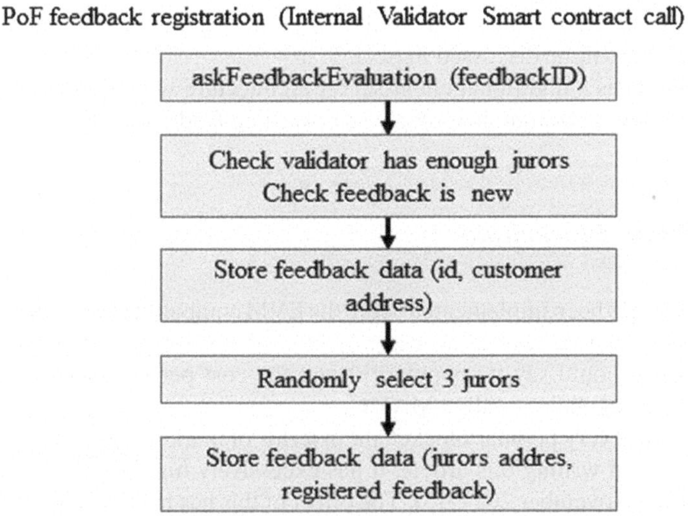

Fig. 5 PoF feedback registration smart contract call

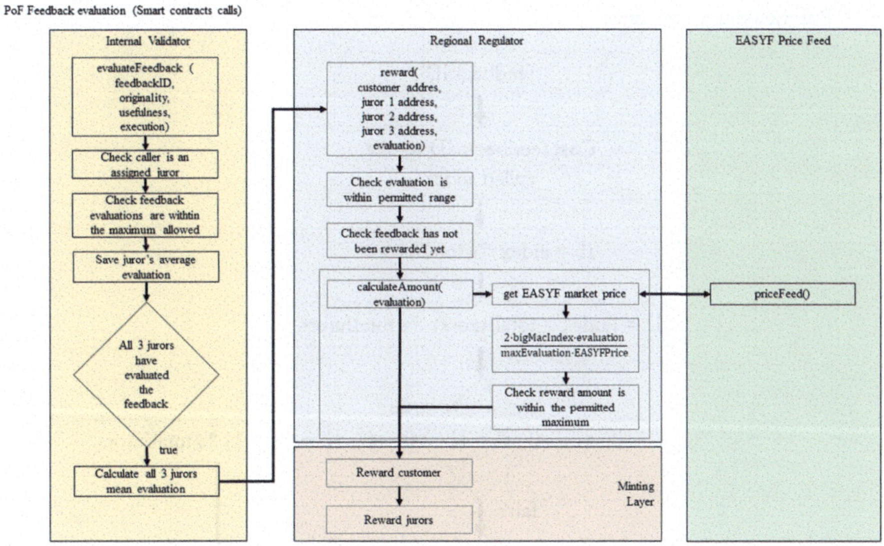

Fig. 6 PoF feedback evaluation smart contract calls

the structure. Then it needs to be evaluated by the jurors. All the feedback registration process happens in an internal validator smart contract.

Once the feedback is stored in the smart contract, the evaluation process can start as shown in Fig. 6. Each juror needs to interact with the smart contract providing the feedback ID and their 3 scores evaluating that feedback. The smart contract first checks that the jurors has been assigned that feedback and that the feedback evaluations are in the correct range. Then it stores the average evaluation of the juror in the structure of the feedback. Once the last juror inserts his/her evaluation, the function calculates all 3 jurors average and calls the regional regulator for giving the reward. That smart contract checks the evaluation is within the permitted range and makes sure the feedback has not been rewarded yet. Then it calculates the amount to be rewarded using Eq. 1:

$$\frac{2 \cdot \text{bigMacIndex} \cdot \text{evaluation}}{\text{maxEvaluation} \cdot \text{EASYFPrice}} \tag{1}$$

The EASYF price is retrieved by calling the EASYF price feed oracle. If the amount to be rewarded does not exceed the maximum set in the smart contract, then it proceeds to reward the user and the jurors calling the ERC-20 smart contract functions.

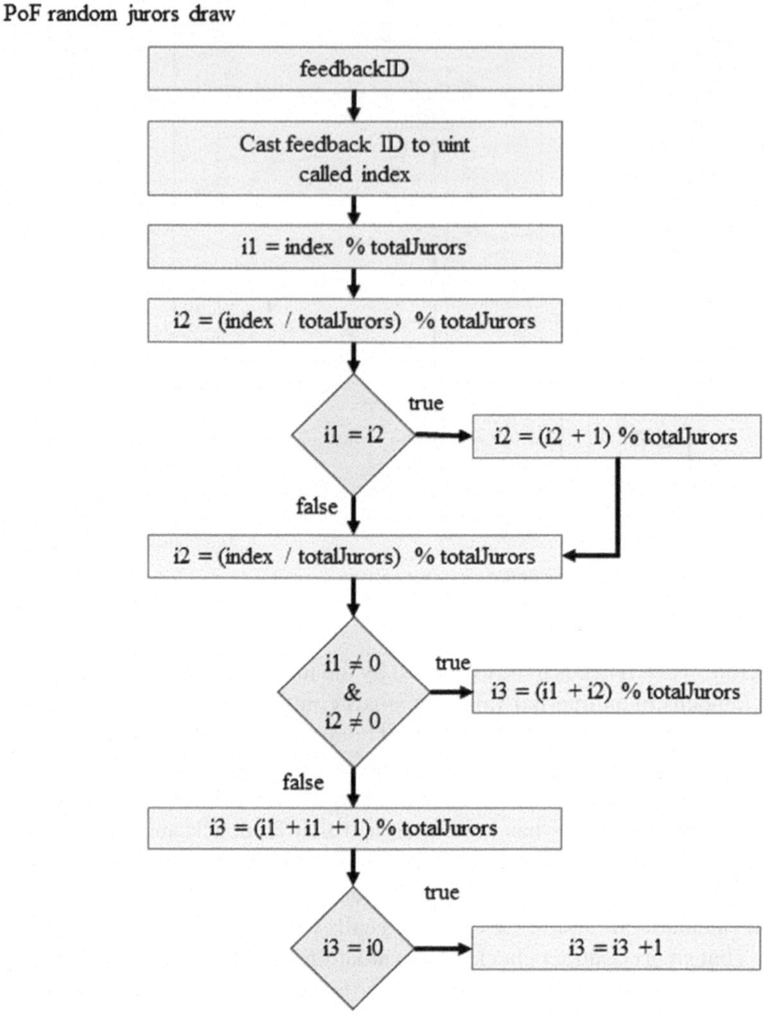

Fig. 7 PoF random jurors draw algorithm

5.2.1 Random Jurors Draw

To avoid an attack like the one to the Fomo3d game [10] where the seed for generating a random number was generated from data publicly available from the blockchain, it needs to be given from the outside real world.

As previously mentioned in Sect. 4.6.1, once a feedback is registered it needs to be randomly assigned to 3 different jurors. This poses the problem of generating randomness on-chain. But this may be solved by having a feedback ID that is random (the ID can be generated outside the blockchain and input as a parameter). Feedback

ID should be in the format of a valid 20-byte address [26]. For that, the feedback could be contained in a structure and ciphered using sha256 algorithm and take the right most 160-bits. This creates a practically collision free ID that is almost impossible to predict unless the user inserting the feedback knows exactly the structure of the feedback content used to be hashed and add words getting a favorable hash. A simple fix would be to add a random number (nonce) to that content or a timestamp in the back-end making it practically impossible for the user to predict the outcome.

Then in the smart contract can be used a modulo of the hash with the total number of jurors. The jurors' wallets are stored in an array, and the function that assigns jurors randomly takes as a parameter a feedback ID (160-bit address) that is first cast to an unsigned integer to draw the indexes as it is explained in Fig. 7 describing the algorithm.

The more jurors the less often collisions. With 7 jurors around 49% of the times there is some collision. With 17 there is already only 18% collisions. This has been empirically tested. It has also been proved that the jurors are uniformly assign for evaluation, meaning that there is no favourite index with this algorithm, all indexes are drawn with the same frequency.

6 Validation of the Proposed System

There are mainly two aspects analyzed for the validation of the system: a general audit of the protocol and the interface in Sect. 7 and to what degree the whole system really is decentralized in Sect. 8. At the end of the validation section, some possible improvements are presented in Sect. 9.

7 Smart Contracts Audit

Like for any system involving ownership of some asset, in this case digital assets in the form of cryptocurrencies, there are standard mechanisms for auditing smart contracts and different tools to perform an automated audit [10, 23] as well as specialized companies like Certik [3]. The use of such tools has been left out of the scope of this paper.

In this paper, we have summarized what seems to be the general practice in the blockchain industry [24], since no unified standards exist in the literature for auditing smart contracts. Firstly the cost of deploying and operating the system is analysed in Sect. 7.1, then the different possible vulnerabilities of the smart contracts are defined in Sect. 7.2 and finally what can be the platform security flaws in Sect. 7.3.

Table 2 Table the gas cost of the different smart contract functions

Solc version: 0.8.12		Optimizer enabled		Runs: 200	Block limit: 30000000 gas	
Methods		21 gwei/gas			1.20 eur/matic	
Contract	Method	Min	Max	Avg	# calls	eur (avg)
EasyFeedBackToken	mint	38651	55775	47213	32	0.00
EASYFPriceFeed	setPriceFeed	24799	46711	29736	100	0.00
PoFInternalValidator	addJuror	79701	113901	86032	54	0.00
PoFInternalValidator	askFeedbackEvaluation	148018	148419	148084	204	0.00
PoFInternalValidator	evaluateFeedback	61845	311161	132236	303	0.00
PoFRegionalRegulator	addInternalNode	74580	91680	90823	20	0.00
PoFRegionalRegulator	setMaxEvaluation	51203	59603	55403	2	0.00
PoFRegionalRegulator	withdraw	–	–	57276	1	0.00
Deployments					% of limit	
EasyFeedBackToken		–	–	1044625	3.5 %	0.03
EASYFPriceFeed		–	–	268824	0.9 %	0.01
PoFInternalValidator		1189237	1189249	1189248	4 %	0.03
PoFRegionalRegulator		1625688	1625700	1625699	5.4 %	0.04

7.1 Gas Costs Analysis

The first analysis to validate the system is the cost of deploying and using such system. For that, the different operations' costs have been evaluated using the developed prototype.

In Table 2 are presented the gas cost of each function that have been measured while running the tests locally. It is important to take on account that the gas cost may vary a bit depending on the moment usage of the network. Also, when translating that cost to euros must be reminded that the conversion from the Polygon native cryptocurrency to euros might change overtime.

We can observe that the cost of deployment is very small, a maximum of 4 euro cents for the regional regulator. But then we need to process what is the cost each time a feedback is registered and evaluated. To have an idea of what that cost means, the cryptocurrency units will be converted to the local fiat, euros. But the price of MATIC in relation to euros is volatile along time. Table 3 shows the variation in price during the last year (from April 2021 to April 2022). That data will be used to analyzed the possible operational variable cost.

The cost of registering a feedback in the blockchain is: $(148084/10^9) * 21 * 1,20 = 0,0037317168€$

While the average cost for a juror to evaluate a feedback is: $(132236/10^9) * 21 * 1,20 = 0,0033323472€$

So the total cost per feedback is $0,0037317168 + 3 * 0,0033323472 = 0,0137287584€$

If the yearly prognosticated amount of feedbacks is 100 000 the cost per year would be 1 372,88€.

Table 3 April 2021–April 2022 historical MATIC price

	MATIC/€
Minimum	0,587305667
Maximum	2,543323738
Average	1,350823265
Standard deviation	0,371608122

Table 4 Future cost estimation in euros

Function	Gas cost	Ether/Gwei	Gas	Matic/Eur			Cost		
			Multiplier	Min	Avg	Max	Min	Avg	Max
AskFeedbackEvaluation	148084	1000000000	21	0,58	1,35	2,54	0,0018	0,0042	0,0079
EvaluateFeedback	132236	1000000000	21	0,58	1,35	2,54	0,0016	0,0038	0,0071
Total cost per feedback							0,0067	0,0155	0,0291
Total cost per 100 000 feedback							671,91	1545,40	2909,70

Table 5 Generated value by feedback getting rewarded and jurors evaluating feedback

Generated value (€)	Customer	Jurors	Total
Min	0,00	7,69	7,69
Avg	3,84	7,69	11,53
Max	7,69	7,69	15,38

From the previous cost tables we can make a prediction of the operational cost for the Proof of Feedback protocol as shown in Table 4

But more interesting than the total cost is the relative cost of the value generated versus the cost of generating that value. The value generated is more stable since it has established an amount of 2 BigMacs to reward feedback an another 2 to reward the jurors. That amount is updated once or twice per year, allowing to do a more precise analysis. Table 5 shows the generated variable generated value (maximum, minimum and the average). The estimation has been done using the last BigMac index for Estonia 4,43 and the average conversion from USD to EUR during April 2021 to April 2022: 0,867916.

And Table 6 presents the variable relation of the generated value versus cost. It can be observed that even in the least favorable scenario when the generated value is the minimum (7,69€) and the cost is the maximum (0,0279€) the generated value is still more than 264 times bigger, which seems to justify the cost. In the best case scenario that ratio is even bigger with the generated value being more than 2288 times bigger than the cost.

Table 6 Relation of generated value versus cost

Total			**Cost**		
			Min	Avg	Max
			0,0067	0,0155	0,0291
Generated	Min	7,69	1144,46	497,59	264,28
	Avg	11,53	1716,69	746,38	396,42
	Max	15,38	2288,92	995,18	528,56

7.2 *Contract Vulnerabilities*

There are mainly three types of vulnerabilities analysed:

1. *Reentrancy issues* [21]: it refers to issues that arise when a smart contract makes an external call to another smart contract function before the effects of the original smart contract call are resolved. Because the original contract's balance has not yet been changed, the external contract can then recursively call the original smart contract and interact with it in ways it should not be allowed to. There have been several serious and big attacks of this kind in terms of capital in DeFi projects ($11.7M in the case of Agave DAO and Hundred Finance [17] and $2M in Revest Finance [19] and even bigger ones [9]).
2. *Integer overflows and underflows*: since the blockchain is a deterministic machine and smart contracts must be too, they work only with integers, and when a smart contract performs an arithmetic operation but for instance the result exceeds the storage limit of the smart contract (usually 18 decimal places) it leads to leaks or errors. As a result, inaccurate numbers may be estimated, or function calls being wasted.
3. *Front running opportunities*: Market purchases or sales can be predicted by poorly designed programming. As a result, others may be able to use and trade on the knowledge for their own gain. This affects mostly DeFi projects, but this prototype is not left out of this vulnerability.

7.2.1 Reentrancy Issues

In the current prototype there are no calls to external untrusted contracts at the moment, so this kind of attack is more theoretic. But analyzing the points where there are calls to external contracts, there is one reentrancy vulnerability when the last juror has evaluated the feedback. There is one requirement to check if the feedback has been rewarded or not to avoid double rewarding but, that flag is set after the customers and the jurors have been rewarded (a call to the ERC-20 token contract), so that could be exploited but only by designated jurors. In theory all jurors are externally owned accounts (EOA), that means the callers of the functions are not contracts, and therefore they cannot have a `fallback` function that exploits that

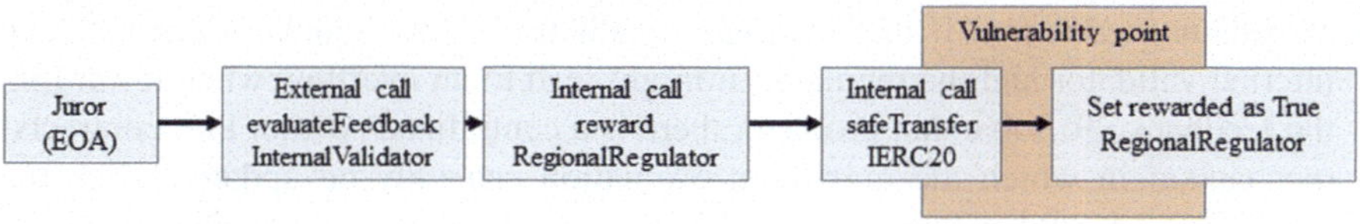

Fig. 8 PoF reentrancy vulnerability point

vulnerability. Figure 8 shows the point within calls where the reentrancy vulnerability can happen.

To avoid any possibility of that exploit, for instance a juror is registered with a smart contract he/she controls, adding a `noReentrant` modifier that locks reentrancy calls would prevent that call. Though that kind of juror should not be allowed. The proposed modifier for avoiding that attack is shown in Listing 1.

```
// attribute to be added in the contract
bool internal locked;

modifier noReentrant() {
    require(!locked, "Locked: ongoing contract
        call");
    locked = true;
    _;
    locked = false;
}
```

Listing 1 Lock modifier

There are also several modifiers: in the internal validator they make sure only jurors can evaluate and they are required to have that feedback assigned (this is checked within the function of the smart contract. There is also a modifier `onlyInternalValidator` that prevents anyone from calling the reward function in the regulator.

7.2.2 Integer Overflows and Underflows

The basic arithmetic operation security is handled by the standard safeMath library that most smart contract use in EVM blockchains.

In the proposed system, there are several points where overflows or underflows might happen:

- *Evaluation number*: if the jurors were to enter an excessively high evaluation, there could be a situation where the tokens rewarded would be unfairly high. Of course, the front-end does not allow so, but the juror could still register the feedback directly using the block explorer of Polygon or through some node. To prevent such a scenario, the smart contracts have a `maxEvaluation` parameter that cannot be exceeded.

- *Maximum evaluation*: if the maximum evaluation is not synchronized between internal validator and the regulator, it might lead to an overflow when rewarding the feedback. To solve that problem, there is a centralized (within PoF contracts) mechanism in which the maximum evaluation can only be updated using the regulator node and is that contract that updates the maximum evaluation in the internal nodes.
- *Regulator liquidity*: if the regulator does not have enough liquidity, it reverts the call preventing the situation were feedback has been evaluated and cannot be rewarded due to lack of liquidity and a bug in the code so when the internal validator rewards within the regulator, one of the things it requires is for the smart contract to have enough balance to proceed with the call.
- There is one "for" loop in the code for the internal validator that could be problematic if there would be millions of jurors registered in a smart contract, which should never be the case. This could only be exploited by hacking the owner's wallet and registering millions of jurors, which would be absurdly expensive for the hacker in terms of gas fees.

7.2.3 Front Running Opportunities

There are no sales in this project but there are a couple of points in the prototype that could become front running opportunities:

1. A customer and one or more evaluators agreeing on a certain feedback rewards. For that they would need to generate a favourable feedback ID that would draw the evaluators that agreed on the process. With few jurors this would not be very hard to achieve since you could brute force a favourable hash by changing the feedback content with some number used as nonce. This vulnerability is stopped in the server by adding a secret nonce.
   Of course, if the administrators of the system got to know about it, they can remove the jurors from the smart contract and the user can also be banned in the DApp (that functionality is added in the server code and there is the administrator role that can execute such action). That does not stop any user from registering any feedback ID in the validator, but that could not get rewarded since there is no register of that feedback. This is still a vulnerability, and that is why a possible improvement could be verifying users in a whitelist that the internal validators can check against.
   There is another requirement for using any internal validator: there have to be at least 7 jurors registered in the contract to start operating. If there are not enough jurors, then the call is reverted.
2. The price feed for EASYF that would come from an external oracle, if it got hacked like it happened with DeusDao [18] it could be used to manipulate the amount of tokens rewarded and therefore drain funds from the contract. In the current prototype this is not a problem since that oracle is controlled by the administrators of the system. But even if that oracle got hacked or there are

some problems with an external oracle the architecture of the system would not allow draining more funds that the limited amount allocated to the regulator node. And at the same time there is an extra attribute in the contract (`maxReward`) against which the calculated amount to be rewarded is checked, and if it exceeds that maximum the call is reverted.

7.3 Platform Security Flaws

When a protocol or platform is ready for production and professional audits are performed, they look at the network that hosts the contracts, as well as the API that is used to interface with the DApp. Users could connect their wallets to fraudulent blockchain applications if a project is subject to a defacement attack where the UI hijacked. Also, the system could be subject of DDoS (Distributed Denial of Service) attack and the interface would become useless.

The prototype developed during the course of this paper is no yet at a sufficient technological readiness level for production and does not include protection for DDoS and defacement attacks. However, sufficient security for the prototype has been implemented, e.g. an authenticated connection between the front-end and the back-end storing all the private data. No external users can access the feedback data, only assigned evaluators and owners of the feedback.

8 Decentralization Analysis

When analyzing how decentralized the proposed prototype is, the first task is finding out how many single failure points there are. Meaning, points that if they are exploited, the whole system would not work. The first obvious answer would be the point where any of the smart contracts does not work, but that has been previously covered, and if it gets hacked most of the token funds still remain secured as long as the EASYF ERC-20 smart contract remains unhacked. Also, the previously mentioned situation is more than one failure point. Each of the contracts has an owner that at the beginning would be the administrator of the system, but each of those contract owners (in the case they would become several different) would be a single failure point (excluding the EASYF token contract since the protocol needs tokens but not the contract). Here, the interface part for the PoF is not considered a single failure point since, the system can still work without it.

But there is indeed a bottleneck in the interface part, a point where one entity has control over that flow diminishing decentralization. If the server component stops working, the system can be used, but the jurors have no way to know the contents of the feedback they need to evaluate.

Another point for analyzing decentralization is transparency of the protocol. The users' personal data is kept private for obvious reasons: complying with data protection laws, as well as the fact that there is no need for that data to be public. The content of the feedbacks is also kept private.

8.1 Preserving Feedback Requirements

As described at the beginning of this paper, the feedback must be private and non-anonymous. Previously was discussed how the privacy of the feedback is ensured by the authentication mechanism which allows to securely store the data and provide it only to the jurors that need to evaluate it, as well as to the entity receiving the feedback.

The prototype presented in this work does not fulfill the non anonymity requirement, since the user can log in with a blockchain wallet and there is no further verification required, as well as no credentials are asked to provide feedback. But it is easy to restrict the registering of feedback until the user has provided personal information, the same way that when a juror that has not been assigned feedback cannot retrieve information about it.

9 Possible Improvements

One concern is the determinism of smart contracts, once it is deployed it cannot be changed. That is why for the prototype of this system there are mechanisms that prevent different exploits like funds getting locked (for that there is withdrawing function that allows to return funds to the owner of the contracts); deprecated internal nodes with access to the regulator node (this is done by adding or removing linked validators to the regulator node); and so on. But not all possible exploits or vulnerabilities arise from the beginning, as well as improvements to the contracts. That is why there exist mechanisms to upgrade contracts that have been developed in EVM compatible blockchains. Further research would be needed, but it is possible to create smart contracts that are upgradable, secure and scalable [2].

The implementation of a DAO is out of the scope of this research, since it is a whole topic and implementation process on its own. But this would be one of the next improvements towards decentralization of the system.

The system is currently designed in a way, that it can be accessed only by people having an EVM compatible wallet and that has certain knowledge on how to use Web3 applications. As per today, there are still many people that are not so familiar with these kinds of systems, and therefore would be left out from this system. This is the main obstacle in the use of the proposed system, since it is meant to be for everyone.

There are at least two different possible solutions for people without a blockchain wallet:

1. Qredo [15]: is a layer 2 protocol for managing crypto assets. Where dozens of wallets can be open and assigned to different customers, so the customer without his/her own wallet can still get the rewards and claim them in the future once has opened a blockchain wallet. The main disadvantage of this solution is that in the ledger it can be seen as a many rewards going to one address, that is the bridge to the Qredo layer 2 solution, giving the impression that someone is unfairly benefiting from the system.
2. Having a smart contract with a list of clients, where the correspondent EASYF tokens are deposited. This is a bit more expensive solution but more transparent.

In both solutions, the management of the funds would still be centralized. These solutions have been thought and discovered in the course of this paper and are a feasible future implementation.

In the current implementation there is lack of connection between the server side and the blockchain in the registering process. The authenticity of the data relies on the secure connection between client and server. It is true that the server provides the feedback ID to the client, which is mandatory to proceed. But it is lacking the mechanism to check if the feedback has been indeed registered. The server does verify the jurors assignation and for that operation reads data from the blockchain.

10 Conclusions

The main contributions of this paper were:

- Several architecture possibilities for an EVM blockchain protocol to reward valuable feedback along with different proposals on how to store feedback with different levels of decentralization;
- An actual prototype using smart contracts in an EVM blockchain for the Proof of Feedback protocol;
- A DApp prototype (client and server type application) to function as an interface for interacting with the proposed protocol smart contracts.
- A new algorithm for drawing three numbers on-chain at random given an external seed in the form of a valid EVM address.
- An audit report analysing the prototype to validate the protocol and the initial user interface. The gas cost analysis in Sect. 7.1 shows that the value generated each time feedback is provided justifies by far the cost of operating this system.
- An analysis of the degree of decentralization. Additional future decentralization steps to be taken are described in Sect. 8;
- An initial draft on how a DAO could be designed to govern the protocol: what functions in managing the protocol it could fulfill Sect. 4.9.1 and by what mechanisms could it be ruled Sect. 4.9.2.

This paper presented a novel approach for behaviormetrics in blockchains to record, process, and reward valuable behavior. The choice of using blockchain technology has the following positive properties:

- Transparency: by having a public record of transactions with a set of addresses.
- Tokenization: monetizing systems that were impossible or extremely cumbersome to do before.
- Data verification: Everybody can check what has happened in each transaction and know that it has not been tampered. Likewise, this is an opportunity for research since it is public data to be analyzed.

The proposed protocol is technologically feasible, secure and economically viable. Such an approach and technology encourages novel aspects of transparency and reproducibility for the wider behaviormetrics community as well.

References

1. Breidenbach, L., Cachin, C., Chan, B., Coventry, A., Ellis, S., Juels, A., Koushanfar, F., Miller, A., Magauran, B., Moroz D., et al. (2021). Chainlink 2.0: Next steps in the evolution of decentralized oracle networks.
2. Bui, V. C., Wen, S., Yu, J., Xia, X., Haghighi, M. S., & Xiang, Y. (2021). Evaluating upgradable smart contract. In *Proceedings—2021 IEEE International Conference on Blockchain, Blockchain 2021* (pp. 252–256).
3. Certik blockchain security. (2022). https://www.certik.com/
4. Chondrogiannis, E., Andronikou, V., Karanastasis, E., Litke, A., & Varvarigou, T. (2022). Using blockchain and semantic web technologies for the implementation of smart contracts between individuals and health insurance organizations. *Blockchain: Research and Applications,3*, 100049.
5. Data retrieved from sanbase analysis platform. (2022). https://app.santiment.net/charts?settings=%7B%22slug%22%3A%22bitcoin%22%2C%22ticker%22%3A%22BTC%22%2C%22from%22%3A%222023-02-14T18%3A29%3A59.815Z%22%2C%22to%22%3A%222023-08-14T18%3A29%3A59.815Z%22%7D&widgets=%5B%7B%22widget%22%3A%22ChartWidget%22%2C%22wm%22%3A%5B%22price_usd%22%5D%2C%22whm%22%3A%5B%5D%2C%22wax%22%3A%5B0%5D%2C%22wpax%22%3A%5B%5D%2C%22wc%22%3A%5B%22%2326C953%22%5D%7D%5D
6. Easy Feedback Token Whitepaper. (2022). https://easyfeedbacktoken.io/wp-content/uploads/2019/06/Easy_Feedback_Token_EFT_WhitePaper_en.pdf
7. Faqir-Rhazoui, Y., Ariza-Garzón, M. J., Arroyo, J., & Hassan, S. (2021). Effect of the gas price surges on user activity in the daos of the ethereum blockchain. In *Conference on Human Factors in Computing Systems—Proceedings* (p. 5).
8. Forever Isn't Free: The Cost of Storage on a Blockchain Database | by Jamila Omaar | IPDB Blog | Medium. (2017). https://medium.com/ipdb-blog/forever-isnt-free-the-cost-of-storage-on-a-blockchain-database-59003f63e01
9. Hack solidity: Reentrancy attack | hackernoon. (2022). https://hackernoon.com/hack-solidity-reentrancy-attack
10. He, D., Deng, Z., Zhang, Y., Chan, S., Cheng, Y., & Guizani, N. (2020). Smart contract vulnerability analysis and security audit. *IEEE Network,34*, 276–282.
11. Introduction to Chainlink VRF | Chainlink Documentation. (2022). https://docs.chain.link/docs/chainlink-vrf/
12. Introduction—The Graph Docs. (2022). https://thegraph.com/docs/en/about/introduction/
13. IPFS Powers the Distributed Web. (2021). https://ipfs.io/
14. Kaleem, M., & Shi, W. (2021). Demystifying pythia: A survey of chainlink oracles usage on ethereum. In *Lecture notes in computer science (including subseries Lecture Notes in Artificial Intelligence and Lecture Notes in Bioinformatics)* (Vol. 12676 LNCS, pp. 115–123).

15. Mccusker, K., & Spector, B. (2020). Qredo network—yellow paper. https://milagro.apache.org
16. Reichheld, F. F. (2003). The one number you need to grow. *Harvard Business Review, 81*(12), 46–55.
17. Rekt—agave dao, hundred finance - rekt. (2022). https://rekt.news/agave-hundred-rekt/
18. Rekt—deus dao—rekt. (2021). https://rekt.news/deus-dao-rekt/
19. Rekt—revest finance—rekt. (2022). https://rekt.news/revest-finance-rekt/
20. Rooksby, J., & Dimitrov, K. (2017). Trustless education? a blockchain system for university grades. In *New value transactions: understanding and designing for distributed autonomous organisations*.
21. Samreen, N. F., & Alalfi, M. H. (2020). Reentrancy vulnerability identification in ethereum smart contracts. In *IWBOSE 2020—Proceedings of the 2020 IEEE 3rd International Workshop on Blockchain Oriented Software Engineering* (pp. 22–29).
22. Talukder, A. K., Chaitanya, M., Arnold, D., & Sakurai, K. (2018). Proof of Disease: A Blockchain Consensus Protocol for Accurate Medical Decisions and Reducing the Disease Burden. In *2018 IEEE smartworld, ubiquitous intelligence computing, advanced trusted computing, scalable computing communications, cloud big data computing, internet of people and smart city innovation (SmartWorld/SCALCOM/UIC/ATC/CBDCom/IOP/SCI)* (pp. 257–262).
23. Tsankov, P., Zurich, E., Dan, A., Drachsler-Cohen, D., Gervais, A., Bünzli, F., & Vechev, M. (2018). Securify: Practical security analysis of smart contracts. In Proceedings of the 2018 ACM SIGSAC Conference on Computer and Communications Security (Vol. 18).
24. What is a smart contract security audit? | binance academy. (2022). https://academy.binance.com/en/articles/what-is-a-smart-contract-security-audit
25. What is ipfs? | ipfs docs. (2022). https://docs.ipfs.io/concepts/what-is-ipfs/
26. Wood, G., et al. (2014). Ethereum: A secure decentralised generalised transaction ledger. *Ethereum Project Yellow Paper, 151*(2014), 1–32.
27. Woodall, P. (1986). The big mac index. *The Economist*.

Propositions for Quantification Theory

Shizuhiko Nishisato

Abstract Over the past 50 years, behaviormetrics has been promoted for interdisciplinary research, and we see enormous progress in this endeavor. The current paper contains one researcher's observations on some problems which, the author considers, are worth further investigations. The problems are all related to quantification theory (e.g., Hayashi's quantification theory), and as presumptuous as they may sound, they are presented here as one researcher's propositions for further work. Some of them are only commentary in nature but some are very crucial for further development of quantification theory.

1 To Begin With

The research environment changes rapidly and makes a number of giant leaps in theory and scope from time to time. Our 50-year history of the Behaviormetric Society attests this observation.

When I left Japan in 1961 and started my graduate work at the Psychometric Laboratory of the University of North Carolina (UNC), Chapel Hill, I was more impressed than anything else by my observation that the academia in the United States was filled with interdisciplinary collaborations. As a good example, I can mention that UNC had joint seminars of departments of *Statistics* (e.g., Hotelling, Madow, Roy, Bose, Hoeffding and Chakravarti), *Biostatistics* (e.g., Grizzle, Sen, Koch, Greenberg, Glasser, Gabriel, Quade and Donnely) and *Psychometric Laboratory* (e.g., Jones, Bock, Thurstone, Shuford, Adkins-Wood, Fillenbaum and such visiting scholars as Kaiser, Schönemann, Ekman and Toda).

The interdisciplinary education in the US provided a stark contrast to the typical education of Japanese universities in the 1950s. This was an inspirational reckoning of what an ideal academia should be like.

When I returned to Japan in September 1965, I was invited to a one-week research camp on *Factor Analysis* at Kowaki-en, near Mr. Fuji, organized by the Union of

S. Nishisato (✉)
University of Toronto, Toronto, Ontario, Canada
e-mail: shizuhiko.nishisato@utoronto.ca

© The Author(s), under exclusive license to Springer Nature Singapore Pte Ltd. 2023
A. Okada et al. (eds.), *Facets of Behaviormetrics*, Behaviormetrics: Quantitative Approaches to Human Behavior 4, https://doi.org/10.1007/978-981-99-2240-6_8

Japanese Scientists and Engineers. To my pleasant surprise, the participants were from diverse areas of social, natural, and medical sciences. This gathering was my first Japanese experience of interdisciplinary gathering.

A remarkable event ensued shortly after that, namely the birth of our Behaviormetric Society, a truly interdisciplinary academic society. Statistics then gained a momentum to become an indispensable part of university education in Japan. This was a stark contrast to my 1950s education: As a psychology student, I was totally deprived of the opportunity to take courses in statistics and mathematics during my 6-year undergraduate and graduate education.

Thanks to the birth of the Behaviormetric Society, interdisciplinary research gained a critical momentum towards the advancement of science.

Now at the juncture of its 50th anniversary, let us pose a question if the current interdisciplinary research is effective enough towards far-reaching collaborative outputs. Ideally, each participant in interdisciplinary projects should be fully committed to the projects, not as a casual consultant, with their unique knowledge and skills. This is one query in promoting our mission of the Behaviormetric Society for the next fifty years.

As a researcher trained under the pre-Behaviormetric Society education, my work on quantification theory has been filled with illustrative examples and redundant expositions. This became my *modus operandi* to convince colleagues with backgrounds similar to mine. I am now a full-fledged senior member of this interdisciplinary society, and on this occasion, I would like to present several propositions for the next generation of researchers.

Proposition 1 *Promote effective interdisciplinary research*

2 Names and Classification of Quantification Theory

There are more than 50 aliases for quantification theory (see Nishisato [26]). This suggests a very unique and unusual aspect of quantification theory: the method can be derived by optimizing any one of a large number of objective functions. Some of the currently used names are:

(1) **Hayashi's theory of quantification**: This name is very general and rightly reflects Hayashi's contributions. However, it does not specify a particular domain of data analysis.
(2) **Optimal scaling**: This is an excellent name by Bock [3], but as J. Zinnes objected to it during the 1976 Psychometric Society meeting, many other methods are also optimal.
(3) **Correspondence analysis**: This is an English translation of the original French name *analyse des correspondances* by Benzécri and others [2], and it refers to the analysis of contingency information, and as such it is also an excellent name to characterize quantification theory

(4) **Homogeneity analysis**: This is a name proposed by the Dutch group (de Leeuw, Heiser, Meulman, Kroonenberg, van de Burg, van der Heijden, van Rijckevorsel and others) and is another excellent name since the technique is based on maximizing the homogeneity coefficient.
(5) **Dual scaling**: This was an outcome of the 1976 symposium at the Psychometric Society in Murray Hill, N.J. and first appeared in the title of Nishisato's 1980 book. Since then the name dual scaling has often been justified as probably the most appropriate one out of many aliases (see Nishisato [28]).

Admittedly, it is unlikely that the researchers on quantification theory would ever agree with a single name, but this is an intellectual exercise apart from its adoption possibility. When we look at the current stage of progress, it looks as though a better naming of quantification theory is to classify it into two categories.

(1) Symmetric Analysis: Nishisato [29, 30] discussed the role of symmetry in quantification and extensively illustrated quantification processes through expanding dual space to multidimensional symmetric space. To accommodate both bimodal and multi-modal symmetry, the name dual scaling is not general enough. In contrast, *symmetric analysis* seems to be general enough to cover the main spectrum of the traditional quantification. Under this name, both simple and multiple correspondence analyses are included, so is dual scaling of different types of categorical data.
(2) Non-Symmetric Analysis: As is thoroughly reviewed by Beh and Lombardo [1], non-symmetric correspondence analysis was proposed and developed by Lauro and D'Ambra [17], D'Ambra and Lauro [9, 10] and others. This important group of contributions will continue to be further developed, and all the work related to this approach should be categorized into *non-symmetric analysis*.

This dichotomy of quantification theory into symmetric and non-symmetric analyses will continue to be further investigated and generalized into a wider range of data types than we currently deal with. This is definitely better than simple versus multiple correspondence analysis or dual scaling of contingency tables versus dual scaling of multiple-choice data.

Proposition 2 *Quantification theory is encompassed with symmetric and non-symmetric analyses of categorical data.*

3 Two-Stage Quantification Analysis

It has been almost like an established practice to treat quantification analysis of contingency tables (simple correspondence analysis) as different from that of multiple-choice data (multiple correspondence analysis). Somehow this tradition of dichotomy has influenced the way in which quantification problems have been investigated, the most noted case being the CGS scaling controversies to be discussed later (Carroll, Green & Schaffer [6–8], Greenacre [13]).

One of the distinct contributions of Nishisato [21] was his treatment of the above two types of data under *a single umbrella*, which made it clear how the so-called perennial problem of joint graphical display can be solved. His handling of both types of data together was finally formalized by Nishisato [30]. We will use a numerical example to illustrate the essence of this amalgamation proposal.

Suppose we ask two multiple-choice questions:

Q1: Which do you like better? [1 = coffee, 2 = tea]
Q2: Are you (1) right-handed, (2) left-handed, or (3) ambidextrous?

The data can be represented in three formats (Tables 1, 2, and 3).

Nishisato [21] has shown that the response-pattern table and the condensed response-pattern table have identical singular structure and recommends to analyze the condensed response-pattern table which is often much smaller in size than the response-pattern table. The three data formats and the results of quantification analysis are presented in Tables 1, 2, 3, 4, and 5.

It is important to note that the data set in any one of the three formats can be reproduced from the other formats. This inter-replacement property, however, does not mean that the quantification results of the three formats are identical. See, for example, the contingency table yielded one component, while the other two pro-

Table 1 Contingency table format **F**

	Right-handed	Left-handed	Ambidextrous
Coffee	4	2	1
Tea	1	2	1

Table 2 Response-pattern Table $\mathbf{F_p}$

Subject	Coffee	Tea	Right-handed	Left-handed	Ambidextrous
1	1	0	1	0	0
2	1	0	1	0	0
3	1	0	1	0	0
4	1	0	1	0	0
5	0	0	1	0	0
6	1	0	0	1	0
7	1	0	0	1	0
8	1	0	0	0	1
9	0	1	1	0	0
10	0	1	0	1	0
11	0	1	0	1	0
12	0	1	0	0	1

Table 3 Condensed response-pattern Table **F***

Pattern	Coffee	Tea	Right-handed	Left-handed	Ambidextrous
I	4	0	4	0	0
II	2	0	0	2	0
III	1	0	0	0	1
IV	0	1	1	0	0
V	0	2	0	2	0
VI	0	1	0	0	1

Table 4 Principal coordinates of contingency Table **F**

C*	ρ^2	ρ	Coffee	Tea	Right-h	Left-h	Ambi
1	0.0964	0.3105	–0.2347	0.4108	–0.3402	0.2835	0.2835

Notes C*= Component; ρ^2 = eigenvalue; ρ = singular value
Right-h = right-handed, Left-h = left-handed, Ambi = ambidextrous

Table 5 Principal coordinates of response-pattern Table **F***

C*	ρ^2	ρ	Coffee	Tea	Right-h	Left-h	Ambi
1	0.6553	0.8095	−06119	1.0708	−0.8867	0.7390	0.7390
2	0.5000	0.7071	0.0000	0.0000	0.0000	−0.9574	1.9149
3	0.3447	0.5871	0.4438	−0.7767	−0.6432	0.5360	0.5360

Notes C* = Component; ρ^2 = eigenvalue; ρ = singular value
Right-h = right-handed, Left-h = left-handed, Ambi = ambidextrous

duced three components each (Tables 4 and 5). As for the response-pattern tables, we should use the smaller data set, the *condensed* response-pattern table **F***, rather than the larger response-pattern table $\mathbf{F_p}$.

The condensed response-pattern format yields more than twice the number of components associated with the contingency table **F**. In terms of our example, we note the following:

(1) Although we obtained a single component from the contingency table, the drinks and the hand dexterity do not span the same space, but they require two-dimensional space, where the two row and column axes cross at the origin with the angle of $cos^{-1}\rho = cos^{-1}0.3105 = 71.9°$ (see Nishisato & Clavel [33]).
(2) The above is a well-known fact that the two non-perfectly correlated variables cannot be graphed on one axis, but require a two-dimensional graph.
(3) The eigenvalue of 0.5 from the response-pattern table (component 2) corresponds to the correlation 0 of the contingency table; therefore, this component does not carry any joint information, and should therefore be discarded from analysis (Look at the second component of the response-pattern table with the eigenvalue of 0.5 and 0 weights of the row categories, meaning that this component does not contain any information about the row variable).

(4) The remaining two components provide us with the two-dimensional principal coordinates for this data set. It is important to note that one component has eigenvalue greater than 0.5 and the other smaller than 0.5 and that the sum of the two eigenvalues is 1.
(5) Following Nishisato's space theory (Nishisato [28, 30, 32]), these two components of (4) constitute *dual subspace*.

Therefore, the plot of the results from the response-pattern table provides the correct graph with the two-dimensional principal coordinates. If, however, we wish to adhere to the contingency table analysis and obtain only one component for our example, we can still construct a two-dimensional Euclidean graph, using Nishisato's formula to generate coordinates for the second dimension (e.g., see Nishisato [32]). This means that the total information of the contingency table is embedded in the table, and one must unfold the total information by an additional procedure.

Note that in the response-pattern format, both rows and columns of the contingency table appear in the columns, which guarantees that the resultant principal coordinates of both rows and columns of the contingency table span the same space (Young & Householder [43]).

In reflection of the aforementioned studies, Nishisato [30] finally defines quantification theory of the frequency data as consisting of two stages: Stage 1 is the analysis of the contingency table to investigate the correlational structure of the rows and the columns, and Stage 2 is the analysis of the corresponding response-pattern table to identify principal coordinates of both row and column variables in the common Euclidean space. This two-stage analysis appears to provide a complete picture of how analysis of association between categorical variables should be analyzed.

Proposition 3 *Quantification of contingency table and its response-pattern table should comprise a complete set of association analysis*

4 Space Theory for Two Variables

On the basis of his 1980 book, Nishisato later formulated his theory of doubled dimensionality (Nishisato [28, 30]; see also Nishisato [32]). He defines:

Contingency Space as the space used in the traditional quantification analysis of contingency tables (*i.e.*, simple correspondence analysis, dual scaling of the contingency table). The dimensionality of contingency space is given by the smaller of the number of rows minus 1 and that of columns minus 1. When the response-pattern format of the contingency table is used, all the eigenvalues of components in contingency space are greater than 0.5.

Dual Space is the space with twice the dimensions of the contingency space. One half of the total number of components in dual space have eigenvalues greater than 0.5 and the other half the number of components have eigenvalues smaller than 0.5. The traditional joint graphical display ignores those components with eigenvalues smaller than 0.5.

Table 6 Decomposition of quantification space

Space	ρ^2	Coffee	Tea	Right-h	Left-h	Ambi
CS	0.6553	−06119	1.0708	−0.8867	0.7390	0.7390
DS1	0.6553	−06119	1.0708	−0.8867	0.7390	0.7390
DS2	0.3447	0.4438	−0.7767	−0.6432	0.5360	0.5360
RS	0.5000	0.0000	0.0000	0.0000	−0.9574	1.9149

Notes CS = contingency space; DS = dual space; RS-residual space

Dual Subspace is defined for a pair of sets of coordinates when the dimension of dual space is greater than 2, and each dual subspace has the property that the sum of the eigenvalues of the pair is 1. This pair provides the two-dimensional space for each component of contingency space. In other words, each component in contingency space should be two-dimensional and the two-dimensional coordinates are given by these paired components in each dual subspace.

Residual Space is the space that shows the contributions of either rows or columns, and not both, of the contingency table. Each component in residual space has eigenvalue equal to 0.5, which corresponds to the zero correlation between the rows and the columns of the contingency table. When the number of rows of the contingency table is equal to that of columns, the dimensionality of residual space is nil.

Total Space is the sum of dual space and residual space. See more discussion of space theory in Nishisato [28–30] and Nishisato et al. [32].

For the current example, the principal coordinates of contingency space, dual space, and residual space are as shown in Table 6. Since there are only two components in dual space, they also constitute dual subspace (Note that the sum of the two eigenvalues is 1).

The complete information therefore can be captured by the two-dimensional graph, obtained from those coordinates in dual space.

When the data set is larger than our current example, we obtain many pairs of components which constitute a collection of dual subspace. See examples of space decomposition of a larger data set in Nishisato [29, 30] and Nishisato et al. [32].

Proposition 4 *Use a joint graph in dual space.*

5 Data Format for More than Two Variables

When we extend the bilinear expansion of a two-way table to the tri-linear expansion of a three-way contingency table, we can represent it as follows:

$$f_{ijk} = \frac{f_{i..}f_{.j.}f_{..k}}{f_{...}}[1 + \rho_1 x_{1jk} y_{i1k} z_{ij1} + \rho_2 x_{2jk} y_{i2k} z_{ij2} + \cdot + \rho_K x_{Kjk} y_{iKk} z_{ijK}$$

This decomposition corresponds to the expansion we use for quantification of the contingency table. As before, this starting point will yield a number of intangible difficult problems if we are to represent the outcomes in joint graphical display. When we extend the three variable case to the n variable case, we can conjecture easily how difficult it would be to derive optimal orthogonal coordinates of all the variables in common space.

To avoid this problem, our proposition is again to use the condensed response-pattern table for the case of n variables, where n is greater than 1. Since the response-pattern table is obtained by placing all the options of all the variables in the columns of the data table, this format enables the application of the Young–Householder theorem [43], and we can obtain orthogonal principal coordinates of all the variables in common space.

So, this is nothing but our familiar multiple correspondence analysis or dual scaling of multiple-choice data.

Proposition 5 *Analyze response-pattern tables, not corresponding multidimensional contingency tables*

6 CGS-Scaling Controversies

Carroll et al. [6] proposed the so-called CGS scaling in order to solve the perennial problem of joint graphical display (i.e., the group of French researchers developed the so-called French plot, now called correspondence plot, which is nothing but to plot row weights and column weights of the contingency table in common space, with the proviso that the joint plot of rows and columns does not give accurate descriptions of distances between a row and a column (Lebart, Morineau & Tabard [18]; see also Lebart, Morineau & Warwick [19]).

From the aforementioned theory of quantification space, we can say that the French plot is a graphical method to plot the row weights and the column weights in the same space (We now know that this corresponds to plotting only those components with the eigenvalues of the corresponding response-pattern table being greater than 0.5). For instance, suppose the quantification of the corresponding response-pattern table yields two components in dual subspace with the eigenvalues of 0.7 and 0.3 (note the sum is 1), the French plot adopts only the dominant component in this dual subspace, thus ignoring the minor component. The fact is that each dual subspace is two-dimensional, thus requires a two-dimensional graph. Yet, the French plot represents a unidimensional plot, using the only dominant component of the corresponding response-pattern table.

The CGS scaling was proposed to remedy this situation by analyzing the response-pattern table. The problem of the CGS scaling, however, was the fact that they considered only the contingency space. Recall that whether one analyzed the contingency table or the response-pattern table, those components in contingency space are proportional, thus no significant differences between the two data formats. The *real advantage of the response-pattern table is that it will produce additional components which fill common space.*

It is well known that the CGS scaling was severely criticized by Greenacre [13] and further clarifications were presented by the proponents (Carroll, Green & Schaffer [6–8]). But, their arguments were meaningless because Greenacre and the proponents were discussing only those components in contingency space (just the first component in our small example), where the quantification results from the contingency table are *proportional* to those from the response-pattern table. Their heated arguments were therefore completely out of focus. The relevant information to help the CGS scaling resides in dual space, which has twice the dimensionality of contingency space. That we need to double the space of the contingency table for the response-pattern table, however, was well known (Nishisato [21]) before the 1986 CGS scaling was proposed.

In summary, the traditional space used by the contingency table must be doubled to accommodate both rows and columns in common space, a crucial point that has never been brought up by either the proponents or the opponent of the CGS scaling. The perennial problem of joint graphical display is now a remnant of the history.

Proposition 6 *The controversies were not controversial*

7 Multi-modal Symmetry

Nishisato [29] has explored many symmetric aspects of quantification theory. One of them was a long-forgotten study by Nishisato and Sheu [35], in which they demonstrated multi-modal symmetry embedded in optimal quantification. Because of its importance, we will revisit the study here.

Nishisato and Sheu [35] developed a method, called the piece-wise method of reciprocal averages. To explain their method, let us borrow their example.

$$
\mathbf{F} = \left[\begin{array}{c|ccc|ccc|ccc}
Question & & 1 & & & 2 & & & 3 & \\
Option & 1 & 2 & 3 & 1 & 2 & 3 & 1 & 2 & 3 \\
\hline
Subject\,1 & 1 & 0 & 0 & 1 & 0 & 0 & 1 & 0 & 0 \\
2 & 1 & 0 & 0 & 1 & 0 & 0 & 0 & 1 & 0 \\
3 & 1 & 0 & 0 & 0 & 1 & 0 & 1 & 0 & 0 \\
4 & 0 & 1 & 0 & 0 & 1 & 0 & 0 & 1 & 0 \\
5 & 0 & 1 & 0 & 0 & 1 & 0 & 0 & 0 & 1 \\
6 & 0 & 1 & 0 & 0 & 0 & 1 & 1 & 0 & 0 \\
7 & 0 & 0 & 1 & 0 & 1 & 0 & 0 & 1 & 0 \\
8 & 0 & 0 & 1 & 0 & 0 & 1 & 1 & 0 & 0 \\
9 & 0 & 0 & 1 & 0 & 0 & 1 & 0 & 0 & 1
\end{array}\right]
$$

Since the response-pattern table can often be too large to handle, they considered the cross-product matrix. In our example, this matrix is as follows:

$$\mathbf{F}'\mathbf{F} = \left[\begin{array}{ccc|ccc|ccc} 3 & 0 & 0 & 2 & 1 & 0 & 2 & 1 & 0 \\ 0 & 3 & 0 & 0 & 2 & 1 & 1 & 1 & 1 \\ 0 & 0 & 3 & 0 & 1 & 2 & 1 & 1 & 1 \\ \hline 2 & 0 & 0 & 2 & 0 & 0 & 1 & 1 & 0 \\ 1 & 2 & 1 & 0 & 4 & 0 & 1 & 2 & 1 \\ 0 & 1 & 1 & 0 & 0 & 3 & 2 & 0 & 1 \\ \hline 2 & 1 & 1 & 1 & 1 & 2 & 4 & 0 & 0 \\ 1 & 1 & 1 & 1 & 2 & 0 & 0 & 3 & 0 \\ 0 & 1 & 1 & 0 & 1 & 1 & 0 & 0 & 2 \end{array}\right] = \begin{bmatrix} \mathbf{D}_1 & \mathbf{C}_{12} & \mathbf{C}_{13} \\ \mathbf{C}_{21} & \mathbf{D}_2 & \mathbf{C}_{23} \\ \mathbf{C}_{31} & \mathbf{C}_{32} & \mathbf{D}_3 \end{bmatrix}$$

where

$$\mathbf{D}_1 = \begin{bmatrix} 3 & 0 & 0 \\ 0 & 3 & 0 \\ 0 & 0 & 3 \end{bmatrix}; \mathbf{C}_{12} = \begin{bmatrix} 2 & 1 & 0 \\ 0 & 2 & 1 \\ 0 & 1 & 2 \end{bmatrix} = \mathbf{C}'_{21}$$

$$\mathbf{C}_{13} = \begin{bmatrix} 2 & 1 & 0 \\ 1 & 1 & 1 \\ 1 & 1 & 1 \end{bmatrix} = \mathbf{C}'_{31}\mathbf{C}_{23} = \begin{bmatrix} 1 & 1 & 0 \\ 1 & 2 & 1 \\ 2 & 0 & 1 \end{bmatrix} = \mathbf{C}'_{32}$$

$$\mathbf{D}_2 = \begin{bmatrix} 2 & 0 & 0 \\ 0 & 4 & 0 \\ 0 & 0 & 3 \end{bmatrix}; \mathbf{D}_3 = \begin{bmatrix} 4 & 0 & 0 \\ 0 & 3 & 0 \\ 0 & 0 & 2 \end{bmatrix}$$

Then the correlation ratio or the eigenvalue for the current example can be expressed as

$$\rho^2 = \frac{\mathbf{x}'\mathbf{F}'\mathbf{F}\mathbf{x}}{3\mathbf{x}'\mathbf{D}\mathbf{x}}$$

To determine $\mathbf{x}$ so as to maximize ρ^2, the partial derivative of ρ^2 with respect to $\mathbf{x}$ is set equal to zero, resulting in

$$\mathbf{F}'\mathbf{F}\mathbf{x} = 3\rho^2\mathbf{D}_c\mathbf{x}$$

Namely,

$$\begin{aligned} \mathbf{D}_1\mathbf{x}_1 + \mathbf{C}_{12}\mathbf{x}_2 + \mathbf{C}_{13}\mathbf{x}_3 &= 3\rho^2\mathbf{D}_1\mathbf{x}_1 \\ \mathbf{C}_{21}\mathbf{x}_1 + \mathbf{D}_2\mathbf{x}_2 + \mathbf{C}_{23}\mathbf{x}_3 &= 3\rho^2\mathbf{D}_2\mathbf{x}_2 \\ \mathbf{C}_{31}\mathbf{x}_1 + \mathbf{C}_{32}\mathbf{x}_2 + \mathbf{D}_3\mathbf{x}_3 &= 3\rho^2\mathbf{D}_3\mathbf{x}_3 \end{aligned}$$

By setting

$$3\rho^2 - 1 = \lambda$$

the three equations can be written as

$$\mathbf{C_{12}x_2} + \mathbf{C_{13}x_3} = \lambda\mathbf{D_1x_1}$$
$$\mathbf{C_{21}x_1} + \mathbf{C_{23}x_3} = \lambda\mathbf{D_2x_2}$$
$$\mathbf{C_{31}x_1} + \mathbf{C_{32}x_2} = \lambda\mathbf{D_3x_3}$$

If we set $\lambda = 1$, the above formulas can be written as

$$\mathbf{x_1} = \mathbf{D_1^{-1}}(\mathbf{C_{12}x_2} + \mathbf{C_{13}x_3})$$
$$\mathbf{x_2} = \mathbf{D_2^{-1}}(\mathbf{C_{21}x_1} + \mathbf{C_{23}x_3})$$
$$\mathbf{x_3} = \mathbf{D_3^{-1}}(\mathbf{C_{31}x_1} + \mathbf{C_{32}x_2})$$

To avoid the trivial component (see Nishisato & Sheu [35]), use the following sub-matrices:

$$\mathbf{C_{jk}} = \mathbf{F'_jF_k} - \frac{\mathbf{F'_j11'F_k}}{N}$$

The iterative scheme can then be written as:

Step 1: Assign arbitrary vectors to $\mathbf{x_1}, \mathbf{x_2}, \mathbf{x_3}$ for the right-handed sides of the above formulas.
Step 2: Calculate the new values of $\mathbf{x_1}, \mathbf{x_2}, \mathbf{x_3}$ by the above formula.
Step 3: Calculate p_1, p_2, p_3 by the formula

$$p_j = \frac{\mathbf{x'_jD_jx_j}}{3N}$$

where we use the new $\mathbf{x_j}$, j=1,2,3.
Step 4: Using the new vectors, calculate weighted vectors $\beta_j\mathbf{x_j}$, where

$$\beta_j = \sqrt{\frac{nNp_j}{\mathbf{x_jD_jx_j}}}$$

Indicate these weighted vectors by $\mathbf{x_j}$
Step 5: If the discrepancy between $\mathbf{x_j}$ of the previous iteration and the new ones are all less than say 0.0001, go to Step 6. Otherwise, go to Step 2 and use the new $\mathbf{x_j}$ for the right-hand side of the reciprocal averaging formulas.
Step 6: Calculate λ and ρ^2 by the formula.

The score for a subject is the average of the final weights of those options chosen by the subject, divided by ρ.

The quantification of the original response-pattern table **F** yields the optimal weights,

x' = [(1.56,–0.63,–0.93), (1.93,–0.16,–1.07), (0.29.0.52,–1.37)]

$\rho^2 = 0.6742$.

Table 7 Constancy of correlation ratio

Sum of Squares	1	2	3	Total
$SS(T_j)$	11.14	10.96	4.90	27.00
$SS(B_j)$	7.51	7.39	3.30	16.26
$SS(W_j)$	3.63	3.57	1.59	8.80
$\frac{SS(B_j)}{SS(T_j)}$	**0.6742**	**0.6742**	**0.6742**	**0.6742**

From the piecewise method, we obtain exactly the same scale values for these categories and the eigenvalue,

$$\mathbf{x}_1 = \begin{bmatrix} 1.56 \\ -0.63 \\ -0.93 \end{bmatrix}, \mathbf{x}_2 = \begin{bmatrix} 1.93 \\ -0.16 \\ -1.07 \end{bmatrix}, \mathbf{x}_3 = \begin{bmatrix} 0.29 \\ 0.52 \\ -1.37 \end{bmatrix}, \rho^2 = 0.6742.$$

Calculate the total sum of squares of each of the three items, $SS(T_j)$, the between-group sum of squares, $SS(B_j)$ and the within-group sum of squares, $SS(W_j)$, $j = 1, 2, 3$ under PMRA (Table 7).

Note the last lines of the above table: the *piecewise ratios are constant*. This is an example of *multi-modal symmetry*. Thus, these consistent ratios are embedded in the optimization of multiple-choice data. This is amazing!

The extension of the above procedure to data of n multiple-choice questions is discussed in Nishisato and Sheu [35] and Nishisato [29]. The existence of this kind of internal symmetry was an exciting discovery and suggests that much more work is needed for further exploration of symmetry in quantification.

Proposition 7 *Explore much further into complex symmetry as embedded in optimal quantification.*

8 Quantifying Ordered Categories

Bradley et al. [4] proposed a method of quantification of a contingency table where the rows or the columns have ordered categories, and another study with a different algorithm was also proposed (Nishisato & Ari [31]). These papers can be regarded as prototype procedures for handling order constraints, such as *never* $\leq$ *sometimes* $\leq$ *often* $\leq$ *always*.

However, from the quantification point of view, these studies created an impression that quantification problem with ordered response categories could be handled by imposing the order constraint on categories. This was a wrong message, and it must be rectified here.

As the term nonlinear multidimensional analysis (Nishisato [26]) implies, we are typically looking for both linear and nonlinear relations embedded in data. In this type of search, there is no justifiable room for introducing ordered categories as constraints. Rather we must ignore order constraints completely, and look for whatever relations the data would tell us, linear or nonlinear.

For instance, we may be dealing with data where an infant is weak, a young boy is stronger than an infant, a 20-year-old is stronger than a young child, a 30-year-old person is even stronger than a 20-year old, but 40-year old may be weaker than a 30-year old and an 80-year old may be even weaker than a young child. In this type of data, the order constraint on ages will never be able to describe what the data will tell us. Therefore, the categories of the data collection instrument are irrelevant to the assignment to the categories of the instrument. The above example of body strength suggests that if we are trying to assess the effects of age on memory, wisdom, and running speed, the optimal weights for age groups will likely show different optimal weights for the age categories.

Thus, the optimal weights for ordered categories of age depend on what phenomena we are trying to measure as a function of age. Therefore, in quantification theory, the ordered categories (e.g., ages) do not mean that we must impose any order constraints on categories, but rather we should leave them constraint-free and let the quantification procedure reveal their optimal values.

Proposition 8 *Ordered categories do not warrant order constraints*

9 Transformation of Input Data

Proposition 9 *Direct our attention to input transformations*

(9.1): Investigate whether or not input data should be standardized

Quantification theory was once called *principal component analysis of categorical data* (Torgerson [39]). In principal component analysis, it is well known that principal components associated with original data are very different from those with standardized data. For example, Nishisato and Yamauchi [36] demonstrated using artificial data how principal component structures of the same data are different between those of non-standardized variables and those of standardized variables.

Their findings can easily be understood once we know that principal axes will change once we change the measurement units. Then, we must answer the question of which structure is what we are looking for. Unfortunately, there does not seem to be a definite answer to the question: *what is the true structure of data?*.

For quantification theory, perhaps the first study relevant to this question was reported by Nishisato [24], in which he standardized the contributions of (1) the number of categories of each variable and (2) the frequencies of categories. As reported in the original paper and in Nishisato [26], the *standardization* has decisive influences on the outcomes of quantification.

Our problem then is whether or not we should *standardize* the input data. This question, raised for principal component analysis, should also be raised for quantification theory. So far, there does not seem to be any definite answers to the question either in principal component analysis or quantification theory. Is this because there is no right answer to the question? We must continue to look for an appropriate answer to the question.

(9.2): Investigate how many categories are useful.

Quantification theory deals with categorical data, but there does not seem to be much discussion on how many categories each variable should have, except when the number of categories is intrinsically fixed (i.e., +, –; gender). In practice, the researchers analyze whatever categorical data they have without much discussion on how many categories each variable should have (e.g., always, never; always, often, sometimes, never).

As Nishisato [21] demonstrated, the number of categories of each variable plays an important role in the optimal outputs.

Example 1 For instance, when we have two categorical variables, no matter how many categories of one variable may have, the number of components is limited by the other variable with a smaller number of categories. Suppose that the other variable has only two categories (e.g., yes, no; agree, disagree' left, right; urban, rural). In this case, contingency space is one-dimensional and dual space is two-dimensional, irrespective of the number of categories of the other variable.

Example 2 With the intention of capturing detailed information, one may decide to use many categories for all the variables (e.g., extremely disagree, strongly disagree, slightly disagree, neutral, slightly agree, strongly agree, extremely agree). Then, one consequence may be that there may be many categories which are not chosen by anyone, or the results of quantification may not lead to any more detailed analysis than one anticipates. Of course, if the number of respondents is very large, most categories are likely to be checked or chosen, but it is then possible that there are too many components to interpret. Sometimes it may be more useful to obtain only a few components that are easy to interpret than too many components to compare.

From the information retrieval point of view, it is ideal that all the variables have the equal number of categories (response options). In practice, however, there are a number of variables with the fixed numbers of categories (e.g., winners versus losers; rural versus urban; alive–moribund–dead; right-handed–left-handed–ambidextrous) and we cannot control these numbers.

Thus, we are talking about controllable categorical variables. For example, given the original response categories of (definite disagreement, strong disagreement, moderate disagreement, moderate agreement, strong agreement, definite agreement), can we change them into (disagreement, agreement), or (strong disagreement, moderate disagreement, moderate agreement, strong agreement), without changing the conclusion?

Although the purpose of quantification is to extract *meaningful* information, this is not a straightforward task. No matter how we may define the word meaningful, the number of response categories is one of the key ingredients of the rightful definition of *meaningful*.

(9.3): Investigate how to categorize continuous variables.

There are occasions in which the investigators want to categorize continuous data so that quantification of categorical data can be carried out. There are also occasions in which the data set consists of categorical data and continuous data. Our question then is how to categorize continuous variables. The main object is to extract as much information as possible. As Eouanzoui's study [12] shows, there are a large number of approaches to the problem of optimal categorizations. This is an important topic for quantification, and needs to be left for future work. See also Kim and Frisby [16].

10 Quantification of Dominance Data

The most typical example of dominance data is a set of ranks, for instance, Movie A is rank 3, Movie B is rank 1 and Movie C is rank 2. As may be obvious, rank numbers are not amenable to the operations of addition and subtraction, for those rank numbers do not have units. For instance, the rank 1 movie may be much better than the rank 2 movie, and the rank 2 movies may be only slightly better than the rank 3 movie. It should be obvious then that one must somehow find a way to assign reasonable numbers to those ranks before the data may be subjected to statistical analysis.

There is one unique aspect of dominance data, called *ipsativity*. Note that rank orders of the movies are meaningful only within subjects, for rank 1 by subject 1 cannot be equated to rank 1 by subject 2 or 3. Thus, the comparison over the rows (different subjects) of the data matrix is not meaningful, and this property is called *row-conditional*. But, we can compare the numbers within each row (subject).

The data with this property are called *ipsative data*. If there are two ipsative variables (e.g., movies), the correlation between the variables is always -1, irrespective of the number of subjects. How does ipsativity of data affect quantification outcomes? Row conditionality of the data implies, for example, that our quantification analysis can be carried out even when the number of subjects is 1. How the conditionality affects our analysis must be much further investigated than what we now know.

Quantification of dominance number was pioneered by Guttman [14], followed by a number of studies (e.g., Carroll [5], de Leeuw [11], Hojo [15], Nishisato [20, 26], Okamoto [37], Slater [38], Torres-Lacomba & Greenacre [40], Tucker [41], van de Velden [42]).

Nishisato [20] proposed an alternative formulation of Guttman's study and extended to the tied ranks as well as paired comparison data. He defined the following response variable.

$$
{}_i f_{jk} = \begin{cases} 1 & \text{if Subject i judges } X_j > X_k \\ 0 & \text{if Subject i judges } X_j = X_k \\ -1 & \text{if Subject i judges } X_j < X_k \end{cases} \tag{1}
$$

$i = 1, 2, ..., N$; $j, k = 1, 2, ..., n$ $(j \neq k)$. He then defined the *dominance number* for Subject i and Object j by

$$
e_{ij} = \sum_{k=1}^{n} {}_i f_{jk} \tag{2}
$$

The judge-by-object table of e_{ij} is called *dominance matrix* and is indicated by **E**. The eigenequation to be solved is given by

$$
(\mathbf{H}_n - \lambda \mathbf{I})\mathbf{x} = \mathbf{0}, \tag{3}
$$

where

$$
\mathbf{H}_n = \frac{1}{Nn(n-1)^2}\mathbf{E}'\mathbf{E} \tag{4}
$$

He then demonstrated his formulation is equivalent to Guttman's [14], Slater's [38] and Tucker-Carroll's [41] formulations. For the details of comparisons, see Nishisato [20, 26].

The quantification is carried out by solving the eigenequation for $\mathbf{H}_n$, but the dominance data have peculiar problems. For example, the sum of the eigenvalues T(inf) is given by

$$
T(inf) = tr(\mathbf{H_n}) = \frac{1}{Nn(n-1)^2} tr(\mathbf{E}'\mathbf{E}) = \frac{n+1}{3(n-1)}.
$$

The total information is therefore bounded by

$$
1 \geq T(inf) \geq \frac{1}{3}.
$$

$T(inf)$ converges to the minimum $\frac{1}{3}$ as n goes to infinity and the maximum of 1 when $n = 2$.

The total number of components T(sol) can be inferred from the fact that the sum of each row of the dominance table is zero. Thus, if the number of subjects N is larger than the number of objects n minus one, then $T(sol) = n - 1$. Otherwise, T(sol) = N. Note that there is *no trivial solution* in dominance data.

The above is only a surface difference of dominance data from the response-pattern data, which affects the quantification outcomes. The interpretation of the outcome is another matter because the object of the quantification is to reproduce not the input data, but the rankings of the input data for each subject. This creates a totally different problem when we want to develop a theory of quantification space for ordinal data and graphical display of the outputs, the reason for the next proposition.

Proposition 10 *Investigate the ipsative nature of dominant data*

11 Projection Pursuit

Nishisato [22] launched the pursuit of quantification of a mosaic structure of categorical data under the name of forced classification, and extended it to a more general procedure for projecting data structure onto substructure of data under the name of generalized forced classification (Nishisato [23]). In terms of projection operators, Nishisato and Lawrence [34] proposed perhaps the most general expression of a two-way table of data: Given the categorical data matrix **F**, express the row structure **P** and the column structure **Q**, in the following way:

$$\mathbf{F} = \left(\sum_i \mathbf{P_i}\right)\mathbf{F}\left(\sum_j \mathbf{Q_j}\right)$$

such that

$$\sum_i \mathbf{P_i} = \mathbf{I_p}$$

and

$$\sum_j \mathbf{Q_j} = \mathbf{I_q}$$

Using this scheme, one can carry out quantification in the chosen subspace, which we can call quantification focused on a chosen subspace. There are an infinite number of applications of this focused quantification, and this will be a promising horizon of quantification applications.

Proposition 11 *Explore the goldmine of categorical data*

Finally, there are two papers by Nishisato [25, 27], in which he discussed *Gleaning in the Field of Dual Scaling*, a collection of left-over problems in quantification theory. The papers may be of some interest to the readers.

References

1. Beh, E. J., & Lombardo, R. (2014). *Correspondence analysis: Theory, practice and new strategies*. Wiley.
2. Benzécri, J. P., et al. (1973). *L'analyse des données: II. L'analyse des correspondances*. Dunod.
3. Bock, R. D. (1960). *Methods and applications of optimal scaling* (p. 25). No: The University of North Carolina Psychometric Laboratory Research Memorandum.
4. Bradley, R. A., Katti, S., & Coons, I. J. (1962). Optimal scaling of ordered categories. *Psychometrika, 27*, 355–374.

5. Carroll, J. D. (1972). Individual differences and multidimensional scaling. In R. N. Shepard, A. K. Romney & S. B. Nerlove (Eds.), *Multidimensional scaling: Theory and applications in the behavioral sciences*, (Vol. I). Seminar Press.
6. Carroll, J. D., Green, P. E., & Schaffer, C. M. (1986). Interpoint distance comparisons in correspondence analysis. *Journal of Marketing Research, 23*, 271–280.
7. Carroll, J. D., Green, P. E., & Schaffer, C. M. (1987). Comparing interpoint distances in correspondence analysis: A clarification. *Journal of Marketing Research, 24*, 445–450.
8. Carroll, J. D., Green, P. E., & Schaffer, C. M. (1989). Reply to Greenacre's commentary on the Carroll-Green-Schaffer scaling of two-way correspondence analysis solutions. *Journal of Marketing Research, 26*, 366–368.
9. D'Ambra, L. & Lauro, N. C. (1989). Non-symmetrical correspondence analysis for three-way correspondence analysis. In R. Coppi & S. Bolasco (eds.), *Multiway Data Analysis* (pp. 301–315). Elsevier.
10. D'Ambra, L. & Lauro, N. C. (1992). Non-symmetric exploratory analysis. *Statistica Aplicata, 4*, 511–529.
11. de Leeuw, J. (1984). *Canonical analysis of categorical data*. Leiden University: DSWO Press.
12. Eouanzoui, K. B. (2004). On desensitizing data from interval to nominal measurement with minimum information loss. The University of Toronto Ph.D. thesis.
13. Greenacre, M. J. (1989). The Carroll-Green-Schaffer scaling in correspondence analysis: A theoretical and empirical appraisal. *Journal of Marketing Research, 26*, 358–365.
14. Guttman, L. (1946). An approach for quantifying paired comparisons and rank order. *Annals of Mathematical Statistics, 17*, 144–163.
15. Hojo, H. (1994). A new method for multidimensional unfolding. *Behaviormetrika, 21*, 131–147.
16. Kim, S. K., & Frisby, C. L. (2019). Gains from discretization of continuous data: The correspondence analysis biplot approach. *Behavior Research Methods, 51*(2), 589–601.
17. Lauro, N. C. & D'Ambra, L. (1984) L'analyse non symetrique des correspondences. In E. Diday (Ed.), *Data Analysis and Informatics* (pp. 433–446). Elsevier.
18. Lebart, L., Morineau, A., & Tabard, N. (1977). *Techniques de la description statistique: Méthodes et logiciels pour l'analyse des grands tableaux*. Dunod.
19. Lebart, L., Morineau, A., & Warwick, K. M. (1984). *Multivariate descriptive statistical analysis*. Wiley.
20. Nishisato, S. (1978). Optimal scaling of paired comparison and rank order data: An alternative to Guttman's formulation. *Psychometrika, 43*, 263–271.
21. Nishisato, S. (1980). *Analysis of categorical data: Dual scaling and its applications*. The University of Toronto Press.
22. Nishisato, S. (1984). Forced classification: A simple application of a quantification method. *Psychometrika, 49*(25), 36.
23. Nishisato, S. (1986). Generalized forced classification for quantifying categorical data. In Diday, E. et al. (Eds.), *Data analysis and informatics, IV* (pp. 351–362). Amsterdam: Elsevier Science Publishers B. V.
24. Nishisato, S. (1991). Standardizing multidimensional space for dual scaling. In *Proceedings of the 20th Annual Meeting of the German Operations Research Society* (pp. 584–591). Hohenheim University.
25. Nishisato, S. (1996). Gleaning in the field of dual scaling. *Psychometrika, 61*, 559–599.
26. Nishisato, S. (2007). *Multidimensional nonlinear descriptive analysis*. Chapman-Hall/CRC.
27. Nishisato, S. (2016). Dual scaling: Revisit to 'gleaning of the field'. *Theory and Applications of Data Analysis, 5*, 1–9. (in Japanese).
28. Nishisato, S. (2019). Reminiscence: Quantification theory and graphs. *Theory and Applications of Data Analysis, 8*, 47–57. (in Japanese).
29. Nishisato, S. (2022). *Optimal quantification and symmetry*. Springer Nature.
30. Nishisato, S. (2023). *Measurement, mathematics and quantification theory*. Springer Nature. in press.

31. Nishisato, S., & Ari, P. S. (1975). Nonlinear programming approach to optimal scaling of partially ordered categories. *Psyhcometrika, 40*, 525–548.
32. Nishisato, S., Beh E. J., Lombardo, R. & Clavel, J. G.(2021). *Modern quantification theory: Joint graphical display, biplots, and alternatives*. Springer Nature.
33. Nishisato, S., & Clavel, J. G. (2003). A note on between-set distances in dual scaling and correspondence analysis. *Behaviormetrika, 30*, 87–98.
34. Nishisato, S., & Lawrence, D. R. (1989). Dual scaling of multiway data matrices: Several variants. In R. Coppi & S. Bolasco (Eds.), *Multiway data analysis* (pp. 317–326). Elsevier Science Publishers.
35. Nishisato, S., Sheu, W., & J. (1980). Piecewise reciprocal averages for dual scaling of multiple-choice data. *Psychometrika, 45*, 467–478.
36. Nishisato, S., & Yamauchi, H. (1974). Principal components of deviation scores and standardized scores. *Japanese Psychological Research, 16*, 162–170.
37. Okamoto, Y. (1995). Unfolding by the criterion of the fourth quantification method. *Journal of Behaviormetrics, 22*, 126–134. (In Japanese with English abstract).
38. Slater, P. (1960). Analysis of personal preferences. *British Journal of Statistical Psychology, 3*, 119–135.
39. Torgerson, W. S. (1958). *Theory and methods of scaling*. Wiley.
40. Torres-Lacomba, A., & Greenacre, M. J. (2002). Dual scaling and correspondence analysis of preference, paired comparisons and ratings. *International Journal of Research in Marketing, 19*, 401–405.
41. Tucker, L. R. (1960). Intra-individual and inter-individual multidimensionality. In H. Gulliksen & S. Messick (Eds.), *Psychological scaling*. Wiley.
42. van de Velden, M. (2000). Dual scaling and correspondence analysis of rank order data. *Innovations in Multivariate Statistical Analysis, 12*, 87–99.
43. Young, G., & Householder, A. A. (1938). Discussion of a set of points in terms of their mutual distances. *Psychometrika, 3*, 19–22.

Exploring Heterogeneity in Happiness: Evidence from a Japanese Longitudinal Survey

Fulvia Pennoni and Miki Nakai

Abstract The present research explores the heterogeneity in trajectories of subjective well-being of Japanese citizens using longitudinal data collected with the Preference Parameters Study from 2003 to 2018. The analysis is carried out through the hidden Markov model that assumes a latent process underlying the individual perception of happiness observed through a categorical response variable. The first-order Markov chain is parameterized in terms of initial and transition probabilities depending on time-constant and time-varying socioeconomic and demographic variables. Maximum likelihood estimation of model parameters accounts for longitudinal sampling weights and missing responses under the Missing-At-Random assumption. Through this model-based clustering approach, we discover three clusters of individuals showing different dynamics across the life course, each of which represents, namely "not so happy", "moderately happy", and "very happy" individuals. We find that males tend to be less happy than females, and a U-shaped association between happiness and age is not detected. Each state has a substantial persistence over time, meaning that initial happiness perceptions play an important role in the life course inequalities of subjective well-being.

F. Pennoni (✉)
Department of Statistics and Quantitative Methods, University of Milano-Bicocca, Via Bicocca degli Arcimboldi 8, 20126 Milan, Italy
e-mail: fulvia.pennoni@unimib.it

M. Nakai
College of Social Sciences, Ritsumeikan University, 56-1 Tojiinkitamachi Kita-ku, 603-8577 Kyoto, Japan
e-mail: mnakai@ss.ritsumei.ac.jp

© The Author(s), under exclusive license to Springer Nature Singapore Pte Ltd. 2023
A. Okada et al. (eds.), *Facets of Behaviormetrics*, Behaviormetrics: Quantitative Approaches to Human Behavior 4, https://doi.org/10.1007/978-981-99-2240-6_9

1 Introduction

In the following, we first review the main literature on research related to subjective well-being (SWB), and we describe the Japanese context from which the analyzed data are derived. Then, we illustrate the main research questions and introduce the proposed methodological approach.

1.1 Literature Review and Features of the Japanese Context

Notions of SWB have a long tradition as central elements of people's quality of life. SWB is measured according to how people evaluate their lives, and it includes self-esteem, joy, lack of depression and anxiety, and positive moods and emotions [20]. In empirical studies, a variety of features have been used in order to measure SWB. Individual happiness is one definition of SWB. Another component may be satisfaction or feelings of fulfillment. Some scholars take a position that happiness and life satisfaction can be used interchangeably, see, among others, [26], while other scholars take opposing sides, suggesting significant differences between these two concepts, see also [28]. In general, happiness is used to measure SWB most frequently.

A better understanding of social mechanisms that produce inequalities in SWB has recently become a more serious subject in philosophy, ethics, psychology, economics, political sciences, and sociology. Interest in the topic among policymakers has also increased markedly during the past decade because of its potential to shed light on citizens' economic, social, and health conditions and informing policy decisions, see also [29].

Much work has been made to evaluate the correlates of SWB with the determinants of a happy and fulfilling life for people in a society, see, among others, [19, 25], and an interesting discussion on positive phenotypes in [24]. On the one hand, a large body of research on SWB has demonstrated that well-being remains relatively stable over time, see also [26], meaning that people cannot do much to change their long-term average levels of happiness throughout adulthood and old age. On the other hand, evidence suggests that self-report measures of SWB are sensitive to external circumstances and responsive to change [23, 36].

Differentials or heterogeneity among individuals in the levels of SWB and its dynamics or transition among different levels are not yet fully understood. Some people may have a consistently higher level of SWB and lead a happy life, whereas others may experience prolonged ill-being. Meanwhile, some people may remain stable in terms of their perception of well-being, while others may experience changes in the levels of SWB in response to the person's contextual changes. The mechanisms of distinct patterns over time also have not been thoroughly examined.

In addition to the above issues, since a large body of previous literature on the inequalities in SWB or happiness comes from Western cultures, the question of whether findings are generalizable in different cultural contexts remains unaddressed. Japan has been regarded as one of the industrialized countries with the longest average life expectancy. However, according to the World Values Survey [33], its average happiness score is low compared to developed and high-income countries. The World Happiness Report [30], based on a series of 2021 surveys, also showed that Japan is ranked 54th in the world, which is a relatively low position compared to other countries. Finland is ranked the highest, followed by Denmark, Iceland, and Switzerland. Japan results as the least happy of all the G7 nations. Germany, for example, is ranked the 14th, the United States the 16th, and Italy the 31st.

A more comprehensive understanding of how happiness changes as people age and how social factors affect happiness over time or across multiple life course stages may be crucial for Japanese policymakers, especially because Japan is at the forefront of aging societies. Aging well, living happy and healthy in middle and late adulthood has been a key public health policy strategy, especially in Japan, and nowadays, it is also one of the targets of the third goal of the Sustainable Development Goals [58] adopted by the United Nations in 2015, which have to be reached by 2030.

A lot of work has argued that SWB is U-shaped in age: well-being is highest for people in their 20s, decreases to its nadir in midlife around the age late 40s to 50, and then rises into old age [4, 5, 56]. At the same time, findings from the US suggest that the overall happiness levels increase with age and support the role theory of the aging process [62]. However, in 2008, White Paper on the National Lifestyle [8] reported that happiness of Japanese people does not increase with age.

Moreover, there has been much empirical research on SWB [31, 41, 44, 52, 59], most of which has examined data from cross-sectional or repeated cross-sectional surveys to investigate the underlying mechanisms that may affect happiness and SWB of Japanese people. Cross-sectional studies are not necessarily suitable to address dynamics like expanding divergence in the socioeconomic statuses (SES) gap in well-being over the life course. For the past couple of decades, the deep gaps in SES and their effects have been gaining increased attention in Japanese society. Inequalities in early life, such as educational attainment, which are explained substantially by family background, can have a lasting cumulative impact on people's quality of life across the life span, which implies that the influences of prior events grow over time. The cumulative advantages or disadvantages have been applied to the study of social inequality [21, 49]. They predict that initial well-being advantages for the well educated and disadvantages for the poorly educated diverge over time, producing more heterogeneity and socioeconomic-based inequalities with time.

Furthermore, Japanese society lags far behind in gender equality compared to other industrialized societies. Traditional norms about gender roles still prevail, in which men work extended hours as breadwinners, while housework and childrearing fall mainly on women [45]. In such societies, circumstances that make lives happier may differ by gender. Analyzing empirical data by gender permits a clearer understanding of culture and social norms.

1.2 Research Questions and Proposed Approach

To date, researchers have analyzed the survey data collected for exploring this topic, primarily using cross-sectional data to identify factors that correlate with happiness and SWB. As the importance of the usage of panel data for analyzing happiness has been recognized, more and more researchers have turned to panel studies that allow for longitudinal analysis to evaluate changes in SWB over time and assess the causal influence of people's experiences of life events on SWB. However, there is still relatively little sociological understanding of the heterogeneity in life course patterns and dynamics of SWB. Is there SWB differential over the life course between people from different social groups? How do the SWB trajectories differ among social groups? How does the SWB differential between individuals from different social groups develop throughout their lifespan? Does the U-shaped happiness curve hold for people in Japan? Are there any different factors driven by gender determining SWB?

To address the above-mentioned research questions, we consider longitudinal data collected through a Japanese representative survey repeated over sixteen years from 2003 to 2018 carried out by the Osaka University [43]. Since happiness may be conceived as an unobservable psychological process of each person that fluctuates over time, we analyze these data through a sequence of latent variables following a Markov process and assuming a discrete distribution. A hidden Markov (HM) model allows us to perform a dynamic model-based clustering [6] with hidden states representing heterogeneous subgroups of individuals in the population with individuals in the same state sharing similar degrees of happiness. The model allows people to move between clusters across time and time-varying covariates affect the latent variables through suitable parameterizations [1]. In this way, the influence of socioeconomic factors can be estimated on the probability of belonging to each cluster for the first time and at the following time occasions. A weighted maximum likelihood estimation approach is considered to account for longitudinal sampling weights provided within the multiple rounds survey so that the sample is representative of the population in each year. Missing responses are addressed under the Missing-At-Random (MAR) assumption [11, 35], according to which the missing pattern is independent of the missing responses given the observed data. Missing data referred to the covariates are accounted for through dummy variables as proposed by [12].

The remainder of the paper is structured as follows. Section 2 details the available data. Section 3 illustrates the methodology. Section 4 reports the results, and Sect. 5 presents some conclusions.

2 Data Description

The Preference Parameters Study [40, 43] collects data from a nationally representative sample of Japanese individuals. We consider data referring to surveys conducted annually from 2003 to 2013 and 2016 to 2018 since the survey was not run during 2014 and 2015. Therefore, the study tracks a large sample of individuals (N = 9,680) from 14 waves over a period of up to 16 years and allows for longitudinal analysis of change in SWB before and after significant life events. Participants were selected through a two-stage stratified random sample, including men and women whose birth years ranged from 1933 to 1989. The panel survey was conducted by a self-administered placement and pick-up visits method. In the first wave 1,418 individuals completed the questionnaire. Additional participants were selected and added to the study during the second wave in 2004, and they were 4,600 individuals; 2,000 individuals were added at the fourth wave in 2006, and 6,000 individuals were added at the seventh wave in 2009. Response rates for each year of the survey are reported in the Appendix along with additional information on the data. The sample used in this study represents 50,421 person-year observations aged between 20 and 85. The respondents were asked questions about their happiness, life satisfaction, sense of fulfillment, and sociodemographic characteristics. Demographic features such as sex, age group, and geographic region (prefecture) have been considered for sampling weights as auxiliary variables to correct any imbalances between survey sample and population.

2.1 *Measurement of Subjective Well-being*

In the following, we first describe the response variable related to happiness and then show some features of the observed socioeconomic variables which are listed in the Appendix.

2.1.1 Happiness

Respondents were asked to rate their level of happiness with a single item having categories ranging from 0 to 10. The question reads: "*Overall, how happy would you say you are currently? Using a scale from 0 to 10 where 10 is very happy and 0 is very unhappy, how would you rate your current level of happiness?*". A visual aid was also used in the administration of the question, which involved a pictorial representation of the scale with the extreme values labeled "very unhappy" and "very happy". This self-reported measure is regarded to be a reliable and valid measure to assess SWB [15]. The original categories are merged by joining adjacent categories to provide a more straightforward interpretation of the results and avoid sparseness [55]. Response categories from 0 to 2, 3 to 4, 6 to 7, and 8 to 10 are aggregated so that a five-point scale of "very unhappy", "unhappy", "neither happy nor unhappy", "happy", and "very happy" results. Figure 1 shows the observed responses of each individual over time. Looking at Fig. 1 clear patterns seem to emerge: a few individuals consider themselves unhappy throughout the survey, and most respondents expressed high levels of well-being over the years. In fact, the colors blue and green, referring to a feeling of happiness, are predominant in the figure.

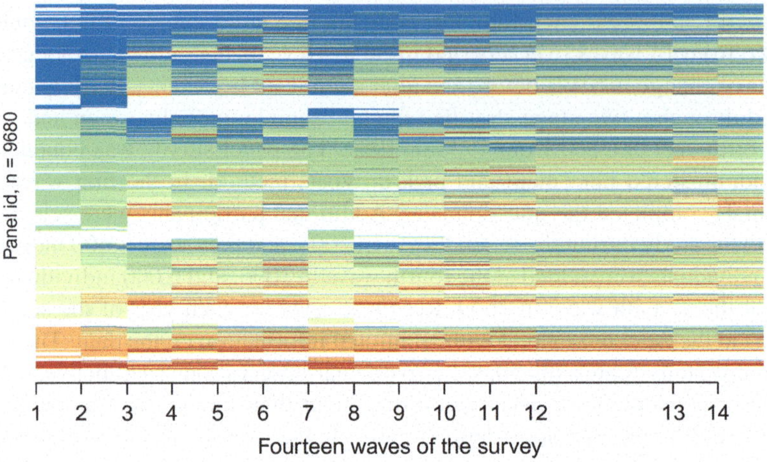

Fig. 1 Lasagna plot according to the fourteen years of the survey referred to the categories of the response variable happiness ranging from: "very unhappy" (red), "unhappy" (orange), "neither happy nor unhappy" (yellow), "happy" (green), "very happy" (blue), missing responses are shown in white

This tendency of Japanese individuals is in line with previous findings showing that most people reported positive levels of SWB throughout the world [17, 41].

Table 1 shows the observed proportions of each response category at every survey. In 2009, happy individuals were 67% of the sample; in subsequent years, this percentage increased to reach 70%. The percentage of individuals who consider themselves extremely unhappy is always not higher than 2%. The percentage of individuals who consider themselves more unhappy than happy remains constant over time at around 10%. The percentage of individuals who perceive themselves as neither happy nor unhappy in 2003 is 30%. This percentage decreases over time and settles around 19% at the end of the survey period.

2.1.2 Covariates

Researchers have examined individual and societal influences on enhancing happiness and SWB in the extant empirical literature. They have shown the effects of demographic factors such as age, gender, socioeconomic status, health, social support, religion, culture, and geographical variables [16]. Happiness and well-being are also associated with the social-institutional context and personality traits. What makes people happy in one society may not be the same in another when there are vast differences in the social norms surrounding individuals and institutions [42]. In the present study, the available independent variables (covariates) include indicators of respondents' socioeconomic resources and demographic characteristics that are considered to enhance or worsen SWB.

Table 1 Proportion of each category of the response variable happiness, by year

Response category		Years						
		2003	2004	2005	2006	2007	2008	2009
1	Very unhappy	0.028	0.022	0.017	0.022	0.024	0.023	0.021
2	Unhappy	0.080	0.107	0.107	0.105	0.116	0.115	0.094
3	Neither happy or unhappy	0.301	0.239	0.235	0.236	0.231	0.204	0.199
4	Happy	0.260	0.328	0.371	0.359	0.365	0.383	0.369
5	Very happy	0.287	0.296	0.262	0.263	0.255	0.265	0.296
	DK, NA	0.044	0.009	0.007	0.015	0.010	0.010	0.021
Number of observations		1416	4217	2980	3765	3107	2733	6184
Response category		Years						
		2010	2011	2012	2013	2016	2017	2018
1	Very unhappy	0.020	0.021	0.024	0.019	0.017	0.019	0.018
2	Unhappy	0.103	0.097	0.102	0.090	0.095	0.099	0.089
3	Neither happy or unhappy	0.192	0.178	0.169	0.162	0.162	0.181	0.187
4	Happy	0.385	0.411	0.392	0.420	0.407	0.401	0.441
5	Very happy	0.285	0.273	0.300	0.292	0.308	0.294	0.252
	DK, NA	0.015	0.019	0.013	0.017	0.011	0.006	0.013
Number of observations		5384	4936	4589	4343	2951	2119	1692

DK for do not know, and NA for no answer

Table 2 shows summary statistics of the time-varying covariates referred to the first, intermediate, and last waves of the survey by gender weighted with longitudinal sampling weights. There are significant differences in educational attainment and employment statuses between females and males. Nowadays, women have improved their educational attainment in many countries, but Japan still has a substantial male-female educational gap. Women tend to finish a two-year college rather than a four-year college. We also find sizable gender differences in the workforce. The majority of men work full-time, whereas women are much more likely to work part-time. Moreover, about one third of women are outside of paid work. These gender gaps in educational attainment and employment statuses could explain the observed differences in income distribution. See the Appendix for a detailed description of the categorical covariates.

3 Hidden Markov Model

Let Y_{it} be the ordinal longitudinal response variable measured at time $t, t = 1, \dots, T$, for individual i, $i = 1, \dots, n$, with categories labeled from 0 to l, and let $\boldsymbol{U}_i = (U_{i1}, \dots, U_{iT})'$ be the vector of time-varying discrete latent variables influencing the distribution of the responses. The latent process is defined for each individual

Table 2 Weighted percentages of respondents in each category of the covariates for the first (2003), intermediate (2010), and last (2018) wave of the survey by gender

Covariate	Years					
	2003		2010		2018	
	Female	Male	Female	Male	Female	Male
Education						
Junior high or lower	13.8	14.0	7.6	7.0	5.7	5.8
High school	57.5	56.1	50.8	45.5	48.1	44.9
Junior college	14.1	3.8	23.2	7.2	25.4	10.5
University and above	12.3	25.1	16.2	38.4	18.6	38.1
DK, NA	2.3	1.0	2.3	1.8	2.2	0.7
Employment status						
Working full-time	24.1	62.3	17.4	55.6	22.3	55.2
Working part-time	20.0	2.5	24.4	9.1	30.3	11.8
Self-employed	9.6	14.5	10.8	13.7	9.3	13.0
Unemployed	–	–	2.0	2.2	0.9	1.3
Not in labor force	42.7	17.5	31.3	10.9	30.7	15.5
DK, NA	3.6	3.2	14.0	8.7	6.5	3.2
Respondent's income						
None	39.3	10.2	12.2	4.0	10.7	3.8
Less than 2 million yen	31.3	13.9	38.0	12.6	41.6	12.6
2–4 million yen	10.5	21.3	12.8	25.7	17.6	28.1
4–6 million yen	4.9	23.5	3.1	20.6	7.1	21.1
6 million yen and more	2.5	20.3	2.1	19.7	1.8	19.2
DK, NA	11.5	10.8	31.8	17.4	21.1	15.2
Marital status						
Never married	15.6	22.7	13.2	19.7	13.2	19.9
Married	73.3	73.3	76.8	76.5	73.3	75.9
No longer married	11.0	3.6	9.7	3.5	13.2	3.2
DK, NA	0.1	0.3	0.3	0.4	0.3	1.0
Age class						
Early adulthood (20 and 30s)	34.0	37.1	29.4	31.3	22.1	25.2
Early middle age (40s)	16.7	18.0	16.5	18.6	19.9	21.1
Late middle age (50s)	18.6	19.4	17.6	18.4	15.8	16.0
Later adulthood (60s and older)	30.7	25.5	36.5	31.6	42.3	37.7
Number of family member						
One	7.3	6.3	5.3	5.2	11.5	7.7
Two	25.0	21.9	25.6	23.0	24.7	26.8
Three	21.3	23.2	22.1	24.6	24.1	22.7
Four	24.8	24.4	24.9	25.7	21.3	25.6
Five or more	21.0	23.4	20.5	20.3	17.8	16.4
DK, NA	0.7	0.7	1.7	1.2	0.6	0.9

(continued)

Table 2 (continued)

Covariate	Years					
	2003		2010		2018	
	Female	Male	Female	Male	Female	Male
Region of residence						
Hokkaido, Tohoku	12.5	12.0	12.3	11.7	11.7	11.3
Kanto	31.3	33.4	31.8	33.8	33.3	34.9
Koshinetsu	4.0	4.4	4.1	4.0	4.0	3.8
Hokuriku	2.9	2.5	2.7	2.7	2.4	2.7
Tokai	11.4	11.6	11.5	11.9	11.6	12.0
Kinki	16.6	16.2	16.5	16.0	16.5	15.9
Chugoku	5.5	6.1	6.2	5.7	6.4	5.9
Shikoku	4.1	3.0	3.2	3.3	2.6	2.8
Kyushu	11.7	11.0	11.7	10.9	11.5	10.7
Size of the place of living						
Cabinet order-designated cities	21.7	23.1	24.8	24.8	23.3	25.8
Cities with 100,000 or more population	40.6	37.1	43.1	43.0	44.3	42.0
Cities with less than 100,000 population	17.5	17.5	22.8	24.3	20.1	24.5
Towns or villages	20.2	22.3	9.3	7.8	12.3	7.7
Number of observations	729	685	2782	2604	875	819

DK for do not know, and NA for no answer

as a first-order Markov chain with a finite number of states denoted by k so that a hidden (or latent) Markov model results [1, 3]. Under the local independence assumption, the responses collected into the vector $\boldsymbol{Y}_i = (Y_{i1}, \ldots, Y_{iT})'$ are assumed to be conditionally independent given $\boldsymbol{U}_i$.

The conditional distribution of every Y_{it} for individual i at time t given the corresponding latent variable U_{it} is denoted as

$$\phi_{y|u} = p(Y_{it} = y|U_{it} = u), \quad u = 1, \ldots, k, \ y = 0, \ldots, l. \tag{1}$$

The hidden states may be interpreted as clusters of individuals with different degrees of happiness. Time-varying sociodemographic features are collected into the vector $\boldsymbol{x}_{it}$ for the t-th time occasion and individual i. These may affect the underlying perceived SWB represented by the latent variables.

The parameters of the latent model are the initial probabilities $\pi_{iu|x} = p(U_{i1} = u|\boldsymbol{x}_{i1})$ denoting the class weight for component u, $u = 1, \ldots, k$, and the transition probabilities $\pi_{i,u|\bar{u},x} = p(U_{it} = u|U_{i,t-1} = \bar{u}, \boldsymbol{x}_{it})$, $t = 2, \ldots, T$, $u, \bar{u} = 1, \ldots, k$ which are time homogeneous if the same evolution for all time occasions is conceived. In order to account for the effects of the covariates, we adopt a multinomial logit parameterization of following type

$$\log \frac{\pi_{iu|x}}{\pi_{i1|x}} = \beta_{0u} + \boldsymbol{x}'_{i1}\boldsymbol{\beta}_{1u}, \quad u = 2, \ldots, k, \tag{2}$$

where β_{0u} is the intercept specific to each latent state, and $\boldsymbol{\beta}_{1u} = (\beta_{2u1}, \ldots, \beta_{2up})'$ is a column vector of regression parameters referring to the covariates, with p denoting the number of these covariates. As a reference category, the multinomial logits have the first category corresponding to latent state 1. For the transition probabilities, we consider a multinomial logit parameterization based on the difference between two sets of parameters:

$$\log \frac{\pi_{i,u|\bar{u},x}}{\pi_{i,\bar{u}|\bar{u},x}} = \gamma_{0\bar{u}u} + \boldsymbol{x}'_{it}(\boldsymbol{\gamma}_{1u} - \boldsymbol{\gamma}_{1\bar{u}}), \tag{3}$$

where $\bar{u} = 1, \ldots, k, u = 2, \ldots, k, \bar{u} \neq u, \boldsymbol{\gamma}_{1u} = (\gamma_{1u1}, \ldots, \gamma_{1up})'\boldsymbol{\gamma}_{1\bar{u}}$ are vectors of regression coefficients corresponding to the time-varying covariates.

We account for missing responses due to refusal to respond, deleted data, or absence of contact with the respondent relying on the MAR assumption. In this case, the missing pattern is non-informative [35]. In order to estimate the model parameters, we account for customized individual weights calculated according to the cross-sectional weights provided within each survey round. Maximum likelihood estimation is performed considering to the observed log-likelihood given by

$$\ell(\boldsymbol{\theta}) = \sum_{i=1}^{n} w_i \log f(\boldsymbol{y}_i),$$

where $\boldsymbol{\theta}$ is the vector of all model parameters, $f(\boldsymbol{y}_i)$ is the manifest distribution expressed with reference to the observed data $\boldsymbol{y}_i$, and w_i is the sample weight for sample unit i, $i = 1, \ldots, n$. The EM algorithm [14, 60] is based on the complete-data log-likelihood, and once properly initialized with suitable starting rules [1], it is employed to compute maximum likelihood estimates of the model parameters. It alternates two steps, named E- and M-step, until convergence. With missing observations and under the MAR assumption, the E-step also requires additional computations. Convergence is checked on the basis of the relative log-likelihood, and inference is based on the solution corresponding to the largest value of the log-likelihood at convergence. Model selection is performed according to the information criteria such as the Bayesian Information Criterion [51, BIC]. It is worth mentioning that, with large data, such criteria may lead to selecting a model with many states. In this case, a pragmatic approach accounting for interpretability of the states related to the aims of the study can be considered. Once the parameter estimates are computed, standard errors may be obtained on the basis of the observed or expected information matrix. The estimated posterior probabilities are directly provided by the EM algorithm and are employed to perform *local decoding* so as to obtain a prediction of the latent states for each unit i at every time occasion t, denoted by $\hat{u}$. This prediction is obtained as the value of u which maximizes the posterior expected value of the latent variable given the observed responses and covariates. The entire sequence of predicted latent states resulting from the local decoding is defined as $\hat{u}_{i1}, \ldots, \hat{u}_{iT}$.

Suitable functions to estimate a general class of HM models, including the proposed model, are available in the package `LMest` [2] of the open source software R [46]. The code implemented for this research is available from the authors upon request.

4 Results

First, a model for the whole sample is estimated, and model selection is performed considering the BIC index. We selected a model with three latent states since the BIC index is decreasing, mostly passing from 2 to 3 states with respect to other models with more latent states. The model shows a log-likelihood value at convergence equal to -55,776 with 168 parameters. The model is adequate, and states can be reasonably interpreted as unobserved subgroups of individuals in the population [13]. Furthermore, as described in Sect. 1.2, we also aim to evaluate gender differences, and we estimated separate models using sample data of females and males to explore happiness determinants by gender on the initial and transition probabilities. The HM models for females and males with three latent states show a log-likelihood value at convergence equal to –26,187 and –29,177, with 164 parameters, respectively.

Table 3 presents conditional probabilities of the response as in Eq. (1), and initial probabilities of the latent variable for each of the three models. According to the estimated conditional probabilities, the latent states may be labeled as: “not so happy” (1st state), “moderately happy” (2nd state), and “very happy” (3rd state). The first state is mainly characterized by neutral to slightly lower levels of perceived happiness compared to the other states. The second state is characterized by moderate levels of happiness, and the third state shows the largest probability of responding to be very happy. Referring to the first year of the survey, 34% of the respondents are “not so happy”, 35% are “moderately happy”, and 31% are “very happy”.

4.1 Covariates Effects on the Initial Probabilities

Table 4 displays parameter estimates regarding the multinomial logit model as in Eq. (2) concerning the initial probabilities. First, we consider the results of the model for the whole sample. The first column shows the covariate effects on the logit of being in the second latent state “moderately happy” than in the first state “not so happy”. The second column shows the covariate effects on the logit of being in the third state “very happy” than “not so happy”.

Socioeconomic resources such as education and income have a significant effect. Higher educational levels foster higher SWB when controlling the other sociodemographic variables. For an individual that has not received a high school diploma, the

Table 3 Estimated conditional probabilities of the response variable given the latent states and initial probabilities of each state in 2003 for the HM model with $k = 3$ latent states estimated with the whole sample (All) and with data of females and males separately

	All			Female			Male		
	Latent state								
Response category	$u = 1$	$u = 2$	$u = 3$	$u = 1$	$u = 2$	$u = 3$	$u = 1$	$u = 2$	$u = 3$
1	0.073	0.001	0.002	0.073	0.000	0.002	0.076	0.001	0.001
2	**0.325**	0.022	0.004	**0.316**	0.024	0.004	**0.346**	0.018	0.004
3	**0.475**	0.147	0.020	**0.488**	0.163	0.017	**0.455**	0.151	0.018
4	0.114	**0.714**	0.167	0.113	**0.699**	0.160	0.112	**0.726**	**0.201**
5	0.013	0.116	**0.808**	0.010	0.114	**0.817**	0.012	0.104	**0.776**
	Initial probabilities								
	0.339	0.347	0.314	0.310	0.361	0.329	0.385	0.317	0.298

Higher conditional probabilities in each hidden state are highlighted in bold

estimated odds ratio for the second state "moderately happy" with respect to the first "not so happy" is $\exp(-1.662) = 0.190$ compared to the baseline category represented by individuals with a high school diploma. In contrast, people holding post-secondary education, relative to high school graduates, are more likely to be "moderately happy" or "very happy" rather than "not so happy": the estimated odds ratio for the third state with respect to the first is $\exp(1.109) = 3.031$. We notice that the association between education and SWB has been accounted for a significant part by the effects of socioeconomic status on social-psychological resources and health [48], and non-monetary factors such as interpersonal network [9]. The advantages of the well educated in work and economic circumstances, social-psychological resources, and lifestyle improve health [48, p. 720]. On the other hand, people with fewer years of schooling have more stressors, hardships, and an adverse health lifestyle. Also, the highly educated have more extensive social networks and greater involvement with the wider world, and these living conditions are positively related to happiness. Existing studies suggest that educational inequality leads to differences in health and social network, affecting the individual's quality of life [50]. Our result is consistent with those arguments.

Other characteristics appear to be significant in shaping happiness during the initial period of the survey. In particular, people in the highest income quartile, when compared to the base category (that of people in the lowest income quartile), are more likely to feel moderately happy (0.984) and also very happy (1.229).

Social capital, or the existence of a network of social relationships, has been regarded as another important factor that improves SWB of people. Studies have found that married persons have more extensive social networks, and therefore, greater social support, which in turn were expected to have positive effects on well-being than unmarried persons [32]. Among the measures of social capital, we found significant effects of marital status and the number of family members. As for differ-

Table 4 Estimates of logit regression parameters of the initial probabilities for the other latent states with respect to the first state for the HM model with $k = 3$ latent states estimated with the whole sample (All) and with data of females and males separately, concerning levels of the covariates reported in Table 2 excluding the baseline category (with significance indicated at the †10%, *5%, and **1% level)

	All		Female		Male	
Covariate	u = 1–2	u = 1–3	u =1–2	u = 1–3	u =1–2	u =1–3
Intercept	–0.299	–0.378	–0.128	–0.170	–0.147	1.005
Female	0.565*	0.957**	–	–	–	–
Junior high or lower education	–1.662**	–0.979**	–0.876	–1.149**	–1.465**	–1.847**
Junior college	0.792*	1.181**	1.226**	1.366**	1.399†	1.080
University and above	0.531†	1.109**	1.405*	1.309**	0.461	1.257**
Education DK, NA	–0.899	–0.088	–1.094	0.330	–0.834	–11.564**
Working part-time	–0.558	–0.217	–0.180	0.074	–1.091	–11.079
Self-employed	0.060	0.145	0.495	1.035*	0.151	–0.187
Not in labor force	–0.170	0.233	0.181	0.506	0.368	0.156
Employment status DK, NA	–0.658	0.666	–2.371†	–0.854	1.839†	2.751**
No income	–0.116	–0.479†	–0.050	–0.497	–0.287	–0.463
Second income quartile	0.603†	–0.263	–0.334	–0.589	0.967†	–0.488
Third income quartile	0.503	0.180	0.243	0.190	0.975†	0.227
Highest income quartile	0.984*	1.229**	2.860†	2.037	1.337*	1.321*
Income DK, NA	0.517	0.075	0.702	0.629	0.480	–0.613
Never married	–0.362	–1.267**	0.588	–0.772†	–1.246*	–1.927**
Divorced or widowed	–1.056**	–1.473**	–0.722	–1.635**	–1.679†	–3.396
Marital status DK, NA	1.397	–0.307	14.077	1.738	–4.791**	–6.730**
Aged 40–49	–0.297	–0.733**	0.137	–0.561	–0.628	–1.515**
Aged 50–59	–0.054	–0.975**	0.413	–0.855**	–0.498	–1.296**
Aged 60+	–0.298	–1.167**	–0.554	–1.360**	–0.817	–1.147*
Number of family 1	0.168	0.083	1.170	1.233†	–0.564	–0.886
Number of family 3	–0.083	–0.520†	0.171	–0.038	–0.478	–1.376**
Number of family 4	–0.337	–0.731**	–0.394	–0.984**	–1.201**	–1.341**
Number of family 5+	0.117	–0.443	0.524	–0.223	–0.406	–0.870†
Number of family DK, NA	0.000	0.000	–10.995	–1.601	3.258**	14.943**
Kanto region	–0.155	0.217	–0.129	0.725	0.002	–0.576
Koshinnetsu region	0.375	–0.114	0.882	0.856	0.718	–1.024
Hokuriku region	–0.114	0.156	–0.910	0.633	0.095	–0.041
Tokai region	–0.537	0.341	–0.666	0.715	–0.482	–0.333
Kinki region	0.193	0.447	0.365	1.326**	0.082	–0.547
Chugoku region	–0.043	–0.160	–0.047	–0.017	0.708	–0.210
Shikoku region	0.179	0.281	0.096	1.104	–1.352	–0.488
Kyushu region	–0.080	0.705†	–0.653	1.084*	0.588	0.316
Cities 100,000+	0.224	0.444	–0.230	0.372	0.673	0.826
Cities less than 100,000	–0.153	0.300	–0.499	0.143	0.232	0.785†
Towns and villages	0.751*	0.750**	–0.043	0.560	1.160**	1.042*

Reference categories: male, high school, working full-time, income first quartile, married, age 20–39, number of family members 2, region Hokkaido, Tohoku, cabinet order-designated cities. As for the variable of employment status, the respondent's unemployment category was not included in examining the covariate effects regarding the initial period due to the lack of information on unemployment. In fact, the questionnaire used for the first two waves in 2003 and 2004 did not provide the choice "being unemployed". DK for do not know, and NA for no answer entiation between "not so happy" and "moderately happy", although married persons do not have a decisive well-being advantage over single persons, being divorced or widowed is associated with lower perceived happiness than those married.

On the other hand, concerning the number of family members, respondents of one-person households do not necessarily associate with lower happiness than those living with others. Contrary to findings in other research [18], our findings reveal that people living in households with more than three or four persons are less likely to describe themselves as being very happy compared with the baseline category of those in two-person households. The results suggest that living in a household with multiple members may be more stressful than rewarding for adults. It has been argued that family relationships, such as marital, parent-child, grandparent, and sibling relationships, are significant for an individual's well-being since they may provide an informal intensive social network on the one hand. On the other hand, they may impose more duties such as individual childcare and caregiving for family members [57]. Moreover, how parenthood and caregiving affect adults' SWB depends on gender. In Japan, the idea that the key to a happy and fulfilling life for women is to be good wives and wise mothers is still very predominant [7]. Women are viewed as being responsible for caretaking tasks, and in fact, women experience a greater caretaker burden than men. However, public childcare facilities, as well as social services for the elderly, are still distinctly lacking in Japan [39].

Female respondents are more likely to be classified as "moderately happy" (0.565) and "very happy" (0.957) than "not so happy" compared to their male counterparts, meaning that women are more likely than men to feel happy. In previous research, findings concerning the association between gender and SWB have been inconsistent: some researchers have found that men have significantly higher levels of SWB, while others have shown that women have slightly higher levels of SWB, see, among others [27]. In regard to this, some recent studies have suggested that greater gender equality or high female status does not necessarily lead females to be happier or satisfied with their lives [38, 53, 54]. For example, results based on several countries show that additional involvement in gainful employment and prolonged female schooling are associated with lower happiness and satisfaction levels than males. In contrast, in societies where gender roles are still differentiated, with men earning money and women responsible for domestic chores, women tend to be happier and more satisfied than men [38]. The expansion of opportunities may have come at a cost for women, such as an increase in the total amount of work that women do, thus benefiting men more than women [53]. The theoretical framework of these studies [47] emphasizes the influence of the interactions between societies (such as labor market institutions) and cultures (such as gender-role ideology and expected behavior for men and women). Our results concerning the Japanese society, where gender inequality in the labor market is widely accepted, can be understood following the reasoning in previous research.

Age does not seem to affect the probability of being "moderately happy" rather than "not so happy", at all ages rather than the ages 20–39. For the age 40s or higher, there is a low probability of being "very happy". For example, for the highest age class (aged 60+), the estimated odds ratio for the third state with respect to the first is $\exp(-1.167) = 0.311$. Such a result is not consistent with the statement that happiness is U-shaped in age [4, 5] as illustrated in Sect. 1.1.

Regarding the effects of urbanization, compared with the residents in urbanized cities, those living in towns or villages are more likely to experience higher levels of happiness (the effect is of around 0.750 for both states "moderately happy" and "very happy"). Participation in community service and other social activities, which are found more frequently among rural regions, might be associated with greater happiness of the rural residents.

Below, we shed light on some gender differences, considering the results of the estimated HM models using sample data of females and males separately, which are reported in the last four columns of Table 4. First, we comment on the initial class weights reported in Table 3 noticing some differences between males and females. More females belong to the state labeled as "moderately happy" (36%) compared to males (32%), and more males belong to the state labeled as "not so happy" (39%) compared to females (31%). Women's SWB is more associated with finishing tertiary education. Earning a bachelor's or higher degrees makes both men and women "very happy" rather than "not so happy", but completing two-year post-secondary education does not enlarge the probability of being "very happy" for men. Low education hurts men more than women.

Employment status shows significant gender-specific effects. Engagement in self-employment compared to full-time standard employment exhibits beneficial effects on well-being only for women (1.035). For men, on the other hand, employment status does not matter, but being in the second income quartile or higher matters to be "moderately happy" rather than "not so happy", whereas women benefit from being in the highest income quartile for being "moderately happy".

Concerning age, the coefficients for males referred to the third state compared to the first are always negative and significant in contrast with those estimated for females. This indicates that the negative effect of age on higher happiness levels is more predominant for males. We also notice that Japanese women have the world's longest average life expectancy, but longevity does not necessarily imply living happily.

Being married is more important for men's well-being. By contrast, for females, the beneficial effect of being married is limited: Although married women are significantly more likely to be "very happy" than "not so happy", marital status does not affect the probability of being in the groups of "moderately happy" versus "not so happy".

Moreover, for women, living on one's own is associated with being "very happy" compared with living in two-person households. Living in a household with three or more members is more likely associated with men's negative well-being, meaning that having more family members for men may be a heavier mental burden. It has been expected that women living with many family members are more likely to be exposed to caregiving stressors, but the results suggest that the negative effects of women's caregiving responsibilities on their happiness levels seem to be relatively limited. On the other hand, negative effects for men seem to reflect widely accepted gender norms in Japanese society that men should be breadwinners to provide sufficient income to support a family.

There are some geographical effects on women's SWB. Women in the Kinki and Kyushu regions are comparatively "very happy" rather than "not so happy". An interesting finding is the effects of urbanization and its gender differences. As for the effects of city size on the probability of "moderately happy" or "very happy", the coefficients for female respondents are not significant. By contrast, men living in less urbanized areas are likely happier than those living in large designated cities. For Japanese men, living in rural areas is beneficial for their well-being, which has been consistently reported in most of the previous studies [22]. Undoubtedly, large cities offer higher incomes and employment opportunities, but when one's income, education, and employment statuses are held fixed, men living in very large cities are not as happy as those in smaller cities and rural areas. An adverse effect for those living in bigger cities on SWB might stem from higher costs of living, longer commute time, as well as unhealthy living environment.

4.2 Transitions Among the Different Happiness Levels Across Life Course and Predicted Latent States

Table 5 shows the averaged transition probabilities obtained with the model estimated for all individuals, and Table 6 shows those obtained with the model estimated separately for males and females. Over the sixteen years, the probabilities of switching from one state to another tend to be very small. There is high persistence in the same group thus suggesting that happiness levels remain stable over time. Therefore, as a consequence, initial inequalities of happiness play an important role in the life course inequalities in SWB. Having said that, for the period from 2004 to 2018, we estimate a general tendency to move from the first to the second state, and found that 9.5% of individuals improve their SWB, whereas about 4.4% of the individuals worsen their SWB passing from the second to the first state and 5.6% transit from the third to the second state. Looking at the results of the models estimated for females and males we notice that there is less persistence for males in the first and third state compared to females.

Table 7 shows the estimates of the intercepts of the logit models for the transition probabilities and Table 8 lists the estimated regression parameters as in Eq. (3). Socioeconomic statuses, such as education and respondents' income, significantly improve people's SWB over time. Education has significant effects on the transition from being relatively unhappy to happy, as well as from being relatively unhappy to very happy for both men and women. People not being so happy during the initial period, with a higher probability a person remains unhappy if lower educated, whereas higher educational attainment helps a person to shift out of being unhappy. Educational attainment has been regarded as reflecting human capital resources available to the individual to succeed [37]. In addition, we found that the lack of human

Table 5 Averaged transition probabilities with respect to the HM model with $k = 3$ hidden states

All	$u = 1$	$u = 2$	$u = 3$
$\hat{\pi}_{u\|1,x}$	0.896	0.095	0.009
$\hat{\pi}_{u\|2,x}$	0.044	0.912	0.044
$\hat{\pi}_{u\|3,x}$	0.011	0.056	0.933

Table 6 Averaged transition probabilities with respect to the HM model with $k = 3$ hidden states estimated with data of females and males separately

Females	$u = 1$	$u = 2$	$u = 3$
$\hat{\pi}_{u\|1,x}$	0.901	0.093	0.006
$\hat{\pi}_{u\|2,x}$	0.041	0.908	0.051
$\hat{\pi}_{u\|3,x}$	0.004	0.048	0.948
Males	$u = 1$	$u = 2$	$u = 3$
$\hat{\pi}_{u\|1,x}$	0.885	0.101	0.014
$\hat{\pi}_{u\|2,x}$	0.041	0.929	0.030
$\hat{\pi}_{u\|3,x}$	0.016	0.058	0.926

Table 7 Estimates of the intercepts of the logit models (based on the difference between logits) for the transition probabilities with respect to the HM model with $k = 3$ latent states estimated with the whole sample (All) and with data of females and males separately (with significance indicated at the †10%, *5%, and **1% level)

Intercept						
	All		Female		Male	
Latent state	1	2	1	2	1	2
$u = 1$	−2.585**	−6.017**	−2.257**	−5.727**	−2.279**	−4.682**
$u = 2$	−2.886**	−4.100**	−3.278**	−3.592**	−3.357**	−3.741**
$u = 3$	−3.702**	−1.921**	−5.155**	−2.445**	−4.496**	−2.626**

capital reduces self-efficacy and obstructs happiness. Interestingly, having a junior college education has a significant positive effect on men's initial happiness levels but has a significant negative effect on the transition to being very happy. It means that, men holding a post-secondary two-year educational certificate primarily related to certain vocational skills are more likely to be moderately happy than those who completed only high school, but they are less likely to become very happy as they age. Women benefit from getting an income in the second or higher quartile to improve their happiness levels. Men, however, benefit from getting higher than a median income. Gender pay gap in Japan is large, with women still making 74% of men's income in 2020.

Even though a woman does not earn the same money as a man, reaching the second income quartile helps to promote happiness. And yet, having no income impedes women from living a comfortable life. Women's income has been regarded as one of the determinants of women's autonomy in decision-making, which is an important indicator of women's empowerment. Women with no income may have little bargaining power in the household, and therefore, lack of autonomy is associated with lowering SWB. Conversely, autonomy can enhance SWB.

A self-employed working style is again beneficial for women's SWB. Unemployment has significant negative effects on SWB of men; meaning that the experience of unemployment obstructs men from getting happier. In contrast, the effects of being not in the labor force, compared with full-time regular employees, on the probability of transit from the group of "not so happy" individuals to that of "moderately happy" or "very happy" individuals are positive. Not being in the labor force or being retired seems to promote happiness for both men and women. Working hours in Japan have been among the longest in the OECD areas for many decades. A lack of work-life balance might negatively influence workers' well-being. Retired people, therefore, may start enjoying more leisure and lower stress levels. Contrary to expectations, being in part-time or non-standard employment has a significant positive effect on the transition from being unhappy to very happy only for men.

The effect of age 40s or older is found to be negative compared with those of age 20s and 30s. Therefore, U-shaped age effects are not confirmed for both men and women. As time goes by, although the number of family members has limited effects, being married to someone improves SWB for both men and women. Never married, divorced, or widowed are more likely to remain in the group of "not so happy" individuals than married persons.

Figure 2 shows the relative frequency of the individuals assigned to each state at every time occasion according to the local decoding, as illustrated in Sect. 3, of the whole sample and of males and females separately. The estimated sequences of latent states include those of people who never provided a response and also of individuals providing intermitted responses over time.

Interestingly, the probability of being very happy does not change through the period, and individuals are mainly predicted as "moderately happy" at the end of the period. For males, we find a declining trend in being unhappy and a slight decline in being very happy, although the predicted probability of "moderately happy" substantially increases. The frequency of the predicted probability of "very happy" for females is higher than that predicted for males over the entire period.

5 Concluding Remarks

The present study employs a data-driven approach using a hidden Markov model to examine individual happiness levels during the life course. The observed response variable is assumed as a proxy of latent variables characterizing the hidden process regarding happiness perceptions. We consider discrete latent variables, and we avoid restrictive assumptions on their distribution. We classify individuals into homogeneous subpopulations or clusters and each individual may be assigned to different

Table 8 Estimates of logit regression parameters of the transition probabilities based on the differences between logits with respect to the HM model with $k = 3$ latent states estimated with the whole sample (All) and with data of females and males separately, concerning levels of the covariates reported in Table 2 excluding the baseline category (with significance indicated at the †10%, *5%, and **1% level)

	All		Female		Male	
Covariate	u = 1–2	u = 1–3	u =1–2	u = 1–3	u =1–2	u =1–3
Female	0.596**	1.321**	–	–	–	–
Junior high or lower education	–0.523**	–0.647**	–0.688**	–0.422†	–0.420**	–0.897**
Junior college	0.390**	0.513**	0.459**	0.648**	–0.089	–0.469†
University and above	0.660**	0.966**	0.813**	1.035**	0.314**	0.500**
Education DK, NA	–0.625**	–0.402	–0.450	–0.419	–0.735*	–1.039†
Working part-time	0.007	0.218	–0.093	0.092	0.188	0.614†
Self-employed	–0.021	0.293†	–0.045	0.450†	–0.019	0.003
Unemployed	–1.255**	–0.635	–0.572	0.201	–1.589**	–0.239
Not in labor force	0.223†	0.783**	0.277	1.053**	0.433*	0.807**
Employment status DK, NA	–0.025	0.431†	0.081	0.565†	–0.465*	0.042
No income	–0.245	–0.398†	–0.304	–0.588*	0.205	–0.230
Second income quartile	0.329**	0.710**	0.558**	1.267**	0.234	0.277
Third income quartile	0.851**	1.412**	1.235**	1.834**	0.847**	1.452**
Highest income quartile	1.287**	2.351**	0.926*	1.800**	1.559**	2.534**
Income DK, NA	–0.044	–0.032	–0.094	0.023	0.248	0.347
Never married	–0.854**	–1.640**	–0.596**	–1.346**	–1.111**	–1.970**
Divorced or widowed	–0.653**	–0.999**	–0.577**	–0.841**	–1.423**	–1.849**
Marital status DK, NA	0.982**	1.526**	1.547†	1.547	1.201**	1.519*
Aged 40–49	–0.765**	–1.030**	–0.542**	–0.864**	–1.100**	–1.303**
Aged 50–59	–0.794**	–0.876**	–0.683**	–0.674**	–1.108**	–1.385**
Aged 60+	–0.464**	–0.517**	–0.405**	–0.744**	–0.602**	–0.512†
Number of family 1	–0.207	–0.448†	–0.211	–0.110	–0.038	–0.670†
Number of family 3	–0.025	–0.092	–0.008	0.003	–0.007	–0.157
Number of family 4	–0.172†	–0.154	–0.170	–0.042	–0.060	–0.266
Number of family 5+	–0.067	–0.165	0.041	–0.130	–0.114	–0.272
Number of family DK, NA	0.000	0.000	–1.171**	–0.925	–0.145	–0.327
Kanto region	0.133	0.036	0.161	–0.059	0.005	0.135
Koshinnetsu region	–0.148	–0.069	0.090	–0.120	–0.306	0.265
Hokuriku region	–0.127	–0.315	–0.542*	–0.535	0.143	–0.460
Tokai region	–0.031	–0.141	0.137	0.130	–0.115	–0.263
Kinki region	0.161	0.252	0.210	0.439†	0.018	–0.157
Chugoku region	–0.146	–0.285	–0.063	–0.421	–0.427†	–0.674*
Shikoku region	–0.199	–0.364	–0.004	–0.397	–0.272	–0.320
Kyushu region	0.078	0.090	0.008	–0.062	0.172	0.287
Cities 100,000+	0.212†	0.097	0.297†	0.328	0.129	–0.144
Cities less than 100,000	0.101	0.101	0.194	0.250	0.025	–0.103
Towns and villages	0.178	0.006	0.063	0.047	0.401**	0.105

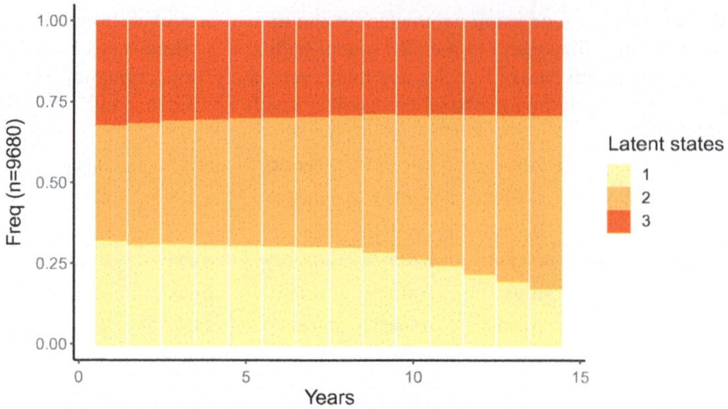

(a) All

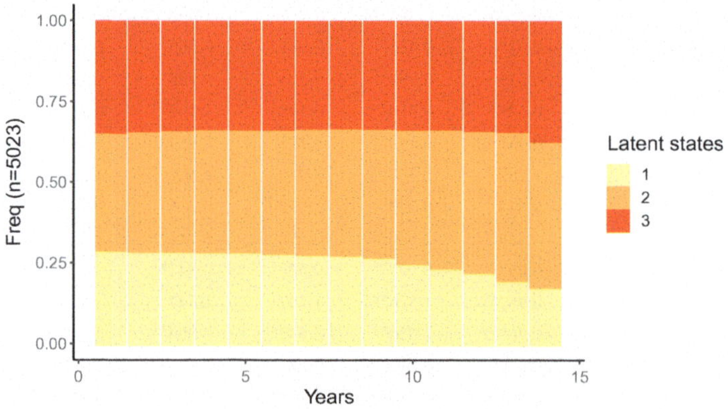

(b) Gender: Female

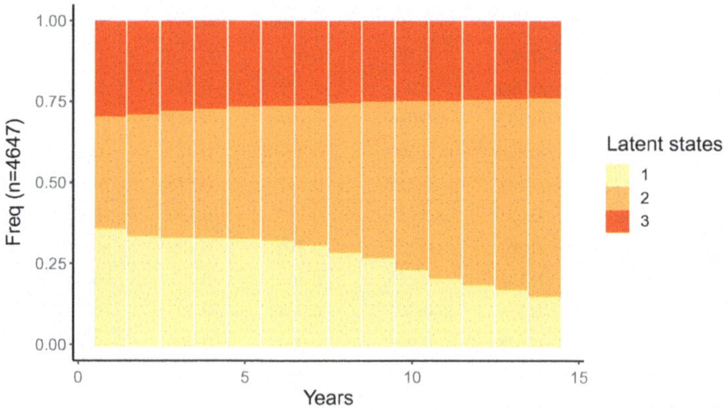

(c) Gender: Male

Fig. 2 Estimated relative frequencies of each state with respect to the HM model with $k = 3$ latent states estimated with all sample data (panel a) and with data of females (panel b) and males (panel c) separately clusters across time. We explore the between-group differences in the happiness trends in Japanese society using fourteen repeated measurements from the Preference Parameters Study. Socioeconomic factors impact subjective well-being (SWB), and we quantify their incidence on the initial and transition probabilities of the Markov chain.

Our findings suggest that the population may be described by three main subgroups of individuals showing different dynamics across the life course. These three clusters are referred to as "not so happy", "moderately happy", and "very happy" individuals. There is a substantial persistence in a certain happiness level over time, meaning that initial inequalities in happiness play an important role in the life course inequalities of SWB.

The major findings of covariates effects are the followings. First, although previous findings based primarily on cross-sectional data have indicated that SWB is U-shaped through the life course, our study found that U-shaped association is not confirmed in Japan. When people reach middle age and late adulthood, they are likely to feel less happy than in early adulthood. Secondly, our study consistently provides evidence that people of higher socioeconomic status have the advantage of having a higher SWB. They feel happy at the initial period, and they also improve their happiness over time. Employment statuses have a limited effect on SWB, whereas educational attainment has a lasting impact on SWB for both men and women. These findings highlight the important role of an earlier stage of the life course, such as education, and its relative advantage/disadvantage resulting in heightened/exacerbated SWB of individuals. Given that the effects of individual's social origins on their educational attainment are unquestioned, lower-background individuals are expected to attain lower levels of education and therefore they are more likely not only to evaluate their lives as dissatisfied in early stage, but also the happiness differentials between individuals with disadvantaged backgrounds and individuals with advantaged backgrounds widen with age. Third, we confirmed significant gender differences in the level of SWB in the initial period and in subsequent periods. Japan is one of the countries where gender equality has lagged far behind. Men get paid more than women for similar work; women are more likely to work in non-standard arrangements with lower earnings. Despite that, we confirmed that women report higher levels of happiness than men and are more likely to become happier over time. The result regarding women's higher contentment suggests that women themselves have lower socioeconomic expectations due to the cultural acceptance of lower status-attainment opportunities as a product of the patriarchal structures, constraints, and social norms in terms of gender that prevail in Japanese society. Nonetheless, the results also suggest that having more balance between work and other parts of life, as well as individual autonomy in decision-making or bargaining position within the household, positively influences women's SWB.

We observed that the present analysis could be enlarged by also considering unique features such as a person's inborn temperament and character, which may contribute to defining a generally positive or negative attitude toward life. A further interest is to investigate the role of happiness with respect to health conditions and enrich the analysis by considering changes over time in other hedonic well-being dimensions, such as satisfaction, and eudaimonic well-being dimensions, such as personal growth.

Concerning the missing values referred to the response variable, we rely on the Missing-At-Random (MAR) assumption, whereas we adopt the missing-indicator approach to handle missing values for the covariates. We mention that the proposed model formulation allows us to relax the MAR assumption when it may be restrictive for valid inference. One way to consider the missing process non-ignorable is to augment data considering a binary indicator equal to one if the individual responds to a certain item at a specific time occasion and zero otherwise. In such cases, using maximum likelihood estimation, the MAR assumption is relaxed further allowing selection to depend directly on the covariates and latent variables.

Acknowledgements This research utilizes microdata from the Preference Parameters Study of Osaka University's 21st Century COE Program 'Behavioral Macro Dynamics Based on Surveys and Experiments', its Global COE project: 'Human Behavior and Socioeconomic Dynamics' and JSPS KAKENHI 15H05728: 'Behavioral Economic Analysis of Long Run Stagnation'. F. Pennoni thanks the financial support from the grant "Statistical models for repeated measures with missing data and dropout" of the University of Milano-Bicocca (2020-ATE-0615). M. Nakai thanks the support from JSPS KAKENHI Grant Number 20K02171.

Appendix

The available data are related to the Preference Parameters Study conducted from 2003 to 2018. The data are freely available for research purposes, upon request, at the following website: https://www.iser.osaka-u.ac.jp/survey_data/top_eng.html. Response rates (%) of each wave of the survey from 2003 to 2018 were: 70.9, 70.4, 72.1, 77.1, 85.0, 89.6, 71.2, 87.8, 92.8, 93.9, 95.5, 92.5, 95.8, and 97.7. The available covariates and their categories are described in the following. As mentioned in the note of the tables do not know, and no answer is denoted with DK and NA, respectively; the categories are:

- Gender: male, female.
- Education is considered with four categories: junior high or lower, high school, junior college, university and above.
- Employment status is measured via the current economic activity status of the respondent with five categories: working full-time, working part-time, self-employed, being unemployed, and not in labor force (student, in retirement, fulfilling domestic tasks, other inactive).
- Respondent's annual income is considered according to five classes corresponding to the observed quantile: none, less than 2,000,000 Yen, from 2,000,000 to 3,999,999 Yen, from 4,000,000 to 5,999,999 Yen, equal or more than 6,000,000 Yen.
- Marital status is a person's current relationship status in terms of a couple's relationship: never married, married, and no longer married.

- Age class is grouped into four categories: early adulthood (20 and 30s), early middle age (40s), late middle age (50s), and late adulthood (60s and older).
- Number of family members: one, two, three, four, and five or more.
- Region of residence corresponds to the Japanese region: Hokkaido and Tohoku, Kanto, Koshinetsu, Hokuriku, Tokai, Kinki, Chugoku, Shikoku, and Kyusyu.
- Size of the place of living is considered according to the size of the population and the degree of urbanization: cabinet order-designated cities, cities with 100,000 or more population, cities with less than 100,000 population, and towns or villages.

References

1. Bartolucci, F., Farcomeni, A., & Pennoni, F. (2013). *Latent Markov models for longitudinal data*. Chapman & Hall/CRC Press, Boca Raton.
2. Bartolucci, F., Pandolfi, S., & Pennoni, F. (2017). `LMest`: An `R` package for latent Markov models for longitudinal categorical data. *Journal of Statistical Software,81*, 1–38.
3. Bartolucci, F., Pandolfi, S., & Pennoni, F. (2022). Discrete latent variable models. *Annual Review of Statistics, 9*, 1–30.
4. Blanchflower, D. G., & Oswald, A. J. (2008). Is well-being U-shaped over the life cycle? *Social Science and Medicine,66*, 1733–1749.
5. Blanchflower, D. G. (2021). Is happiness U-shaped everywhere? Age and subjective well-being in 145 countries. *Journal of Population Economics, 34*, 575–624.
6. Bouveyron, C., Celeux, G., Murphy, T. B., & Raftery, A. E. (2019). *Model-based clustering and classification for data science: With applications in R*. Cambridge University Press, Cambridge.
7. Brinton, M. C. (2017). Happiness at work? Marital happiness among Japanese housewives and employed wives. In *Life course, happiness and well-being in Japan* (pp. 138–157). Routledge, Abingdon, Oxon.
8. Cabinet Office. (2008). *White Paper on the National Lifestyle* 2008. Cabinet Office of Japan.
9. Chen, W. C. (2012). How education enhances happiness: Comparison of mediating factors in four East Asian countries. *Social Indicators Research,106*, 117–131.
10. Clark, A. E., & Oswald, A. J. (1994). Unhappiness and unemployment. *The Economic Journal,104*, 648–659.
11. Daniels, M. J., & Hogan, J. W. (2008). *Missing data in longitudinal studies*. Chapman & Hall/CRC, Boca Raton.
12. Dardanoni, V., Modica, S., & Peracchi, F. (2011). Regression with imputed covariates: A generalized missing-indicator approach. *Journal of Econometrics,162*, 362–368.
13. Davies, P. L. (1995). Data features. *Statistica Neerlandica,49*, 185–245.
14. Dempster, A. P., Laird, N. M., & Rubin, D. B. (1977). Maximum likelihood from incomplete data via the EM algorithm (with discussion). *Journal of the Royal Statistical Society, Series B,39*, 1–38.
15. Diener, E. (1984). Subjective well-being. *Psychological Bulletin,95*, 542–575.
16. Diener, E. (2009). *The science of well-being: The collected works of Ed Diener*. Springer, Dordrecht.
17. Diener, E., & Diener, C. (1996). Most people are happy. *Psychological Science, 7*, 181–185.
18. Diener, E., & Seligman, M. E. (2004). Beyond money: Toward an economy of well-being. *Psychological Science in the Public Interest, 5*, 1–31.
19. Diener, E., Suh, E. M., Lucas, R. E., & Smith, H. L. (1999). Subjective well-being: Three decades of progress. *Psychological Bulletin, 125*, 276–302.
20. Diener, E., Suh, E., & Oishi, S. (1997). Recent findings on subjective well-being. *Indian Journal of Clinical Psychology,24*, 25–41.

21. DiPrete, T. A., & Eirich, G. M. (2006). Cumulative advantage as a mechanism for inequality: A review of theoretical and empirical developments. *Annual Review of Sociology,32*, 271–297.
22. Dolan, P., Peasgood, T., & White, M. (2008). Do we really know what makes us happy? A review of the economic literature on the factors associated with subjective well-being. *Journal of Economic Psychology, 29*, 94–122.
23. Easterlin, R. A. (2001). Life cycle welfare: Trends and differences. *Journal of Happiness Studies, 2*, 1–12.
24. Farrelly, C. (2021). Positive biology and well-ordered science. In *Measuring Well-being*, eds. by M. T. Lee, L. D. Kubzansky & T. J. VanderWeele (pp. 191–215). Oxford University Press, New York.
25. Frey, B. S., & Stutzer, A. (2002). *Happiness and economics*. Princeton University Press, Princeton, N.J.
26. Fujita, F., & Diener, E. (2005). Life satisfaction set point: Stability and change. *Journal of Personality and Social Psychology, 88*, 158–164.
27. Fujita, F., Diener, E., & Sandvik, E. (1991). Gender differences in negative affect and well-being: The case for emotional intensity. *Journal of Personality and Social Psychology, 61*, 427–434.
28. Gundelach, P., & Kreiner, S. (2004). Happiness and life satisfaction in advanced European countries. *Cross-cultural Research, 38*, 359–386.
29. Helliwell, J. F. (2021). Measuring and using happiness to support public policies. In *Measuring well-being*, eds. by M. T. Lee, L. D. Kubzansky & T. J. VanderWeele (pp. 29–49). Oxford University Press.
30. Helliwell, J. F., Layard, R., Sachs, J., De Neve, J.-E., & J.-E., Aknin, L. B., Wang, S. (2022). World Happiness Report 2022. Technical report, Sustainable Development Solutions Network. New York
31. Holthus, B., & Manzenreiter, W. (2017). *Life course, happiness and well-being in Japan*. Routledge, Abingdon, Oxon.
32. Hurlbert, J. S., & Acock, A. C. (1990). The effects of marital status on the form and composition of social networks. *Social Science Quarterly, 71*, 163–174.
33. Inglehart, R. (2004). Subjective well-being rankings of 82 societies (based on combined happiness and life satisfaction scores). World Values Survey Working Paper.
34. Juang, B., & Rabiner, L. (1991). Hidden Markov models for speech recognition. *Technometrics, 33*, 251–272.
35. Little, R. J. A., & Rubin, D. B.(2020). *Statistical analysis with missing data* (3rd ed.). Wiley.
36. Lucas, R. E., & Diener, E. (2009). Personality and subjective well-being. *The science of well-being* (pp. 75–102). Springer, Dordrecht.
37. Mayer, S. E. (1997). *What money can't buy: Family income and children's life chances*. Harvard University Press, Cambridge, MA.
38. Meisenberg, G., & Woodley, M. A. (2015). Gender differences in subjective well-being and their relationships with gender equality. *Journal of Happiness Studies,16*, 1539–1555.
39. Ochiai, E. (2009). Care diamonds and welfare regimes in East and South-East Asian societies: Bridging family and welfare sociology. *International Journal of Japanese Sociology,18*, 60–78.
40. Ohtake, F. (2016). Kurashi no konomi to manzokudo panel chosa (Osaka University) niyoru kenkyu seika. [Research results from the Panel Survey on Lifestyle Preferences and Satisfaction (Osaka University)]. *Chuochosaho,709*, 6207–6213.
41. Ohtake, F., Shiraishi, S., & Tsutsui, Y. (2010). *Nihon no kofukudo: Kakusa, roudou, kazoku [the level of happiness in Japan: Inequality, labor, and family]*. Nippon HyoronSha.
42. Ono, H., & Lee, K. S. (2016). *Redistributing Happiness: How Social Policies Shape Life Satisfaction: How Social Policies Shape Life Satisfaction*. ABC-CLIO.
43. Osaka University Institute of Social and Economic Research. (n.d.). *Overview Preference Parameters Study of Osaka University*, Retrieved December 2022, from https://www.iser.osaka-u.ac.jp/survey_data/doc/japan/surveyoverview/english/SurveyOverviewJapan_eng.pdf

44. Oshio, T., & Urakawa, K. (2012). Relative income and subjective well-being in Japan: Perceived happiness, self-rated health, and trust. *Economic Review, 63*, 42–55.
45. Pennoni, F., Nakai, M. (2019). A latent class analysis towards stability and changes in breadwinning patterns among coupled households. *Dependence Modeling, 7*, 234–246.
46. R Core Team. (2022). *R: A language and environment for statistical computing*. R Foundation for Statistical Computing, Austria.
47. Ridgeway, C. L. (2009). Framed before we know it: How gender shapes social relations. *Gender and Society,23*, 145–160.
48. Ross, C. E., & Wu, C. L. (1995). The links between education and health. *American Sociological Review,60*, 719–745.
49. Ross, C. E., & Wu, C. L. (1996). Education, age, and the cumulative advantage in health. *Journal of Health and Social Behavior,37*, 104–120.
50. Schuller, T., Desjardins, R. (2007). *Understanding the social outcomes of learning*. OECD.
51. Schwarz, G. (1978). Estimating the dimension of a model. *Annals of Statistics, 6*, 461–464.
52. Shishido, K., & Sasaki, T. (2011). Happiness in Japan: A hierarchical age-period-cohort analysis based on JGSS cumulative data 2000–2010. *Japanese Sociological Review, 62*, 336–355.
53. Stevenson, B., & Wolfers, J. (2009). The paradox of declining female happiness. *American Economic Journal: Economic Policy, 1*, 190–225.
54. Tesch-Römer, C., Motel-Klingebiel, A., & Tomasik, M. J. (2008). Gender differences in subjective well-being: Comparing societies with respect to gender equality. *Social Indicators Research, 85*, 329–349.
55. Tutz, G. (2012). *Regression for categorical data*. Cambridge University Press.
56. Ulloa, B. F. L., Møller, V., & Sousa-Poza, A. (2013). How does subjective well-being evolve with age? A literature review. *Journal of Population Ageing,6*, 227–246.
57. Umberson, D., Pudrovska, T., & Reczek, C. (2010). Parenthood, childlessness, and well-being: A life course perspective. *Journal of Marriage and Family,72*, 612–629.
58. United Nations. (2015). Resolution adopted by the General Assembly on 25 September 2015, Transforming our world: the 2030 Agenda for Sustainable Development.
59. Wakita, A. (2017). Regional Disparity and Life Satisfaction. *Nenpoh shakaigaku-ronshu,30*, 98–109.
60. Welch, L. R. (2003). Hidden Markov models and the Baum-Welch algorithm. *IEEE Information Theory Society Newsletter,53*, 1–13.
61. Winkelmann, L., & Winkelmann, R. (1995). Happiness and unemployment: A panel data analysis for Germany. *Applied Economics Quarterly,41*, 293–307.
62. Yang, Y. (2008). Social inequalities in happiness in the United States, 1972 to 2004: An age-period-cohort analysis. *American Sociological Review,73*, 204–226.
63. Zucchini, W., MacDonald, I.L., & Langrock, R. (2017). *Hidden Markov Models for Time Series: An Introduction using R*. Chapman and Hall/CRC, Boca Raton.

On Likelihood Ratio Tests for Dimensionality Selection

Yoshio Takane and Peter G. M. van der Heijden

Abstract Many multivariate statistical models have dimensional structures. Such models typically require judicious choice of dimensionality. Likelihood ratio tests are often used for dimensionality selection. However, to this day there is still a great deal of confusion about the asymptotic distributional properties of the log-likelihood ratio (LR) statistics in some areas of psychometrics. Although in many cases the asymptotic distribution of the LR statistic representing the difference between the correct model (of specific dimensionality) and the saturated model is guaranteed to be chi-square, that of the LR statistic representing the difference between the correct model and the one with one dimension higher than the correct model is not likely to be chi-square due to a violation of one of regularity conditions. In this paper, we attempt to clarify the misunderstanding that the latter is also assured to be asymptotically chi-square. This common misunderstanding has occurred repeatedly in various fields, although in some areas it has been corrected.

Keywords Asymptotic chi-square distribution · Regularity conditions · Canonical correlation analysis · Models of contingency tables · Multidimensional scaling · Factor analysis · Normal mixture models

---

This paper is dedicated to the Memory of the late Professor Michael W. Browne at the Ohio State University. This paper is an expanded version of the paper by Takane, van der Heijden, and Browne (2003).

Y. Takane (✉)
Department of Psychology, University of Victoria, 5173 Del Monte Avenue, Victoria, BC V8Y 1X3, Canada
e-mail: yoshio.takane@mcgill.ca

P. G. M. van der Heijden
Department of Methodology and Statistics, Faculty of Social Sciences, Utrecht University, Padualaan 14, 3584 CH Utrecht, The Netherlands
e-mail: p.g.m.vanderheijden@uu.nl; P.G.Van-Der-Heijden@soton.ac.uk

Southampton Statistical Science Research, University of Southampton, B39 West Highfield Campus, University Road, SO17 1BJ Southampton, UK

© The Author(s), under exclusive license to Springer Nature Singapore Pte Ltd. 2023
A. Okada et al. (eds.), *Facets of Behaviormetrics*, Behaviormetrics: Quantitative Approaches to Human Behavior 4, https://doi.org/10.1007/978-981-99-2240-6_10

1 Introduction

Many multivariate statistical models have dimensional structures. We can think of models for contingency tables such as correspondence analysis or correlation models, association models, and latent class models. We can also think of models for other kinds of multivariate data such as multidimensional scaling, factor analysis, and canonical correlation analysis including reduced rank regression, canonical discriminant analysis, and multivariate analysis of variance as its special cases. Another line of models with dimensionality structures are parametric mixture models such as normal mixture models. These models may be considered generalizations of latent class models to continuous manifest variables.

Such models typically require judicious choice of dimensionality. For example, in correspondence analysis, the correlation model, and the association model, the dependence in the contingency table is represented in a number of dimensions, and an important question is the assessment of the number of dimensions that have to be interpreted. In latent class analysis, the dependence between the manifest variables in a contingency table is explained by a latent variable having a number of categories called classes, and the number of classes necessary to explain the manifest dependence has to be assessed. In multidimensional scaling, the (dis)similarities are displayed in a multidimensional space, and in factor analysis, the covariances among manifest variables are explained by continuous latent variables called factors. In either case, the dimensionality of the space or the number of factors enough to explain the observed (dis)similarities or covariances have to be determined. Similarly, in canonical correlation analysis, reduced rank regression, canonical discriminant analysis, and multivariate analysis of variance, the number of canonical variates to be interpreted has to be determined. For the purpose of this paper, we call all of the above problems the "dimensionality" selection problems.

Likelihood ratio tests are often used for the purpose of dimensionality selection. However, to this day there is still a great deal of confusion about the asymptotic distributional properties of the log-likelihood ratio (LR) statistics in some areas of psychometrics. Although in many cases the asymptotic distribution of the LR statistic representing the difference between the correct model (of specific dimensionality) and the saturated model is guaranteed to be chi-square, that of the LR statistic representing the difference between the correct model and the one with one dimension higher than the correct model is not likely to be chi-square due to a violation of one of regularity conditions. In this paper, we attempt to clarify the misunderstanding that the latter is also assured to be asymptotically chi-square. This common misunderstanding has occurred repeatedly in various fields, although in some areas it has been corrected.

As an illustration of the general problem of dimensionality selection, we will discuss the testing problems arising in the correlation model (CM) proposed by Goodman [20] and Gilula and Haberman [19]. Consider an I by J contingency table, $\mathbf{F} = \{f_{ij}\}$. We denote the corresponding observed proportions by $\mathbf{P} = \{p_{ij}\} = \{f_{ij}/N\}$, where N is the total sample size, with row and column marginals denoted by $p_{i.} = \sum_j p_{ij}$ and $p_{.j} = \sum_i p_{ij}$, respectively. In the correlation model, we fit

$$\pi_{ij} = \alpha_i \beta_j (1 + \sum_{t=1}^{T} \phi_t r_{it} c_{jt}) \tag{1}$$

to **F**, where α_i and β_j are parameters associated with the margins, $\phi_t (t = 1, \cdots, T)$ is the correlation associated with dimension t, and r_{it} and c_{jt} are sets of scores associated with the rows and columns. To remove indeterminacies in the model, we impose the following identification conditions:

$$\sum_i \alpha_i = 1, \quad \sum_j \beta_j = 1, \quad \sum_i p_{i.} r_{it} = 0 = \sum_j p_{.j} c_{jt}, \tag{2}$$

and

$$\sum_i p_{i.} r_{it} r_{it'} = \delta_{tt'} = \sum_j p_{.j} c_{jt} c_{jt'}, \tag{3}$$

where $\delta_{tt'}$ is a Kronecker delta, i.e., $\delta_{tt'} = 1$ if $t = t'$ and $\delta_{tt'} = 0$ if $t \neq t'$. For brevity, we denote CM with T dimensions as CM(T). Let $T^* = \min(I - 1, J - 1)$. Then CM(T^*) is a saturated model: in this case, the probabilities are unrestricted. CM(0) is equivalent to the independence model. CM(T) with $1 \leq T < T^*$ is a reduced rank model with rank equal to $T + 1$ (see [12]).

When CM is estimated by the maximum likelihood method, an obvious candidate for tests is the likelihood ratio (LR) test λ, defined as

$$\lambda = L_0 / L_1, \tag{4}$$

where L_0 is the maximum likelihood of the model under H_0 and L_1 is the maximum likelihood under H_1, and H_0 is nested within H_1. If the model under H_0 is true and certain regularity conditions are satisfied, $-2 \log \lambda$ is known to be asymptotically distributed as a chi-square variate with degrees of freedom equal to the difference in the number of parameters under the two hypotheses.

We consider three LR tests. To simplify our presentation, we distinguish two kinds of LR tests, depending on the model under H_1. If the model under H_1 is saturated, we have an unconditional LR test (statistic) denoted by λ_s, and if the model under H_1 is unsaturated, we have a conditional LR test denoted by λ_c. The LR statistic $-2 \log \lambda_s$ is asymptotically chi-square distributed if H_0 is CM(0), H_1 is the saturated model, and CM(0) is true. However, when H_0 is CM(1), H_1 is the saturated model, and CM(1) is true, $-2 \log \lambda_s$ is asymptotically chi-square distributed only if CM(0) is not true. It is no longer assured to be asymptotically chi-square distributed if not only CM(1) but also CM(0) is true. Similarly, $-2 \log \lambda_c$ is not assured to be asymptotically chi-square distributed if H_0 is CM(0) and H_1 is CM(1), and CM(0) is true. Notice that CM(0) being true implies CM(1) is also true. Note that in this particular case we know a much stronger result that $-2 \log \lambda_c$ is indeed not asymptotically distributed according to chi-square, but instead according to the distribution of the largest eigenvalue from a certain central Wishart matrix [24].

These situations are illustrated in slightly more general terms in panels (b') and (c') of Fig. 1, where $\chi^2_m(\delta)$ indicates the (asymptotic) noncentral chi-square distribution with m degrees of freedom and noncentrality parameter δ. The $\delta = 0$ indicates a central chi-square distribution. In panel (b') where the regularity condition (to be defined in the next section) holds, the reproducibility of the chi-square distribution with respect to both the degrees of freedom and the noncentrality parameters holds for nested models. However, in panel (c') where the regularity condition is violated, this is no longer true. Only the LR statistic representing the difference between the saturated model and the T-dimensional model is assured to be asymptotically (central) chi-square distributed, assuming that $T-1$ is not true, or $T = 0$ (the independence model). Asymptotic distributions of the remaining two LR statistics are generally unknown except for a few special cases. These situations are contrasted with the "normal" situations in panels (b) and (c) of Fig. 1, where model B is assumed to be nested within model A, which in turn is nested within the saturated model, and where the regularity conditions are assumed to hold. In these cases, all three LR statistics are asymptotically central or noncentral chi-square distributed depending on which models are assumed true, and the reproducibility of the chi-square distribution holds perfectly. (Panels (a) and (a') of Fig. 1 will be explained in Sect. 4.6.)

The reason that $-2\log\lambda_c$ is not assured to be asymptotically chi-square distributed, when both CM(0) and CM(1) are true, is usually attributed to nonuniqueness of parameters and/or parameters lying on the boundary of the parameter space. That is, we put $\phi_1 = 0$ in CM(1) to obtain CM(0), but $\phi_1 = 0$ puts the parameter vector on the boundary of the parameter space. It also makes r_{i1} and c_{j1} unidentifiable. The problem is that these conditions depend on parametrization in the model. For example, if we reparameterize the correlation model by

$$\pi_{ij} = \alpha_i \beta_j \left(1 + \sum_{t=1}^{T} r^*_{it} c^*_{jt}\right), \tag{5}$$

where $r^*_{it} = \phi_t^{1/2} r_{it}$ and $c^*_{jt} = \phi_t^{1/2} c_{jt}$, there will be no boundary point or unidentifiability problem, and yet the same anomalous condition persists. Note that some of the identification restrictions in (2) and (3) are translated into $\sum_i p_{i.} r^*_{it} = 0 = \sum_j p_{.j} c^*_{jt}$, and $\sum_i p_{i.} r^*_{it} r^*_{it'} = \phi_t \delta_{tt'} = \sum_j p_{.j} c^*_{jt} c^*_{jt'}$. In the next section, we discuss a more fundamental notion of "regularity" due to Shapiro [54]. This is one of the sufficient set of conditions for $-2\log\lambda$ to be asymptotically chi-square distributed, and in the dimensionality selection problems discussed above, this condition is not fulfilled.

In the next section, this notion will be worked out in the context of maximum likelihood multidimensional scaling (MLMDS). We will show an example where, although there is no boundary point or unidentifiability problem in the usual parametrization of the model after removing rotational and translational indeterminacies, there is no assurance that the asymptotic distribution of $-2\log\lambda_c$ is chi-square. (Throughout the rest of this paper we use $-2\log\lambda_c$ to refer to the LR statistic to test the

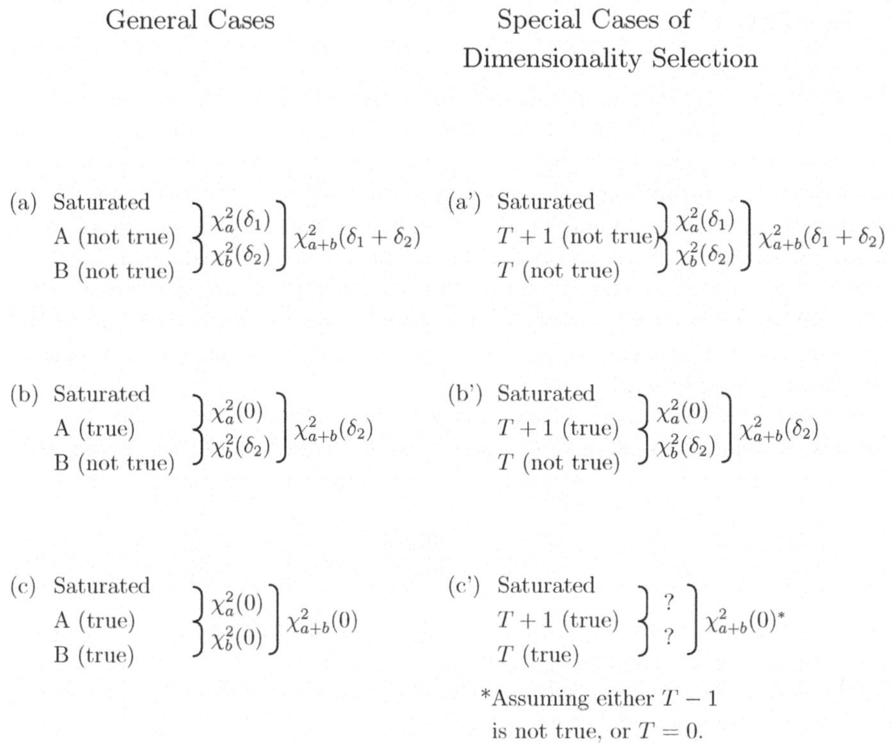

Fig. 1 The LR tests for nested models. In (a) neither model A nor B is assumed true. In (b) model A is assumed true but not B. In (c) both A and B are assumed true. Regularity is assumed to hold in all of these cases. The right column depicts the special cases of dimensionality selection, where in (a') neither the $T+1$- nor the T-dimensional model is assumed true, in (b') $T+1$ is assumed true but not T, and in (c') both $T+1$ and T are assumed true. Regularity is known to hold in (a') and (b'), but not in (c'). In the figure $\chi^2_m(\delta)$ indicates the noncentral chi-square distribution with m degrees of freedom and with noncentrality parameter δ, and $\chi^2_m(0)$ refers to the corresponding central chi-square distribution

T-dimensional model (H_0) against the $T+1$-dimensional model (H_1), and $-2\log\lambda_s$ to refer to the LR statistic to test the T-dimensional model (H_0) against the saturated model (H_1), unless otherwise specified.) In Sect. 3, we will investigate the distribution of $-2\log\lambda_c$ and $-2\log\lambda_s$ using Monte Carlo studies and demonstrate that the former is indeed not asymptotically chi-square in the particular context of MLMDS examined. In Sect. 4, we will discuss similar problems in other models mentioned at the outset of this introduction. In discussing these problems we also point out instances of the incorrect claim that $-2\log\lambda_c$ is assured to be asymptotically chi-square distributed, made in various fields of statistics in the past. We end with a discussion of our results.

2 Regularity

The likelihood ratio statistic is asymptotically distributed as a chi-square variate if certain conditions are fulfilled. In the context of moment structural models, some conditions are specified by Steiger et al. [57]. Some of the well-known conditions, which together constitute part of a sufficient (but not necessary) set of conditions, are that the parameter vector should be identified, that the parameter vector should be in the interior of the parameter space, that the Jacobian matrix (to be defined shortly) is nonsingular, and that the Hessian matrix (the matrix of the second derivatives of the log-likelihood function with respect to the model evaluated at the true parameters) is nonsingular. In this paper, we focus on the concepts of local and global regularity, introduced by Shapiro [54].

We start with the local regularity. Let $\boldsymbol{\theta}$ be the parameter vector. Let $\boldsymbol{\xi}$ be the vector with elements of the model, each of which is a function of the elements in $\boldsymbol{\theta}$. That is, $\boldsymbol{\xi} = \mathbf{g}(\boldsymbol{\theta})$ for some function $\mathbf{g}$. Let $\mathbf{J}(\boldsymbol{\theta})$ be the Jacobian matrix defined by

$$\mathbf{J}(\boldsymbol{\theta}) = \frac{\partial \mathbf{g}(\boldsymbol{\theta})}{\partial \boldsymbol{\theta}'}. \tag{6}$$

This is the matrix of the first derivatives of the model with respect to the model parameters. Then local regularity is defined as follows:

A point $\boldsymbol{\theta}_0$ in the parameter space Θ is locally regular if $\text{rank}(\mathbf{J}(\boldsymbol{\theta})) = \text{rank}(\mathbf{J}(\boldsymbol{\theta}_0))$ for all $\boldsymbol{\theta}$ in a neighbourhood of $\boldsymbol{\theta}_0$.

That is, for the true parameter vector $\boldsymbol{\theta}_0$ to be a locally regular point the rank of the Jacobian matrix, even if it is singular, should be constant in a neighborhood of $\boldsymbol{\theta}_0$. Local regularity is a necessary condition for global regularity (see below). Violation of local regularity implies violation of global regularity, which is one of a set of sufficient conditions for $-2\log\lambda$ to be asymptotically chi-squared distributed. This does not mean that violation of the local (and hence global) regularity condition implies that $-2\log\lambda$ is never asymptotically distributed according to chi-square. It may well be chi-square, but it is not guaranteed, and often it is not chi-square. If the regularity condition does not hold, the asymptotic distribution has to be determined by other means (see the discussion of this paper).

Global regularity is formally defined as follows ([54], Definition 2.3). Let $\boldsymbol{\xi}_0 = \mathbf{g}(\boldsymbol{\theta}_0)$, and let Ξ denote the image of the mapping $\mathbf{g}(\boldsymbol{\theta})$, i.e., $\Xi = \{\boldsymbol{\xi} : \boldsymbol{\xi} = \mathbf{g}(\boldsymbol{\theta}), \boldsymbol{\theta} \in \Theta\}$. Then

A point $\boldsymbol{\theta}_0 \in \Theta$ is (globally) regular if $\boldsymbol{\theta}_0$ is locally regular and there exist neighbourhoods U and V of $\boldsymbol{\theta}_0$ and $\boldsymbol{\xi}_0$, respectively, such that $\Xi \cap V = \mathbf{g}(U)$.

The first part of the above definition is clear, but the second part requires some explanation. First of all, the second part rules out boundary points, since any neighborhood of a boundary point contains a point outside the parameter space for which $\mathbf{g}(.)$ is not in the model space. (That is, no boundary points are globally regular.) When $\boldsymbol{\theta}_0$ is not a boundary point, and the model is an analytic function of its parameters (i.e., $\mathbf{g}$ is an analytic function of $\boldsymbol{\theta}$; a function is analytic on an open set if it can be expanded in power series in a neighborhood of every point of this set [54] and is globally identified (i.e., $\boldsymbol{\theta}^* \in \Theta$ and $\mathbf{g}(\boldsymbol{\theta}^*) = \mathbf{g}(\boldsymbol{\theta}_0)$ implies $\boldsymbol{\theta}^* = \boldsymbol{\theta}_0$), the second part of the above definition is automatically satisfied. However, a bit of care is necessary, if the model is not globally identified. For example, in multidimensional scaling (MDS) to be discussed shortly, the model is not globally identified with respect to reflection and permutation of dimensions even after the rotational and translational indeterminacies are removed. However, local structure of $\mathbf{g}(\boldsymbol{\theta}_0)$ remains essentially the same for all $\boldsymbol{\theta}^* \in \Theta$ obtained by reflection and permutation of dimensions. The second part of the global regularity condition is, therefore, still satisfied. Similarly, it is satisfied by most of the models discussed in this paper. It is thus important to examine where and when local regularity is violated.

In the introduction, we mentioned many models in which dimensionality selection played an important role. For these models, we can use the sufficiency of the regularity condition to assess that under certain circumstances $-2\log\lambda_s$ is distributed as chi-square, whereas $-2\log\lambda_c$ may not be so. We will give three examples to illustrate local regularity. In the first example, local regularity does not hold, and in the second and the third, it holds.

Example 1 Consider the context of MDS. We assume that there are four stimuli numbered 1 to 4. H_1 is the unconstrained two-dimensional solution, and H_0 is the unconstrained one-dimensional solution. We assume that H_0 is true. The question is whether $-2\log\lambda_c$ is asymptotically chi-square distributed if H_0 is true, given that H_1 is true. We collect the coordinates of the four points in a matrix $\mathbf{X}$. We remove the translational indeterminacy in the model by fixing $x_{11} = x_{12} = 0$, and we remove the rotational indeterminacy in the model by fixing $x_{22} = 0$. These are analogous to the identification restrictions (2) and (3) in the correlation model discussed in the introduction. For the model under H_1 the matrix of stimulus coordinates $\mathbf{X}$, and the parameter vector $\boldsymbol{\theta}$ and the model vector $\boldsymbol{\xi}$ are

$$\mathbf{X} = \begin{bmatrix} 0 & 0 \\ x_{21} & 0 \\ x_{31} & x_{32} \\ x_{41} & x_{42} \end{bmatrix},$$

$\boldsymbol{\theta} = (x_{21}, x_{31}, x_{41}, x_{32}, x_{42})'$, and $\boldsymbol{\xi} = (d_{12}, d_{13}, d_{14}, d_{23}, d_{24}, d_{34})'$. The Euclidean distance d_{ij} is related to the coordinates x_{it} by

$$d_{ij} = \{\sum_{t=1}^{2}(x_{it} - x_{jt})^2\}^{1/2}. \tag{7}$$

For this model, the Jacobian matrix is

$$\mathbf{J}(\boldsymbol{\theta}) = \begin{array}{c} d_{12} \\ d_{13} \\ d_{14} \\ d_{23} \\ d_{24} \\ d_{34} \end{array} \begin{bmatrix} -\delta_{121} & 0 & 0 & 0 & 0 \\ 0 & -\delta_{131} & 0 & -\delta_{132} & 0 \\ 0 & 0 & -\delta_{141} & 0 & -\delta_{142} \\ \delta_{231} & -\delta_{231} & 0 & -\delta_{232} & 0 \\ \delta_{121} & 0 & -\delta_{241} & 0 & -\delta_{242} \\ 0 & \delta_{341} & -\delta_{341} & \delta_{342} & -\delta_{342} \end{bmatrix},$$

$$\begin{array}{ccccc} x_{21} & x_{31} & x_{41} & x_{32} & x_{42} \end{array}$$

where $\delta_{ijt} = (x_{it} - x_{jt})/d_{ij}$. It is assumed that $d_{ij} \neq 0$ for all i and j (i.e., no two points coincide). This Jacobian matrix for the two-dimensional model has $\text{rank}(\mathbf{J}(\boldsymbol{\theta})) = 5$. For the one-dimensional model we obtain $\mathbf{J}(\boldsymbol{\theta}_0)$ by putting $x_{32} = x_{42} = 0$ in δ_{ijt} above. In this case $\delta_{132} = \delta_{232} = \delta_{342} = \delta_{142} = \delta_{242} = 0$. It turns out that the $\text{rank}(\mathbf{J}(\boldsymbol{\theta}_0)) = 3$. This shows that $\boldsymbol{\theta}_0$ is not locally regular, since arbitrary perturbations of x_{32} and x_{42} increase the rank of the Jacobian matrix.

This example emphasizes the importance of the notion of local regularity: for this example, there is no unidentifiability problem or boundary point problem after removing rotational and translational indeterminacies, and yet there is no assurance that the asymptotic distribution of $-2 \log \lambda$ is chi-square. This example corresponds with the situation described in panel (c') of Fig. 1.

Note that if we reparameterize the model into $d_{ij} = \{\sum_{t=1}^{2} w_t (x^*_{it} - x^*_{jt})^2\}^{1/2}$, where $w_t \geq 0$, $\sum_i x^*_{it} = 0$ and $\sum_i x^*_{it} x^*_{it'} = \delta_{tt'}$ for all t and t', we face the boundary problem as well as the nonuniqueness problem similar to that encountered in the correlation model discussed in the introduction section. The fact that the boundary problem and the nonuniqueness problem appear or disappear (on the surface) depending on the parametrization of the model points that the nonuniqueness created under H_0 is not an definitive criterion for detecting the anomalous condition.

Example 2 As in example 1, H_1 is the unconstrained two-dimensional solution, so again $\text{rank}(\mathbf{J}(\boldsymbol{\theta})) = 5$. Now H_0 is the two-dimensional solution with the additional restriction $x_{32} = 0$. This will make $\delta_{132} = \delta_{232} = 0$, but $\delta_{342} \neq 0$, so $\text{rank}(\mathbf{J}(\boldsymbol{\theta}_0)) = 5$. This shows that $\boldsymbol{\theta}_0$ is locally regular. Therefore, $-2 \log \lambda$ is asymptotically chi-square distributed under H_0.

Example 3 As in examples 1 and 2, H_1 is the unconstrained two-dimensional solution, where $\text{rank}(\mathbf{J}(\boldsymbol{\theta})) = 5$. Now H_0 is the two-dimensional solution with the additional restriction $x_{32} = x_{42} \neq 0$. This will make $\delta_{342} = 0$, but $\delta_{132}, \delta_{232}, \delta_{142}, \delta_{242}$ are nonzero, so $\text{rank}(\mathbf{J}(\boldsymbol{\theta}_0)) = 5$. This shows that $\boldsymbol{\theta}_0$ is locally regular. Therefore, $-2 \log \lambda$ is asymptotically chi-square distributed under H_0. Examples 2 and 3 correspond with the situation described in panel (c) of Fig. 1.

3 Monte Carlo Studies

The results for the examples in the previous section leave open two questions. First, when the regularity condition is violated, the asymptotic distribution of $-2\log\lambda$ may or may not be chi-square. There are situations in which it is definitely not chi-square, and in some cases, alternative distributions are known (see the discussion section). When the distribution is not known, what is the distribution, and in particular does this distribution depart significantly from chi-square? Second, the regularity condition is one of the sufficient sets of conditions for the asymptotic chi-square distribution, but how fast does the actual distribution of $-2\log\lambda$ converge to chi-square? In this section, we present a few Monte Carlo studies that show how such questions can typically be answered. These studies are performed in the context of multidimensional scaling. We use the program MAXSCAL-4 [60].

The Monte Carlo study is set up as follows. For a fixed configuration of 10 stimulus points the set of Euclidean distances is calculated. For a prescribed set of 45 pairs of distances (tetrads of stimuli which are all distinct) the true probabilities (p) of one distance judged to be larger than the other are calculated by Luce's choice model, i.e., $p(d_{ij}$ judged larger than $d_{kl}) = [1 + \exp\{-(d_{ij} - d_{kl})\}]^{-1}$. For each tetrad N (number of replications) uniform random numbers between 0 and 1 were generated, and the frequency of the numbers larger than p was counted, which constitute the basic data for MAXSCAL-4. This was repeated 100 times for the Monte Carlo study. (In some cases, we tried 1000 Monte Carlo samples, but the results were essentially the same.) The prescribed coordinates of 10 stimulus points are given by

$$\mathbf{X} = \begin{bmatrix} 2.3057 & -3.1959 \\ 1.5734 & -0.8130 \\ 2.8506 & 1.1299 \\ -0.2942 & -1.2028 \\ -0.1792 & 1.5090 \\ 0.8144 & 3.2396 \\ -1.7368 & 0.0227 \\ -2.3505 & 1.0822 \\ -2.9876 & -1.7701 \\ 0.0000 & 0.0000 \end{bmatrix}. \tag{8}$$

MAXSCAL-4 uses an iterative optimization procedure to obtain ML solutions, which is susceptible to convergence to nonglobal maxima, depending on initial estimates of parameters. To ensure getting globally optimal solutions, a general tactic is to obtain solutions from as many different initial estimates as possible. We therefore obtained, for each Monte Carlo sample, two-dimensional and three-dimensional solutions starting from 101 and 151 different initial configurations, respectively, including one rational start, and the remaining partially or completely random initial starts. In partially random initial starts, only the coordinates on a certain dimension or dimensions were randomly generated. We think that the problem of ending up in nonglobal instead of global maxima is negligible in all cases, due to the large numbers of initial starts.

In study 1, the observed distribution of $-2\log\lambda_c$ was investigated between the two-dimensional model (the true model) and the three-dimensional model using 100 Monte Carlo samples. Four replications were drawn for each tetrad. For this study, the regularity condition is violated. The behavior of $-2\log\lambda_c$ is shown in panel (a) of Fig. 2. Horizontally quantiles are set out for the theoretical (central) chi-square distribution with 7 degrees of freedom (χ_7^2) associated with the statistic, and vertically the ordered observed $-2\log\lambda_c$ are shown. The solid line shows the situation when the observed $-2\log\lambda_c$ would follow the theoretical chi-square distribution. The dots present the observed and corresponding theoretical pairs found in the simulation. It shows that, for the situation in this study, the observed distribution of $-2\log\lambda_c$ is approximately 1.5 to 2 times as large as the theoretical chi-square distribution.

The observed behavior of $-2\log\lambda_s$ is investigated between the two-dimensional model (the true model) and the saturated model using the same set of 100 Monte Carlo samples as above. For this LR statistic the regularity condition holds, and therefore, asymptotically, $-2\log\lambda_s$ follows χ_{28}^2 (a central chi-square distribution with 28 degrees of freedom; 28 df = 45 tetrads - 17 parameters). The observed behavior of $-2\log\lambda_s$ is shown in panel (b) of Fig. 2. It shows that quantiles of the observed distribution of $-2\log\lambda_s$ are systematically lower than those of the theoretical chi-square distribution. Four replications do not seem to be sufficient for the statistic to attain the deemed asymptotic distribution.

In study 2, the behavior of the LR statistics is investigated in a similar way to that of study 1. The difference with study 1 is that now 20 replications are drawn for each tetrad. The observed distribution of $-2\log\lambda_c$ seems closer to the theoretical chi-square distribution than in study 1, but somewhat higher than the theoretical chi-square distribution (see panel (c) of Fig. 2). The observed distribution of $-2\log\lambda_s$, on the other hand, is already very close to the theoretical chi-square distribution (see panel (d) of Fig. 2).

In studies 3 and 4, the number of replications per tetrad was increased to 500 and 2000, respectively. Panels (b) and (d) of Fig. 3 confirm that the asymptotic distribution of $-2\log\lambda_s$ is indeed approximately χ_{28}^2. On the contrary, there is no indication in panels (a) and (c) of Fig. 3 that the observed distribution of $-2\log\lambda_c$ gets closer to χ_7^2 beyond the sample size of 2000.

We conclude that, since the regularity condition holds, $-2\log\lambda_s$ does indeed follow an asymptotic chi-square distribution, but 4 replications are not sufficient to attain the asymptotic distribution. In the particular context of the present study, 20 replications seem sufficient. However, we avoid making any definitive remark on the speed of convergence, because it is affected by various factors including the number of stimuli, true dimensionality, particular stimulus configurations assumed, the form of data, and so on. If the regularity condition is violated, we have found that the asymptotic distribution of $-2\log\lambda_c$ is not chi-square.

To what extent can we generalize the foregoing results? As it seems, we may safely state that they apply at least to all MLMDS designed for discrete data. The claim made in Takane [58, 59] and Takane and Carroll [60] that $-2\log\lambda_c$ is asymp-

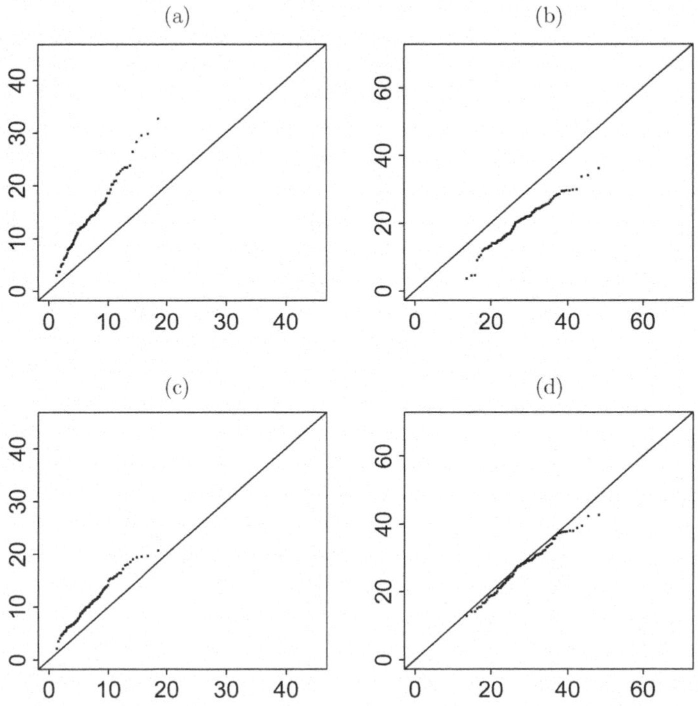

Fig. 2 Q-Q plots of the observed distribution (vertical axis) against the theoretical distribution (χ^2_7 for $-2\log\lambda_c$, and χ^2_{28} for $-2\log\lambda_s$) for relatively small samples. (a) $-2\log\lambda_c$ for $N = 4$, (b) $-2\log\lambda_s$ for $N = 4$, (c) $-2\log\lambda_c$ for $N = 20$, and (d) $-2\log\lambda_s$ for $N = 20$

totically chi-square distributed when the dimensionality is tested is thus incorrect. For these methods, the alternative criterion $-2\log\lambda_s$ should be used for dimensionality selections. In metric MDS [9, 40, 47–49, 62, 68], the situation is less clear. Although any claim that $-2\log\lambda_c$ is assured to follow the asymptotic chi-square distribution is incorrect, actual distributions of this statistic have to be determined in the specific contexts of these models using a similar technique used in the present study.

4 Dimensionality Selection Problems in Various Models

4.1 CANO, RRR, DISC, and MANOVA

We begin with canonical correlation analysis (CANO). Later we discuss how the results in CANO can be extended to reduced rank regression (RRR: [2, 31]), canonical discriminant analysis (DISC), and multivariate analysis of variance (MANOVA).

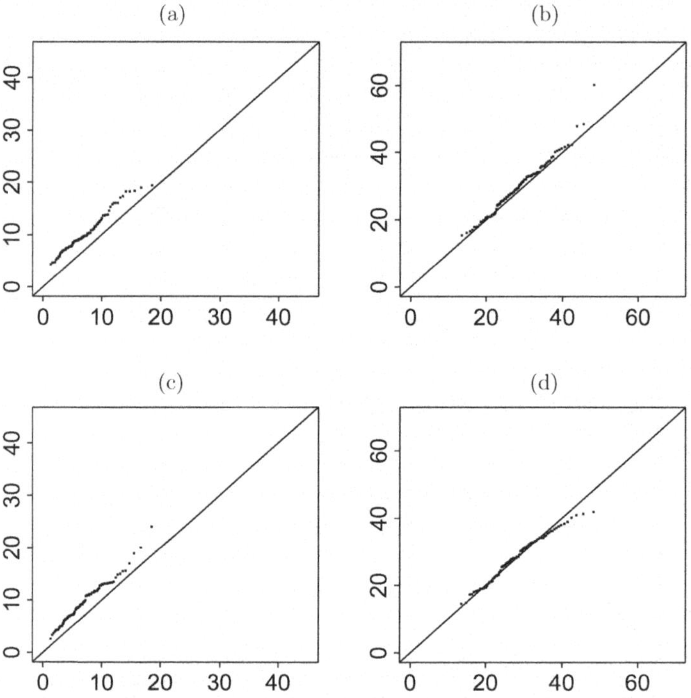

Fig. 3 Q-Q plots of the observed distribution (vertical axis) against the theoretical distribution (χ^2_7 for $-2\log\lambda_c$, and χ^2_{28} for $-2\log\lambda_s$) for relatively large samples. (a) $-2\log\lambda_c$ for $N = 500$, (b) $-2\log\lambda_s$ for $N = 500$, (c) $-2\log\lambda_c$ for $N = 2000$, and (d) $-2\log\lambda_s$ for $N = 2000$

Let $\mathbf{X}$ and $\mathbf{Y}$ be N by I $(I < N)$ and N by J $(J < N)$ matrices, respectively. We denote $T^* = \min(I, J)$. For simplicity, we assume both $\mathbf{X}$ and $\mathbf{Y}$ are columnwise centered and nonsingular. Let $\hat{\mu}_1 > \hat{\mu}_2 > \cdots > \hat{\mu}_{T^*}$ be sample canonical correlations between $\mathbf{X}$ and $\mathbf{Y}$ in descending order of magnitude. They are obtained as singular values of $(\mathbf{X}'\mathbf{X})^{-1/2}\mathbf{X}'\mathbf{Y}(\mathbf{Y}'\mathbf{Y})^{-1/2}$ (or as the square root of eigenvalues of $(\mathbf{Y}'\mathbf{Y})^{-1/2}\mathbf{Y}'\mathbf{X}(\mathbf{X}'\mathbf{X})^{-1}\mathbf{X}'\mathbf{Y}(\mathbf{Y}'\mathbf{Y})^{-1/2}$). For identification, we assume that no $\hat{\mu}^2$'s are identical. We assume that each row of $\mathbf{X}$ and each row of $\mathbf{Y}$ are jointly multivariate normal with zero means and the covariance matrices, $\text{Cov}(\mathbf{x}, \mathbf{x}) = \Sigma_{xx}$, $\text{Cov}(\mathbf{y}, \mathbf{y}) = \Sigma_{yy}$, and $\text{Cov}(\mathbf{x}, \mathbf{y}) = \Sigma_{xy}$. Under this distributional assumption, the LR statistic testing the difference between $H_0 : \Sigma_{xy} = \mathbf{0}$ (independence between $\mathbf{X}$ and $\mathbf{Y}$) and $H_1 : \Sigma_{xy}$ is unrestricted (the saturated model), which is given by

$$\Lambda = 1/\prod_{t=1}^{T^*}(1 - \hat{\mu}_t^2). \tag{9}$$

This is called Wilks' Λ. Note that H_0 is equivalent to $\mu_1 = \mu_2 = \cdots = \mu_{T^*} = 0$, where μ_t's are population canonical correlations. We often denote the number of nonzero population canonical correlations (the dimensionality) by T. Then we can alternatively express $H_0 : T = 0$.

The exact distribution of Wilks Λ is very complicated. Bartlett [3] provided a large sample approximation to this statistic, namely

$$B_1 = M \log \Lambda = -M \sum_{t=1}^{T^*} \log(1 - \hat{\mu}_t^2), \tag{10}$$

where $M = N - (I + J + 3)/2$ is asymptotically chi-square distributed with IJ df. This statistic is analogous to $-2 \log \lambda_s$.

Under the same distributional assumption, the LR statistic for testing H_0 : the independence model ($T = 0$; all μ's are zero) against H_1 : one-dimensional model ($T = 1$; $\mu_1 > \mu_2 = \cdots = \mu_{T^*} = 0$) is given by

$$R_1 = -M \log(1 - \hat{\mu}_1^2). \tag{11}$$

This statistic is $-2 \log \lambda_c$. Some authors claimed that R_1 asymptotically follows the chi-square distribution under H_0 (e.g., [64], p. 88, p. 165, p. 189, [45], pp. 288–289). However, this is generally known to be incorrect [25]. Instead, it follows the distribution of the largest eigenvalue of a certain central Wishart matrix ($W_{T^*}(\max(I, J), \mathbf{I})$).

Both B_1 and R_1 provide consistent and asymptotically unbiased tests of independence between $\mathbf{X}$ and $\mathbf{Y}$. However, asymptotic powers are different. Whereas the test based on R_1 is more powerful when μ_1 is large, the test based on B_1 is more powerful when the departure from independence is sufficiently small.

When the independence model is rejected, Bartlett ([3]; see also [10]) showed that under $H_0 : T = 1$,

$$B_2 = -M \sum_{t=2}^{T^*} \log(1 - \hat{\mu}_t^2) \tag{12}$$

is asymptotically chi-square distributed with $(I - 1)(J - 1)$ df. This statistic can be used to test $H_0 : T = 1$ against H_1 : the saturated model. Unlike R_1 under the independence model, the asymptotic distribution of $R_2 = -M \log(1 - \hat{\mu}_2^2)$ is unknown under $H_0 : T = 1$. Note that R_1 asymptotically follows the noncentral chi-square distribution under $H_0 : T = 1$ (see panel (b') of Fig. 1). However, as $T = 0$ gets closer to the true model (i.e., as δ_2 approaches 0), convergence of the distribution of R_1 to the noncentral chi-square distribution gets slower and slower, until it breaks down when $T = 0$ is exactly true. As noted earlier, R_1 asymptotically follows the distribution of the largest eigenvalue of a central Wishart matrix $W_{T^*}(\max(I, J), \mathbf{I})$ in this case. Similarly, convergence of the asymptotic distribution of B_2 to the (central) chi-square distribution gets slower and slower, as $T = 0$ gets truer, until it breaks down when $T = 0$ is exactly true. The asymptotic distribution of B_2 is unknown under $T = 0$. For the test based on B_2 to achieve a reasonable power with a moderate sample size, $T = 0$ must be substantially incorrect. Some empirical evidence to this effect has been given by Harris [26]. The statistic like B_2 can be defined in general for testing $H_0 : T = t^*(< T^*; \mu_1 > \cdots > \mu_{t^*} > \mu_{t^*+1} = \cdots = \mu_{T^*} = 0)$ against H_1 : the saturated model. We note that

$$B_{t^*+1} = -M \sum_{t=t^*+1}^{T^*} \log(1 - \hat{\mu}_t^2) \tag{13}$$

is asymptotically chi-square distributed with $(I - t^*)(J - t^*)$ df under H_0. Note that $\mu_t^* > 0$ is explicit in H_0.

In RRR we seek to find the subspace in the space of **X** that is best predictive of **Y** in the column metric of $\hat{\Sigma}^{-1} = N(\mathbf{Y}'\mathbf{Q}_X\mathbf{Y})^{-1}$, where $\mathbf{Q}_X = \mathbf{I}_N - \mathbf{X}(\mathbf{X}'\mathbf{X})^{-1}\mathbf{X}'$ is the orthogonal projector onto the space orthogonal to the space of **X**. The solution and the determination of the dimensionality of the subspace remain essentially the same as in CANO ([31]; see also [51]), and the forgoing argument about CANO holds as is except for minor technical details. A special case of RRR is redundancy analysis (RA: [66]), where the column metric matrix in RRR, $\hat{\Sigma}^{-1}$, is replaced by $\mathbf{I}_J$, the identity matrix of order J. In the context of RA, Lazraq and Cléroux [37] made a gross mistake in claiming in effect that $-2 \log \lambda_c$ follows a chi-square distribution, which was incorrect and corrected by Takane and Hwang [61].

In DISC or MANOVA one of two matrices, **X** or **Y**, is a matrix of dummy variables that define subpopulations of samples. However, essentially the same theory as above applies, if the other set still follows the multivariate normal distributions with equal covariance matrices within the subpopulations. The only difference is in the degrees of freedom; the rank of a dummy variable matrix is reduced by 1 by columnwise centering, so that either I has to be replaced by $I - 1$ or J by $J - 1$ depending on which of the two matrices is the dummy variable matrix.

4.2 Correspondence Analysis (CA)

Let **X** and **Y** be two matrices of dummy variables. We form a matrix of observed proportions by $\mathbf{P} = \{p_{ij}\} = \mathbf{X}'\mathbf{Y}/N$. CA is a canonical correlation analysis between columnwise centered matrices of **X** and **Y**. It amounts to the singular value decomposition (SVD) of the matrix with elements $p_{ij}^* = (p_{ij} - p_{i.}p_{.j})/(p_{i.}p_{.j})^{1/2}$. The adjustment made on p_{ij} in this formula is to subtract the effect of the independence model and analyze only the portion of p_{ij} that departs from the independence model. The denominator standardizes the residual. The SVD of the matrix of p_{ij}^* leads to the decomposition of the observed proportions

$$p_{ij} = p_{i.}p_{.j}\left(1 + \sum_{t=1}^{T^*} \hat{\mu}_t \hat{r}_{it} \hat{c}_{jt}\right), \tag{14}$$

where $T^* = \min(I-1, J-1)$, $\hat{\mu}_t$ is the singular value for dimension t, $\hat{r}_{it}$ is the sample score for row i on dimension t, and $\hat{c}_{jt}$ is the sample score for column j on dimension t. The row and column scores are restricted by $\sum_i p_{i.}\hat{r}_{it} = 0, \sum_j p_{.j}\hat{c}_{jt} = 0, \sum_i p_{i.}\hat{r}_{it}\hat{r}_{it'} = \delta_{tt'} = \sum_j p_{.j}\hat{c}_{jt}\hat{c}_{jt'}$, where $\delta_{tt'}$ is a Kronecker delta, $p_{i.} = \sum_j p_{ij}$ and $p_{.j} = \sum_i p_{ij}$. Since CA is a special kind of canonical correlation analysis, the singular value $\hat{\mu}_t$ is the sample canonical correlation for dimension t, and the scores $\hat{r}_{it}$ and $\hat{c}_{jt}$ are the scores that maximize this canonical correlation.

We have the Pearsonian chi-square statistic X^2 for testing independence between rows and columns of contingency tables expressed as a function of $\hat{\mu}$'s, namely

$$X^2 = N\sum_{i,j} p_{ij}^{*2} = N\sum_{t=1}^{T^*} \hat{\mu}_t^2 \tag{15}$$

Lancaster [33]. Because of this relation X^2 is more often used in CA than the corresponding LR statistic ($-2\log\lambda_s = 2N\sum_{i,j} p_{ij}\log(p_{ij}/p_{i.}p_{.j})$). The two statistics are known to be asymptotically equivalent. Both X^2 and $-2\log\lambda_s$ follow the asymptotic chi-square distribution with $(I-1)(J-1)$ df. Since $-2\log\lambda_s$ is approximated by B_1 in (10), this means that X^2 and B_1 are also asymptotically equivalent with both I and J in the latter replaced by $I-1$ and $J-1$ (both sets variables are categorical), respectively.

O'Neill [43, 44], see also Lebart [38] and Corsten [11], showed that under independence $R_1^* = N\hat{\mu}_1^2$ was asymptotically distributed according to the distribution of the largest eigenvalue from a certain central Wishart matrix ($W_{T^*}(\max(I, J) - 1, \mathbf{I})$). This means that under independence $N\hat{\mu}_1^2$ is asymptotically equivalent to R_1 in (11). That is, when the independence model is correct, two tests of independence derived under the multivariate normality assumption (B_1 and R_1) can still be used even when both sets of variables are categorical. They are not exactly the LR tests but are asymptotically equivalent to X^2 and R_1^*. (When each $\hat{\mu}_t^2$ is small, which is likely under independence, $\hat{\mu}_t^2 \approx \log(1-\hat{\mu}_t^2)$ for each t. This relation is often used in establishing the asymptotic equivalence between the LR and the Pearsonian chi-square in contingency table analysis and between Wilks' Λ and Lawley-Hotelling's trace criterion in multivariate normal theory.)

Unfortunately, the above relationships do not extend beyond the independence model. O'Neill [43] showed that when the true dimensionality ($T = t^*$) was greater than 0 in the sense that there were $t^* < \min(I-1, J-1)$ distinct canonical correlations larger than zero, and the remaining canonical correlations were zero, $N\hat{\mu}_{t^*+1}^2$ was asymptotically distributed according to the largest eigenvalue of a certain central Wishart matrix($W_{T^*-t^*-1}(\max(I, J) - t^* - 1, \mathbf{I})$), and $N\sum_{t=t^*+1}^{T^*}\hat{\mu}_t^2$ is asymptotically chi-square distributed with $(I - t^* - 1)((J - t^* - 1)$ df, only when a certain unrealistic condition (Eq. (40) of O'Neill [43]) is satisfied.

In the history of CA, some incorrect assertions regularly appear. A well-known incorrect remark is that of Kendall and Stuart [32], pp. 574–575, who incorrectly claimed that each of the components of the Pearsonian chi-square ($N\hat{\mu}_t^2$) is itself an asymptotically independent chi-square variable. This as well as the same error made by Williams [67] was corrected by Lancaster [34]. However, the same error appeared in van de Geer [65], pp. 50–51, and this time it was corrected by Gilula [18]. A last incorrect result that also appears in Greenacre [23] is a test proposed by Bock [6] and used by Nishisato [42], pp. 41–42. It is claimed that the test statistic for the tth canonical correlation, $-\{N-(I+J+1)/2\}\log(1-\hat{\mu}_t^2)$, follows an asymptotic chi-square distribution. However, this quantity is asymptotically equivalent to $N\hat{\mu}_t^2$ and the error here is of the same nature as the error committed by Kendall and Stuart [32]. Interestingly, Nishisato knew Lancaster's correction of the Kendall and Stuart procedure, and yet he fell into essentially the same pitfall as Kendall and Stuart. The error was corrected by Goodman [20], pp. 301–302. Lebart and Morineau [39] presented a Monte Carlo study on the distribution of the second largest eigenvalue from CA when dependency exists in contingency tables.

4.3 The Correlation Model

The correlation model (CM) was already discussed in the introduction. It was introduced by Goodman [20, 21] and by Gilula and Haberman [19] as a maximum likelihood version of CA. Compare Eqs. (1) and (14). In the introduction, we described the properties of the LR statistics in the context of CM. First, the independence model CM(0) is compared against the saturated model CM(T^*) using $-2\log\lambda_s$, which is asymptotically chi-square distributed under CM(0). If CM(0) is rejected, CM(1) is compared against CM(T^*) using $-2\log\lambda_s$, which is again asymptotically chi-square distributed under CM(1) (with different degrees of freedom). We repeat this process until we find an acceptable model, or we end up with accepting CM(T^*).

4.4 Association Models

The RC(T) association model [19–21] is another model where the assessment of dimensionality is an important issue. The model can apply to multi-way contingency tables. However, we primarily focus on two-way tables here to draw direct comparisons with CA and CM. The model is

$$\pi_{ij} = c\alpha_i^*\beta_j^* \exp\left(\sum_{t=1}^{T^*}\gamma_t u_{it} v_{jt}\right), \tag{16}$$

where α_i^* and β_j^* are parameters associated with the margins such that $\sum_i \log\alpha_i^* = 0$ and $\sum_j \log\beta_j^* = 0$, and $\gamma_t(\geq 0)$ is a parameter associated with the importance of dimension t, and u_{it} and v_{jt} are sets of scores associated with the rows and columns, usually restricted by $\sum_i u_{it}u_{it'} = \delta_{tt'} = \sum_j v_{jt}v_{jt'}$. For brevity, we denote the T-dimensional RC association model as RC(T). RC(T^*) with $T^* = \min(I-1, J-1)$

is a saturated model, in which the probabilities are unrestricted. RC(0) is equivalent to the independence model. RC(T) with $1 \leq T < \min(I-1, J-1)$ is a model where the $I \times J$ matrix of interaction $\sum_t \gamma_t u_{it} v_{jt}$ has a reduced rank. In the RC association model parameters are usually estimated by the maximum likelihood method. The LR test $-2\log\lambda_s$ is asymptotically chi-square distributed if H_0 is RC(0) and H_1 is the saturated model and RC(0) is true. Similarly, $-2\log\lambda_s$ is asymptotically chi-square distributed if H_0 is RC(1) and H_1 is the saturated model and RC(1) is true, but RC(0) is not. However, $-2\log\lambda_c$ is not likely to be asymptotically chi-square distributed if H_0 is RC(0) and H_1 is RC(1) and RC(0) is true. Instead, it is asymptotically equivalent to the $N\hat{\mu}_1^2$, which is asymptotically distributed according to the distribution of the largest eigenvalue from a certain central Wishart matrix [24]. These results are similar to those in the correlation model. Gilula and Haberman [19] have shown that for $1 \leq T \leq T^*$, RC(T) holds approximately if CM(T) holds and ϕ_1 is small, and CM(T) holds if RC(T) holds and γ_1 is small. Thus, essentially the same strategy for dimensionality selection recommended for the correlation model can be followed in the association model as well.

4.5 Latent Class Analysis (LCA)

Testing problems in mixture models are relatively well known. In fact, we initially set out to find why the LR test could not be used to determine the number of components in mixture problems [22, 27, 41, Sect. 1.10]. We found that violation of regularity was the key problem, but then essentially the same difficulty should exist in virtually all other models that require dimensionality selection.

For comparison with other models, we consider latent class analysis (LCA; see, for example, [36]) for a two-way contingency table with row variable A and column variable B, although LCA itself is not restricted to the two variable situations. Let π_{ij} $(i = 1, \cdots, I;\ j = 1, ..., J)$ be the probability for cell (i, j). Then LCA with T latent classes $(k = 1, ..., K)$ is stated as

$$\pi_{ij} = \sum_{t=1}^{T} \pi_t \pi_{i|t} \pi_{j|t}, \tag{17}$$

where π_t is the size of latent class t (with the restriction that $\pi_t \geq 0$ and $\sum_t \pi_t = 1$), $\pi_{i|t}$ is the conditional probability of response i on variable A in class t (with the restriction that $\pi_{i|t} \geq 0$ and $\sum_i \pi_{i|t} = 1$), and $\pi_{j|t}$ is the conditional probability of response j on variable B in class t (with the restriction that $\pi_{j|t} \geq 0$ and $\sum_j \pi_{j|t} = 1$). LCA defines a rank T model (see [12]). For brevity, we denote LCA with T latent classes as LCA(T). Let $T^* = \min(I, J)$. LCA(T^*) is the saturated model. LCA(1) is the independence model. LCA(2) defines a rank 2 model. De Leeuw and van der Heijden [12] show that, when the rank of the matrix is 1, 2, or T^*, LCA is equivalent to CA and CM, but when the rank is larger than 2 but smaller than T^*, LCA implies

CA and CM, but CA and CM do not imply LCA (due to the inherent restrictions on parameters in the latter).

LCA is a kind of reduced rank model, and as such, it is endowed with a similar problem to the LR tests encountered by other reduced rank models. When H_0 is LCA(1) and H_1 is LCA(T^*), $-2\log\lambda_s$ is asymptotically distributed as chi-square with $(I-1)(J-1)$ degrees of freedom if LCA(1) is true. When H_0 is LCA(2) and H_1 is LCA(T^*), $-2\log\lambda_s$ is asymptotically distributed as chi-square with $(I-2)(J-2)$ degrees of freedom if LCA(2) is true, but LCA(1) is not true. When H_0 is LCA(1) and H_1 is LCA(2), the regularity condition is violated, and the conditional test $-2\log\lambda_c$ is not assured to be asymptotically chi-square distributed if LCA(1) is true. This result holds not only for LCA(1) and LCA(2), but more generally for all conditional tests of models with different ranks $T < \min(I, J)$. Using Monte Carlo studies, Everitt [14] and Holt and Macready [29] have shown that $-2\log\lambda_c$ is indeed not asymptotically distributed as a chi-squared variate. However, the exact distribution is unknown. This last finding is well known for LCA, but we occasionally find mistakes similar to those in other fields (e.g., [46]).

McLachlan and Basford [41, Sect. 1.10], explained this state of affairs by nonuniqueness of parameters and by boundary points. The problem of nonuniqueness of parameters refers to the fact that, in going from LCA(2) to LCA(1) by imposing the class size $\pi_2 = 0$, the conditional parameters $\pi_{i|2}$ and $\pi_{j|2}$ are undetermined. However, as has been noted earlier, whether we have indeterminacies in the model often depends on the parametrization of the model. The boundary point problem refers to the fact that LCA(1) can be considered as a restricted version of LCA(2), namely a model where $\pi_2 = 0$ is placed on the boundary of the parameter space rather than in its interior. While no boundary points are regular, whether we have the parameter vector on the boundary of the parameter space also depends on the parametrization of the model. More importantly, however, there is a fundamental difference between setting $\pi_2 = 0$ and setting $\pi_{i|t} = 0$ for some i and t, which also puts the parameter vector on the boundary of the parameter space. While the former incurs additional indeterminacies in the model, the latter does not. Also, while the former creates a locally nonregular point, the latter creates a locally regular but globally nonregular point. In the latter case, $-2\log\lambda_c$ is not likely to be asymptotically chi-square distributed, but its theoretical distribution can often be deduced analytically. See Shapiro [53, 55], Self and Liang [52], and Dijkstra [13] for more details.

In mixture problems where there is no saturated model such as mixtures of normal distributions (e.g., [16]), $-2\log\lambda_s$ cannot be defined. Consequently, testing procedures based on entirely different rationales have to be developed in such cases (e.g., [5]). With the advent of high-speed computers, this kind of model (performing clustering and modeling within each cluster simultaneously) is expected to grow rapidly in number in the near future. It should be kept in mind, however, that such models often require extra efforts in choosing the appropriate number of clusters allowed in the models.

4.6 Factor Analysis (FA) Including Item FA

Consider the factor analysis model for I variables. Let Σ be the $I \times I$ covariance matrix, let $\mathbf{A}$ be the $I \times T$ matrix of factor loadings, and let Ψ be an *nnd* (non-negative definite) diagonal matrix of unique variances. For identification, we require rank($\mathbf{A}$) $= T$ [17, 50], and $\mathbf{A}'\Psi^{-1}\mathbf{A} =$ diagonal. Then the factor analysis model is

$$\Sigma = \mathbf{A}\mathbf{A}' + \Psi. \tag{18}$$

We denote the factor analysis model with T factors as FA(T). If H_0 is FA(T) and H_1 is the saturated model (the covariances are unrestricted), then $-2\log\lambda_s$ is asymptotically chi-square distributed if FA(T) is true but FA($T-1$) is not. However, if H_0 is FA(T) and H_1 is FA($T+1$), $-2\log\lambda_c$ is not assured to be asymptotically chi-square distributed if FA(T) is true. These results are consistent with those in the other models discussed in this paper.

Substantially more general results have been given by Steiger et al. [57] and to some extent by Shapiro [54]. Although their results are presented in the context of structural equation models of which FA is a special case, they apply equally well to all other models discussed in this paper. The first generalization made is with respect to the fitting criterion. It was shown that their asymptotic results could hold for a much wider class of optimization criteria than the maximum likelihood method if certain conditions are satisfied. These criteria are called the minimum discrepancy functions. The second generalization is concerned with the fact that, although they considered a sequence of nested models, they did not assume that any particular one of them was correct. Panels (a) and (a') of Fig. 1 illustrate this. In panel (a), model B is nested within model A, which in turn is nested within the saturated model, but none of these models are assumed correct, except the saturated model which is always assumed correct. Under suitable regularity conditions N times the value of the minimum discrepancy functions are asymptotically noncentral chi-square distributed for all the three possible differences defined among the three models while maintaining the reproducibility of chi-square with respect to degrees of freedom and noncentrality parameters. Under the same regularity conditions, (a) specializes into (b) and (c) depending on whether only model A is assumed correct or both A and B are assumed correct. The situation described in (a) also specializes into (a') for dimensionality selection where neither the $T+1$- nor the T-dimensional model is assumed correct. (a') in turn specializes into (b') and (c'), but when both the T- and the $T+1$-dimensional models are assumed correct, one of the regularity conditions is bound to fail for comparisons between the saturated model and the $T+1$-dimensional model, and between the $T+1$- and the T-dimensional models.

Note that in model (18) $\psi_i = 0$ for any i, where ψ_i is the ith diagonal element of Ψ that puts the parameter vector on the boundary of the parameter space. However, reparameterizing Ψ by Υ^2 will eliminate the boundary problem. Still, the point where $\upsilon_i^2 = 0$ is not regular because the rank of the Jacobian matrix at this point is smaller than in its neighborhood [56].

We did not find many incorrect claims in this context [28]. The only reference deals with full-information item factor analysis, where Bock et al. [7] claim that $-2\log\lambda_c$ is asymptotically chi-square distributed when two models are compared with a different number of dimensions. Note, however, that in this case the saturated model is not a sample covariance matrix, but a set of observed frequencies of response patterns to test items.

5 Discussion

Regularity is one of a sufficient set of conditions for the distribution of $-2\log\lambda$ to be asymptotically chi-square. However, it is not a necessary condition. If the distribution of $-2\log\lambda$ is not assured to be asymptotically chi-square, then there are a few alternative approaches to be taken. First of all, note that there are situations in which the asymptotic distribution of $-2\log\lambda$ is known despite the violation of regularity. There are two representative situations to be distinguished. One is in which a one-dimensional model is compared against the independence model (assumed true) in canonical correlation analysis and related methods. The asymptotic distribution of the LR statistic in this case is that of the largest eigenvalue of a certain central Wishart matrix. The second situation in which the asymptotic distribution is known is when the nonregularity is related to a particular kind of the boundary point problem. When the boundary point causes no additional indeterminacy problems, the asymptotic distribution of $-2\log\lambda$ is often a chi-bar-square ($\bar{\chi}^2$) distribution which is a mixture of chi-square distributions with different degrees of freedom.

If the distribution is unknown, we may investigate the asymptotic behavior of $-2\log\lambda$ using Monte Carlo methods like the ones we conducted earlier. It might turn out that the asymptotic distribution is still chi-square, even though regularity is violated. In our experience, this is not very likely.

A second alternative is to generate the distribution of $-2\log\lambda$ using the so-called parametric bootstrap (see [1, 4, 8, 30, 35]). The procedure followed in the parametric bootstrap is as follows [30].

(i) Evaluate the value of $-2\log\lambda$ for the sample of size N.
(ii) Draw 19 samples of size N from the distribution created from the parameter estimates found in (i) under H_0, and evaluate the 19 values of $-2\log\lambda$.
(iii) If the value of $-2\log\lambda$ for the sample is larger than the 19 values found in (ii), then reject H_0 in the conditional test procedure.

This provides a test with a significance level of 0.05. Hope [30] shows that the power of this test procedure increases when the number of Monte Carlo samples increases to numbers larger than 19. Feng and McCulloch [15] recently presented a rigorous justification for the Monte Carlo significance test procedure.

Acknowledgements The work reported in this paper has been supported by grant A6394 from the Natural Sciences and Engineering Research Council of Canada to the first author. We thank Alexander Shapiro for his valuable help.

References

1. Aitkin, M., Anderson, D., & Hinde, J. (1981). Statistical modelling of data on teaching styles. *Journal of the Royal Statistical Society, Series A, 144,* 419–461.
2. Anderson, T. W. (1951). Estimating linear restrictions on regression coefficients for multivariate normal distribution. *Annals of Mathematical Statistics, 22,* 327–351.
3. Bartlett, M. S. (1938). Further aspects of the theory of multiple regression. *Proceedings of Cambridge Philosophical Society, 34,* 33–40.
4. Barnard, G. A. (1963). Contribution to the discussion of a paper by M. S. Bartlett. *Journal of the Royal Statistical Society, Series B,* **25**, 294.
5. Bock, H. H. (1985). On some significance tests in cluster analysis. *Journal of Classification, 2,* 77–108.
6. Bock, R. D. (1960). *Methods and applications of optimal scaling*. The University of North Carolina Psychometric Laboratory Research Memorandum No. 25.
7. Bock, R. D., Gibbons, R., & Muraki, E. (1988). Full-information item factor analysis. *Applied Psychological Measurement, 12,* 261–280.
8. Bollen, K. A., & Stine, R. A. (1992). Bootstrapping goodness-of-fit measures in the context of structural equation models. *Sociological Methods and Research, 21,* 205–229.
9. Büyükurt, B. K., & Büyükurt, M. D. (1990). Robustness and small-sample properties of the estimators of probabilistic multidimensional sclaing (PROSCAL). *Journal of Marketing Research, 27,* 139–149.
10. Chou, K. J., & Muirhead, R. J. (1979). On some distribution problems in MANOVA and discriminant analysis. *Journal of Multivariate Analysis, 9,* 410–419.
11. Corsten, L. C. A. (1976). Matrix approximation, a key to application of multivariate methods. In *Proceedings of the 9th International Biometric Conference* (Vol. 1, pp. 61–77). Biometric Society.
12. de Leeuw, J., & van der Heijden, P. G. M. (1991). Reduced rank models for contingency tables. *Biometrika, 78,* 229–232.
13. Dijkstra, T. K. (1992). On statistical inference with parameter estimates on the boundary of the parameter space. *British Journal of Mathematical and Statistical Psychology, 45,* 289–309.
14. Everitt, B. S. (1988). A Monte Carlo investigation of the likelihood ratio test for number of classes in latent class analysis. *Multivariate Behavioral Research, 23,* 531–538.
15. Feng, Z. D., & McCulloch, C. E. (1996). Using Bootstrap likelihood ratios in finite mixture models. *Journal of the Royal Statistical Society, Series B, 58,* 609–617.
16. Frayley, C., & Raftery, A. E. (1998). How may clusters? Which clustering method? Answers via model-based cluster analysis. *The Computer Journal, 41,* 578–588.
17. Geweke, J. F., & Singleton, K. J. (1980). Interpreting the likelihood ratio statistic in factor models when sample size is small. *Journal of the American Statistical Association, 75,* 133–137.
18. Gilula, Z. (1994). Book review of "Multivariate analysis of categorical data: applications and theory" by van de Geer. *Journal of the American Statistical Association, 89,* 1140–1141.
19. Gilula, Z., & Haberman, S. J. (1986). Canonical analysis of contingency tables by maximum likelihood. *Journal of the American Statistical Association, 81,* 780–788.
20. Goodman, L. A. (1985). The analysis of cross-classified data having ordered and/or unordered categories: Association models, correlation models and asymmetry models for contingency tables with or without missing entries. *The Annals of Statistics, 13,* 10–69.

21. Goodman, L. A. (1986). Some useful extensions of the usual correspondence analysis approach and the usual log-linear models approach in the analysis of contingency tables. *International Statistical Review, 54,* 243–309.
22. Ghosh, J. K., & Sen, P. K. (1985). On the asymptotic performance of the log likelihood ratio statistic for the mixture model and related results. In L. M. Le Cam & R. A. Olshen (Eds.), *Proceedings of the Berkeley Conference in Honor of Jerzy Neyman and Jack Kiefer, Vol. II* (pp. 789–806). Wadsworth.
23. Greenacre, M. J. (1984). *Theory and applications of correspondence analysis*. Academic.
24. Haberman, S. J. (1981). Tests for independence in two-way contingency tables based on canonical correlation and on linear-by-linear interaction. *Annals of Statistics, 9,* 1178–1186.
25. Harris, R. J. (1976). The invalidity of partitioned-U tests in canonical correlation and multivariate analysis of variance. *Multivariate Behavioral Research, 11,* 353–366.
26. Harris, R. J. (1985). Multivariate statistics: When will experimental psychology catch up? In S. Koch & T. Leary (Eds.), *A century of psychology as a science* (pp. 678–697). McGraw-Hill.
27. Hartigan, J. A. (1985). A failure of likelihood asymptotics for normal mixtures. In L. M. Le Cam & R. A. Olshen (Eds.), *Proceedings of the Berkeley Conference in Honor of Jerzy Neyman and Jack Kiefer, Vol. II* (pp. 807–810). Wadsworth.
28. Hayashi, K., Bentler, P. M., & Yuan, K.-H. (2007). On the likelihood ratio tests for the number of factors in exploratory factor analysis. *Structural Equation Modeling, 14,* 505–526.
29. Holt, J. A., & Macready, G. B. (1989). A simulation study of the difference chi-square for comparing latent class models under violation of regularity conditions. *Applied Psychological Measurement, 13,* 221–231.
30. Hope, A. C. A. (1968). A simplified Monte Carlo significance procedure. *Journal of the Royal Statistical Society, Series B, 30,* 582–598.
31. Izenman, A. J. (1975). Reduced-rank regression for the multivariate linear model. *Journal of Multivariate Analysis, 5,* 248–264.
32. Kendall, M. G., & Stuart, A. (1961). *The advanced theory of statistics. Vol. 2: Inference and relationship*. Griffin.
33. Lancaster, H. O. (1949). The derivation and partition of χ^2 in certain discrete distributions. *Biometrika, 36,* 117–129.
34. Lancaster, H. O. (1963). Canonical correlations and partitions of χ^2. *Quarterly Journal of Mathematics, 14,* 220–224.
35. Langeheine, R., Pannekoek, J., & van de Pol, F. (1996). Bootstrapping goodness-of-fit measures in categorical data analysis. *Sociological Methods and Research, 24*(4), 492–516. https://doi.org/10.1177/0049124196024004004.
36. Lazarsfeld, P. F., & Henry, N. W. (1968). *Latent structure analysis*. Houghton Mifflin.
37. Lazraq, A., & Cléroux, R. (2002). Testing the significance of the successive components in redundancy analysis. *Psychometrika, 67,* 411–419.
38. Lebart, L. (1976). The significancy of eigenvalues issued from correspondence analysis. *Proceedings of Computational Statistics* (pp. 38–45). Physica Verlag.
39. Lebart, L., & Morineau, A. (1981). Statistical significance criteria in multiple-choice data reduction and visualization. A paper presented at the (1981). In *Annual meeting of the psychometric society*. Chapel Hill.
40. MacKay, D. B., & Zinnes, J. L. (1982). PROSCAL: A program for probabilistic scaling. Discussion paper #218. School of Business, Indiana University.
41. McLachlan, G. J., & Basford, K. E. (1988). *Mixture models: Inference and applications to clustering*. Marcel Dekker.
42. Nishisato, S. (1980). *Analysis of categorical data: Dual scaling and its applications*. University of Toronto Press.
43. O'Neill, M. E. (1978). Distributional expansions for canonical correlations from contingency tables. *Journal of the Royal Statistical Society, Series B, 40,* 303–312.
44. O'Neill, M. E. (1981). A note on the canonical correlations from contingency tables. *Australian Journal of Statistics, 23,* 58–66.
45. Overall, J. E., & Klett, C. J. (1972). *Applied multivariate analysis*. McGraw-Hill.

46. Raijmakers, M. E. J., van Koten, S., & Molenaar, P. C. M. (1996). On the validity of simulating stagewise development by means of PDP networks: Application of catastrophe analysis and an experimental test of rule-like network performance. *Cognitive Science, 20*, 101–136.
47. Ramsay, J. O. (1977). Maximum likelihood estimation in multidimensional scaling. *Psychometrika, 42,* 241–266.
48. Ramsay, J. O. (1980). Some small sample results for maximum likelihood estimation in multidimensional scaling. *Psychometrika, 45,* 139–144.
49. Ramsay, J. O. (1981). How to use MULTISCALE. In S. S. Schiffman, M. L. Reynolds, & F. W. Young (Eds.), *Introduction to multidimensional scaling* (pp. 211–235). Academic.
50. Reiersøl, O. (1950). On the identifiability of parameters in Thurstone's multiple factor analysis. *Psychometrika, 15,* 121–149.
51. Reinsel, G. C., & Velu, R. P. (1998). *Multivariate Reduced-rank Regression: Theory anf Applications*. Springer.
52. Self, S. G., & Liang, K.-Y. (1987). Asymptotic properties of maximum likelihood estimators and likelihood ratio tests under nonstandard conditions. *Journal of the American Statistical Association, 82,* 605–610.
53. Shapiro, A. (1985). Asymptotic distribution of test statistics in the analysis of moment structures under inequality constraints. *Biometrika, 72,* 133–144.
54. Shapiro, A. (1986). Asymptotic theory of overparametrized structural models. *Journal of the American Statistical Association, 81,* 142–149.
55. Shapiro, A. (1988). Towards a unified theory of inequality constrained testing in multivariate analysis. *International Statistical Review, 56,* 49–62.
56. Shapiro, A., & Browne, M. W. (1983). On the investigation of local identifiability: A counter example. *Psychometrika, 48,* 303–304.
57. Steiger, J. H., Shapiro, A., & Browne, M. W. (1985). On the multivariate asymptotic distribution of sequential chi-square statistics. *Psychometrika, 50,* 253–264.
58. Takane, Y. (1978). A maximum likelihood method for nonmetric multidimensional scaling: I. The case in which all empirical pairwise orderings are independent - theory. *Japanese Psychological Research, 20,* 7–17.
59. Takane, Y. (1981). Multidimensional successive categories scaling: A maximum likelihood method. *Psychometrika, 46,* 9–28.
60. Takane, Y., & Carroll, J. D. (1981). Nonmetric maximum likelihood multidimensional scaling from directional rankings of similarities. *Psychomerika, 46,* 389–405.
61. Takane, Y., & Hwang, H. (2005). On a test of dimensionality in redundancy analysis. *Psychometrika, 70,* 271–281.
62. Takane, Y., & Sergent, J. (1983). Multidimensional scaling models for reaction times and same-different judgments. *Psychometrika, 48,* 393–423.
63. Takane, Y., van der Heijden, P. G. M., & Browne, M. W. (2003). On likelihood ratio tests for diensionality selection. In T. Higuchi, Y. Iba, M. Ishiguro (Eds.), *Proceedings of Science of modeling: The 30th Anniversary Meeting of the Information Criterion (AIC)* (pp. 348–349). The Institute of Statistical Mathematics.
64. Tatsuoka, M. M. (1971). *Multivariate analysis: Techniques for educational and psychological research*. Wiley.
65. van de Geer, J. P. (1993). *Multivariate analysis of categorical data: applications*. Sage.
66. van den Wollenberg, A. L. (1977). Redundancy analysis: an alternative to canonical correlation analysis. *Psychometrika, 42,* 207–219.
67. Williams, E. J. (1952). Use of scores for the analysis of association in contingency tables. *Biometrika, 39,* 274–309.
68. Winsberg, S., & De Soete, G. (1993). A latent class approach to fitting the weighted euclidean model, CLASCAL. *Psychometrika, 58,* 315–330.

Submitted Articles

Chronological Monitoring of the Cross-Resistance Rate of *Pseudomonas Aeruginosa* Classified by the Radius-Distance Model

Syou Maki, Yasutoshi Hatsuda, Toshihiko Ishizaka, Sachiko Omotani, Naonori Koizumi, Yukako Yasui, Takako Saito, Michiaki Myotoku, and Tadashi Imaizumi

Abstract The approximate situation of the spread of antimicrobial-resistant bacteria can be estimated by the use of the cross-resistance rate (*CRR*). In our previous study, we reported some interesting features of the *CRR* that the pharmaceutical prescription of antibiotic drugs to patients reflects the asymmetric behavior of the pharmacists. Those behaviors can be represented by asymmetric data of the drugs, and we could gain new insights into that matter by applying the radius-distance model (Okada and Imaizumi in Behaviormetrika 14:81–96, 1987). This model was developed for evaluating the similarity of asymmetric data, and the degree of the similarity can be visualized by a configuration with points and the radii of circles in a multidimensional Euclidean space. We adopted the data of *CRR* of *Pseudomonas aeruginosa* as the first attempt to demonstrate the practical utility of this model. As a result, the same strains of antibiotics were distributed closely on the two-dimensional configuration. Even with different groups of antibiotics, the classification was found to be identical to the drugs' mechanism of action. We clarified from the results of one month of chronological monitoring that the radii of the circles were influenced by a change in the environment, i.e., the hospital relocation. In conclusion, the radius-distance model has the potential for discovering new characteristics or specific features of the data from a new perspective in the fields of medical and pharmaceutical research.

S. Maki (✉)
Institute of Frontier Science and Technology, Okayama University of Science, 1-1 Ridai-cho, Kitaku, Okayama 700-0005, Japan
e-mail: makisyou@ifst.ous.ac.jp

S. Maki · Y. Hatsuda · S. Omotani · M. Myotoku
Faculty of Pharmacy, Osaka Ohtani University, 3-11-1, Nishikiori-Kita, Tondabayashi 584-8540, Osaka Pref, Japan

Y. Hatsuda · T. Ishizaka · S. Omotani · N. Koizumi · Y. Yasui · T. Saito
Sakai City Medical Center, 1-1-1 Ebaraji-Cho, Nishi-Ku, Sakai-Shi, Osaka 593-8304, Japan

T. Imaizumi
School of Management and Information Sciences, Tama University, 4-1-1, HijirigaokaTokyo 206-0022, Tama, Japan

© The Author(s), under exclusive license to Springer Nature Singapore Pte Ltd. 2023
A. Okada et al. (eds.), *Facets of Behaviormetrics*, Behaviormetrics: Quantitative Approaches to Human Behavior 4, https://doi.org/10.1007/978-981-99-2240-6_11

Keywords Cross-resistance · Asymmetric multidimensional scaling · Antibiotic drugs · Data mining

1 Introduction

The global spread of antimicrobial-resistant bacteria is a very serious problem. No matter how much effort researchers commit to the development of revolutionary new antibiotics, the problem of *antimicrobial resistance* (AMR) is inevitable as long as they are frequently used. In general, bacteria that are resistant to two or more antimicrobial agents with different mechanisms of action are called *multidrug-resistant bacteria* (MRB). In 2011, the World Health Organization (WHO) issued the imperative announcement of "No action today, no cure tomorrow" [3], stating that AMR and MRB problems should be addressed worldwide. The incidence of resistance, i.e., the resistance rate, varies by country, region, season, and hospital; hence, antimicrobial susceptibility rates should be routinely calculated for each medical institution. When AMR appears in one drug and a similar effect is also present in other drugs with an analogical structure or action to that drug, that case is called "*cross-resistance*" [36]. Since clinical combinations of antimicrobial agents are very complicated, medical staff are under heavy pressure to select from many combinations of drugs so that the effect of cross-resistance will be as small as possible. So far, however, few methods have been able to quickly detect the cross-resistance rate (*CRR*), even though *CRR* provides very important information for the treatment of infectious diseases. In our previous reports [13–15], we proposed a new method to analyze the relationship between *CRR* and antibiotic drugs by applying the method of the radius-distance model [23–25]. This research gave us the crucial insight that pharmaceutical prescription of antibiotic drugs to patients is associated with the "asymmetric" behavior of pharmacists. Such behavioral data can be described by the *CRRs* between various antimicrobial agents and can be summarized by an asymmetric matrix of the *CRRs*. In this study, we visualize the data of *CRR* by a two-dimensional configuration, and we will propose the practical utility of this model by the approach of chronological monitoring of the configurations.

2 Ethical Review of Investigation

This study was approved by the clinical trial review committee of Sakai City Medical Center (approval number: H30-119). Connectable data will be anonymized, and all data will not be personally identifiable and will not be detrimental to the individual.

3 Data

3.1 Cross-Resistance Rate (CRR)

Medical institutions usually conduct many kinds of drug-susceptibility testing and ascertain the susceptible (or resistant) effect of antibiotics on the patients. The *CRR* (%) is calculated by accumulating the testing results and is deduced by the following:

$$CRR_{X \leftarrow A} = \frac{N(R_A \cap R_X)}{N(R_A)} \times 100. \tag{1}$$

The arrow in the *CRR* subscript means the computation results of the *CRR* when it is carried out under the condition of drug X against drug A. $N(R_A)$ is the number of strains that are resistant to drug A against a certain bacteria. $N(R_A \cap R_X)$ is the number of strains that are also resistant to drug X, meaning the number of strains that are resistant to both drug A and X. As shown in Eq. 1, the *CRR* can be thought of as a bivariate function of A and X.

3.2 Data Collection

We utilized the data of the drug-susceptibility test of *Pseudomonas aeruginosa* collected from the Sakai City Medical Center. This hospital is one of the regional base hospitals in Sakai City, has 480 general beds and 7 infectious disease beds, and is equipped with a critical care center. *P. aeruginosa* is a typical bacterium with resistance to many drugs and is also known for nosocomial infection in medical institutions around the world [2, 7, 10, 18, 20, 21, 28, 29, 34, 35, 37, 40]. In the process of data collecting, we used unique software called "Chans", which was developed by Prof. Hatsuda, who is one of the coauthors of the present study [13–15]. In the analyses, we devoted an effort to collecting a large amount of data for 10 or more resistant strains in order to present the effect of cross-resistance. Hence, the interval of the data collection was set to 18 months, and its starting point was moved by one month. To prevent the duplication of calculation, we used only the initial data of patients who were hospitalized for more than 18 months or who underwent multiple drug-susceptibility tests.

In this survey, the data collection was conducted from January 2013 to December 2018. During this period, there was a big event in which the hospital was moved to its current location about 3.5 km away on July 1, 2015. Such hospital relocation was a beneficial event to confirm the validity of our method because we could chronologically investigate the occurrence and propagation of the cross-resistance by comparing the data before and after the hospital relocation. As far as we know, no similar investigation has ever been carried out.

3.3 *Pseudomonas Aeruginosa*

Bacteria that are problematic in clinical practice are broadly divided into two categories of gram-positive cocci (GPC) and gram-negative bacilli (GNR). *P. aeruginosa* belongs to the category of GNR and is one of the well-known opportunistic pathogens. It prefers moist environments and can readily thrive in artificial circumstances such as on or within medical devices, e.g., catheters, injectors, and syringes with a needle. This is one of the major reasons why nosocomial infections due to *P. aeruginosa* are on the rise worldwide. Routine disinfection of such devices exposed to moisture is recommended to prevent infection with *P. aeruginosa*. Encouragement of patient hand washing is also important. However, complete prevention of *P. aeruginosa* is nearly impossible, as some reports have announced drains, sinks, and faucets as sites of nosocomial infection [2, 37]. Nowadays, too many AMR have appeared in *P. aeruginosa*, and there is no practical or clinical countermeasure to deal with the multidrug-resistant *P. aeruginosa* (MDRP) [11, 18]. *P. aeruginosa* is not extremely virulent compared to other major pathogenic bacterial species such as *Staphylococcus aureus*. Instead of blindly administering new antibiotics to patients, we should prescribe proper antibiotics while monitoring the latest emergence situations of resistant bacteria at each hospital [1, 28, 40]. That is, early monitoring of clinical institutions and its feedback are realistically the most effective preventive measures, especially in *P. aeruginosa*.

3.4 *Antibiotic Drugs and the Strain Subgroups*

As shown in Table 1, antibiotic drugs are grouped according to their "mechanism of action" [16]. Furthermore, antibiotics in the same group are subdivided into subgroups depending on the difference in their molecular structure. In this study, the effect of antibiotics on *P. aeruginosa* was analyzed with the following twelve antibiotic drugs: piperacillin (PIPC), piperacillin-tazobactam (PIPC.TAZ), ceftazidime (CAZ), cefepime (CFPM), cefoperazone-sulbactam (CPZ.SBT), imipenem (IPM), meropenem (MEPM), aztreonam (AZT), amikacin (AMK), gentamicin (GM), ciprofloxacin (CPFX), and levofloxacin (LVFX). These drugs ensured that we had 10 or more resistant strains for analyzing the effect of cross-resistance. Other antibiotics were not examined because there was too small a number of strains collected to conduct an analysis.

PIPC and PIPC.TAZ belong to the group of penicillins. This group inhibits cell-wall synthesis [5] and is effective for GPC. Both PIPC and PIPC.TAZ show anti-pseudomonas aeruginosa activity, and these drugs are usually administered intravenously.

CAZ, CFPM, and CPZ.SBT belong to the group of cephems (including cephalosporins). In Japan, CAZ is often classified into the 3.5th-generation cephems [6] and is widely effective for GNRs, including *P. aeruginosa*. CAZ has low effects on

Table 1 Groups of antibiotics and their mechanism of action

Groups of antibiotics	Antibiotic drugs	Abbreviated code[a]	Mechanism of action	Color of circles
Penicillins	Ampicillin	ABPC	Inhibition of cell-wall synthesis	**Green**
	Piperacillin	**PIPC**		
	Piperacillin-Tazobactam	**PIPC.TAZ**		
	Amoxicillin	AMPC		
	Benzylpenicillin	PCG		
Cephems	**Ceftazidime**	**CAZ**	Inhibition of cell-wall synthesis	**Blue**
	Cefepime	**CFPM**		
	Cefmetazole	CMZ		
	Cefetaxis	CTX		
	Cefsulodin	CFS		
	Flomoxef	FMOX		
	Cefoperazone	CPZ		
	Cefoperazone -Sulbactam	**CPZ.SBT**		
Carbapenems	Doripenem	DRPM	Inhibition of cell-wall synthesis	**Purple**
	Imipenem	**IPM**		
	Meropenem	**MEPM**		
Monobactams	**Aztreonam**	**AZT**	Inhibition of cell-wall synthesis	**Water blue**
Aminoglycosides	**Amikacin**	**AMK**	Inhibition of protein synthesis	**Orange**
	Gentamicin	**GM**		
	Isepamicin	ISP		
New-quinolones	**Ciprofloxacin**	**CPFX**	Inhibition of DNA synthesis	**Pink**
	levofloxacin	**LVFX**		

[a]Every antibiotic strain is distinguished with capital letters of the official abbreviated code

GPCs. CFPM is referred to as the 4th-generation cephems [38]. The 4th-generation cephems combine the pharmacological properties of both the 1st-generation GPC-effective and the 3rd-generation GNR-effective drugs [38]. Hence, the 4th-generation cephems are effective against *P. aeruginosa.*

IPM and MEPM belong to the group of carbapenems. Clinically, when an antibiotic drug of carbapenems is prescribed as a first-chosen drug, it is necessary to avoid, as far as possible, using another drug of carbapenems as a second-chosen drug [33]. Thus, a carbapenem is a specific drug for drug-resistant bacteria, but there are many infections for which it does not work (e.g., *MRSA, Legionella, Mycoplasma, Enterococcus, Chlamydophila, and Rickettsia*). However, carbapenems are very effective for GNRs, including *P. aeruginosa.*

AMK and GM belong to the group of aminoglycosides. This group inhibits protein synthesis and is effective for many GNRs, including *P. aeruginosa* [31]. However, since it may cause side effects of kidney or ear symptoms [8], it is indispensable to measure the patient's blood concentration of the drug [17].

CPFX and LVFX belong to the group of new-quinolones. This group has inhibitory effects on nucleic acid synthesis and is effective for GNRs, including *P. aeruginosa* [29]. LVFX can also be used to treat pneumonia [19]. LVFX is currently the only oral drug active against *P. aeruginosa* [19]. Both CPFX and LVFX are known to have low activity on anaerobic bacteria and have side effects such as central nervous system symptoms and inflammation of joints and tendons [9]. New-quinolones have antibacterial activity against mycobacterium tuberculosis [12]. Monotherapy with new-quinolones antibiotics is considered to cause cross-resistance with other new-quinolones antibiotics.

In summary, hereinabove, the groups of penicillins, cephems, carbapenems, and monobactams target the inhibition of cell-wall synthesis, and they have common properties in the aspect of their mechanism of action. On the other hand, the aspects of aminoglycosides and new-quinolones are essentially different from those of penicillins. That is, we knew in advance that there were three groups according to their mechanism of action.

4 Analysis

We calculated the *CRR* for the 12 antibiotics in the first step of the analysis, the results of which are summarized in the table. Those results are presented as a square matrix with the 12 antibiotics symmetrically arranged in rows and columns. As an example, Table 2 exhibits a representative matrix, where the data were collected during the 18 months from January 1, 2013 to June 30, 2014. In this study, twenty-seven similar *CRR* matrixes were prepared in advance, and we shall name them as the *CRR* matrix hereafter. All the cell elements were deduced by Eq. 1 and expressed in percentages. The antibiotic drugs in the rows correspond to the drug A in Eq. 1, and those in the columns correspond to the drug X. The denominator of the element is the number of patients on whom drug A was not effective, and the numerator is the number of the patients on whom both drugs A and X were not effective. Therefore, the cell element becomes asymmetric in terms of the diagonal element, and the diagonal element becomes 100% where it is not used for the analysis.

In other words, the *CRR* matrix can be considered as a kind of conditional probability because the cell elements in the rows are computed under the condition in the columns. The twenty-seven *CRR* matrixes mentioned above are as follows: the data of 2 years before the relocation (BR); the data of 1 year BR; the data of just after the relocation (AR); and twenty-four data of AR from January 2013 to December 2018. They were utilized to investigate longitudinal changes in the *CRR* by comparing the data before and after the hospital relocation.

Table 2 A representative *CRR* matrix of *P. aeruginosa*

		AZT	**AMK**	**GM**	**IPM**	**MEPM**	**CAZ**	**CFPM**	**CPZ SBT**	**CPFX**	**LVFX**	**PIPC**	**PIPC TAZ**
Aztreonam	**AZT**	100.0	16.1	32.3	22.6	19.4	35.5	32.3	54.8	35.5	48.4	41.9	19.4
Amikacin	**AMK**	27.8	100.0	94.4	22.2	22.2	16.7	44.4	16.7	50.0	44.4	16.7	0.0
Gentamicin	**GM**	24.4	41.5	100.0	17.1	17.1	17.1	31.7	19.5	24.4	26.8	19.5	7.3
Imipenem	**IPM**	33.3	19.0	33.3	100.0	52.4	28.6	19.0	28.6	28.6	33.3	33.3	14.3
Meropenem	**MEPM**	40.0	26.7	46.7	73.3	100.0	40.0	33.3	46.7	46.7	46.7	46.7	20.0
Ceftazidime	**CAZ**	73.3	20.0	46.7	40.0	40.0	100.0	53.3	73.3	40.0	60.0	66.7	46.7
Cefepime	**CFPM**	52.6	42.1	68.4	21.1	26.3	42.1	100.0	47.4	63.2	73.7	42.1	26.3
Cefoperazone-sulbactam	**CPZ SBT**	85.0	15.0	40.0	30.0	35.0	55.0	45.0	100.0	40.0	60.0	55.0	25.0
Ciprofloxacin	**CPFX**	37.9	31.0	34.5	20.7	24.1	20.7	41.4	27.6	100.0	86.2	17.2	6.9
Levofloxacin	**LVFX**	51.7	27.6	37.9	24.1	24.1	31.0	48.3	41.4	86.2	100.0	27.6	13.8
Piperacillin	**PIPC**	81.3	18.8	50.0	43.8	43.8	62.5	50.0	68.8	31.3	50.0	100.0	37.5
Piperacillin-tazobactam	**PIPC TAZ**	85.7	0.0	42.9	42.9	42.9	100.0	71.4	71.4	28.6	57.1	85.7	100.0

The cell elements of the matrix represent the CRR, and every element is expressed in percentages.
The CRR of each cell is the CRR of antibiotics in the columns against the rows.

In the next step of the analysis, we used a program for the radius-distance model in R. The program will be released as a Shiny application [39]. We evaluated the relationships between the antibiotics in Table 2 as asymmetric similarities among the 12 antibiotics and inputted them into the program as a data matrix. The asymmetric similarity from an antibiotic j to the other antibiotic k is represented as $m_{jk} = d_{jk} - (r_j - r_k)$, where d_{jk} is the Euclidean distance between two points j and k, and $r_j - r_k$ is the difference of the circle size from the antibiotic j to the antibiotic k. These geometric relations of $m_{jk}, m_{kj}, d_{jk}, d_{jk}, r_j$, and r_k are schematically exhibited in Fig. 1. The degree of similarity between the antibiotics is quantified by m_{jk} and m_{kj}, and the degree of mutual influence on the antibiotics can be evaluated by r_j and r_k. That is, the size and central position of the circles are regarded as important as the circles' placement. In Fig. 1, d_{jk}, d_{jk}, r_j, and r_k are drawn with solid lines, and m_{jk} and m_{kj} are highlighted with dotted lines. To easily visualize the relationship between the *CRR* and the antibiotic drugs, all the results were finally presented in a two-dimensional configuration. In the process of successively compressing the multidimensional data vectors to the final two dimensions, we acted under the condition of selecting the case where the stress of the data was the smallest. When "tying" of the data occurred, we repeated the calculation as changes in the initial conditions. The theoretical details of the radius-distance model and its procedures are described elsewhere [13–15].

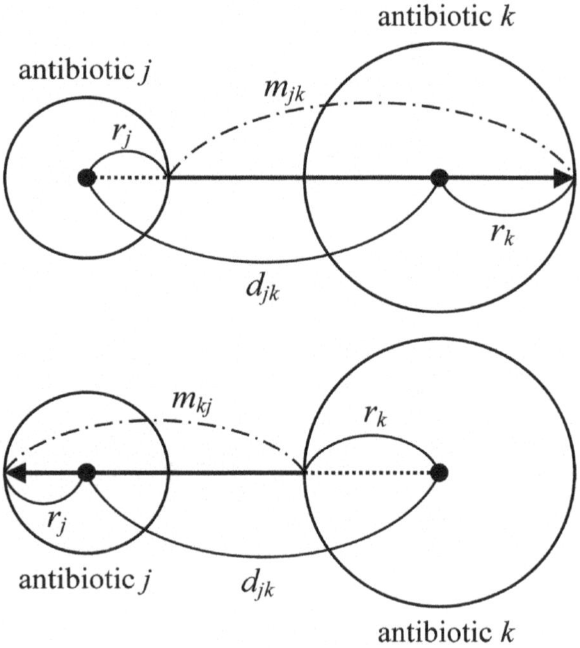

Fig. 1 Schematic illustrations of the geometric relations of $m_{jk}, m_{kj}, d_{jk}, d_{jk}$, r_j, and r_k

5 Results

5.1 Two-Dimensional Configuration

Figure 2a–d exhibits the two-dimensional configurations by the radius-distance model and represents the results before and after the relocation of the hospital. Figure 2a corresponds to the configuration of 2 years BR from 1, Jan. 2013 to 30, June 2014. Figure 2b is the configuration of the 1 year BR from 1, Jan. 2014 to 30, June 2015. Figure 2c is the configuration just AR from 1, July 2015 to 31, Dec. 2016. Figure 2d is the configuration using the last data of AR from 1, July 2017 to 31, Dec. 2018. In a similar way, the monthly change of the configurations is chronologically exhibited in the Appendix. The numbers (1)–(23) in the Appendix correspond to the month after the relocation.

All the configurations are distinguished by colors. PIPC and PIPC.TAZ are drawn in green. CAZ, CFPM, and CPZ.SBT are drawn in blue. IPM and MEPM are drawn in purple. AZT is drawn in light blue. AMK and GM are drawn in orange. CPFX and

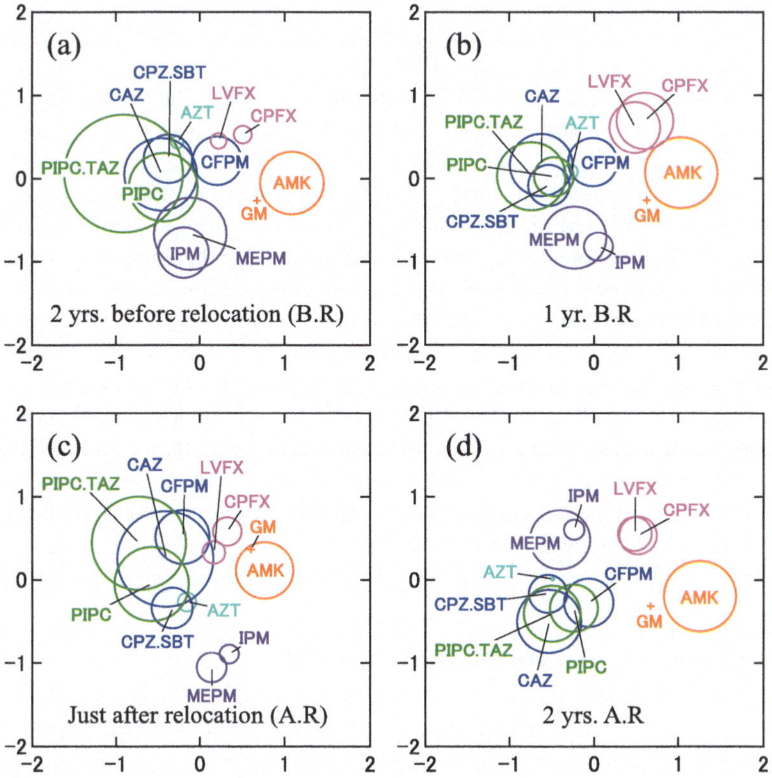

Fig. 2 Two-dimensional configurations using the data of *CRR* of *P. aeruginosa*

LVFX are drawn in pink. We can find that the drugs that inhibit cell-wall synthesis (penicillins, cephems, carbapenems, and monobactams) are visually highlighted in the "cold" colors of green, blue, and purple, respectively. In contrast, the drugs that inhibit protein synthesis (aminoglycosides) and DNA synthesis (new-quinolones) are clearly distinguished by the "warm" colors of orange and pink, respectively.

In the radius-distance model, the result is allowed to rotate around the origin and can also be flipped horizontally as long as the distances among the data are unchanged. Hence, we adjusted the data positions so that the warm color data are placed on the right-hand side and the cold color data are placed on the left-hand side of the figure. As a consequence, aminoglycosides (orange circle) were always on the right side of the figure, and monobactams (light blue circles), penicillins (green circles), and cephems (blue circles) were on the left side of the figure. New-quinolones (pink circles) were placed in the middle, and carbapenems (purple circles) were slightly below the middle.

5.2 Circle Positions and Antibiotic Strains

As shown in Fig. 2a–d, the orange circles (AMK and GM) were located independently to the right, while the green circles (PIPC, PIPC.TAZ), blue circles (CAZ, CFPM, CPZ.SBT), and light blue circle (AZT) were close together. The purple circles (IPM and MEPM) and pink circles (CPFX and LVFX) were close to the green, blue, and light blue circles. From these results, we can find that there were four groups classified in the configurations, and antibiotic drugs that originated from the same strain group were located in close positions. From the chronological results in the Appendix, these circle positions were rather stable during the analyses period. Only in the case of carbapenems (purple circles) did the positions move after the 20th month after the relocation.

As to the circle sizes, the circle of AMK was always far larger than that of GM. The circle of MEPM was constantly larger than that of IPM. The circle of PIPC.TAZ was constantly larger than that of PIPC, except for a temporal disturbance in the 16th month. The circle of CPZ.SBT was constantly smaller than those of CAZ and CFPM. The circles of CPFX and LVFX were of almost equal size during the data collection period. Thus, the circle size relationships in the same strain group were fairly stable during the period of the analyses.

5.3 Influence of the Hospital Relocation

The influence of the hospital relocation was clarified by the monthly change of the circle size in the two-dimensional configurations. Figure 3 represents the changes before and after the relocation and how long the influence of the relocation on the cross-resistance continued. As shown in Fig. 3, we can find that the radius sharply

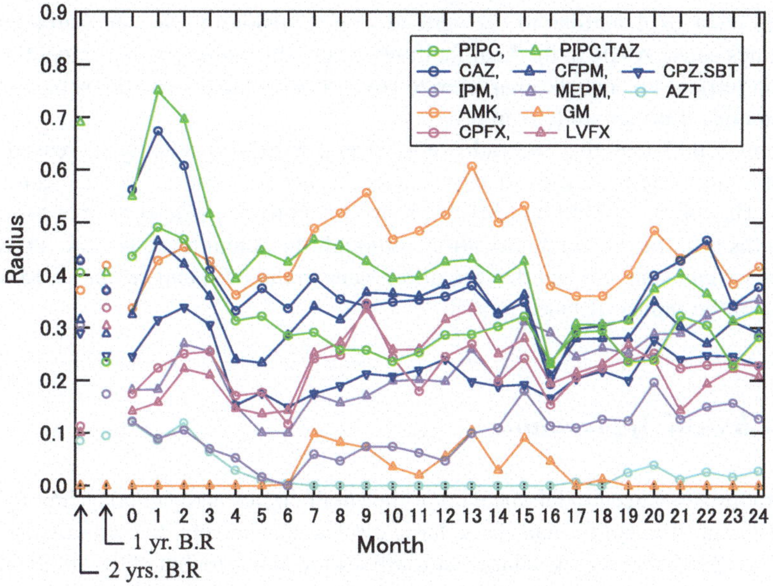

Fig. 3 Monthly radius change of the circles in the two-dimensional configurations

expanded during the 1st and 2nd months of AR. In particular, the increase of penicillins (green circles) and cephems (blue circles) was remarkable. Such drastic changes were temporal, however, and disappeared within 5–6 months. After that, some radii remained gradually increased, and others stayed flat until the 15th month. These results may reflect that the resistant bacteria carried by humans immediately spread in the new hospital and returned to the same initial situation as in the old hospital.

6 Discussion

6.1 Classification by the Circles and Antibiotic Strains.

We expected three groups to be classified by the strains' mechanism of action, but the results showed that there were four groups of circles classified in the configurations. As described in Sect. 3.4, the carbapenems (IPM and MEPM) were placed apart from those of penicillins, cephems, and monobactams, regardless of having the common mechanism of inhibiting cell-wall synthesis. Those groups continued for a long time in keeping almost the same placement throughout the data collection period. Such stable placements suggest that the horizontal axis of the configuration represents the mechanism of action of the antibiotic drugs. On the other hand, it was difficult to apply clear explanations to the vertical axis. Considering that the carbapenems sometimes cause AMR in medical facilities where this antibiotic is frequently used [33], the vertical axis may be characterized by unknown factors caused by the overuse of antibiotics. Further study is in progress.

Figure 3 indicates that the radii of IPM and MEPM gradually increased after the 6th month, the responses of which were in distinct contrast to other antibiotic circles. The circles of IPM and MEPM moved slightly to the upper central location 20 months after the hospital relocation, although the actual reasons were unknown. At least we can state the fact that the differences in the *CRR* can be classified by the method of the radius-distance model.

6.2 Clinical Applications

In the radius-distance model, the data of the small circles is regarded as having little similarity (or affinity) to that of the large circles. Conversely, the data of the large circles is regarded as having a large similarity (or affinity) to that of the small circles. The similar size of circles is regarded as having an equal degree of similarity to each other. Applying these relationships to the antibiotic drugs in this study, the following relationship can be deduced.

When a smaller circle antibiotic is used after larger circle antibiotics, the degree of the emergence of resistant bacteria is expected to be greater than in the case of the opposite procedure in which a larger circle antibiotic is used after smaller circle antibiotics.

Concretely speaking, the following comments (a)–(d) apply:

a. When a smaller circle antibiotic is used after larger circle antibiotics, the smaller circle antibiotic which is used later is more likely to emerge as resistant bacteria.
b. When a larger circle antibiotic is used after smaller circle antibiotics, the larger circle antibiotic which is used later is less likely to emerge as resistant bacteria.
c. When a smaller circle antibiotic is used before larger circle antibiotics, the larger circle antibiotics which are used later are less likely to emerge as resistant bacteria.
d. When a larger circle antibiotic is used before small circle antibiotics, the smaller circle antibiotics which are used later are more likely to emerge as resistant bacteria.

6.3 Limitations of the Analyses

In the early stage when drug-resistant bacteria have not yet diffused so much, the prescription is concentrated on a specific antibiotic drug. As a result, most of the cell components of the *CRR* matrix are often zero, since other antibiotics are rarely used. When analyzed, such a matrix will result in the configuration that many circles overlap in the same location. That is, the radius-distance model is useful only when there are few zero components included in the matrix.

In order to avoid cases in which the matrix component tends to be zero, we should attempt the following two approaches: the collection of data should be set for as long a period as possible, and the collection of data should be carried out at many different medical institutions and hospitals. In the former case, seasonal fluctuations would disappear in the results. In the latter case, it would be impossible to examine the fluctuations in each hospital. These problems are attributable to the data collecting and are essential limitations in this analysis.

Our method cannot examine the details of medical information itself. For example, a temporary change in the radius was found in the 16th month in Fig. 3. This was remarkable in AMK, PIPC, PIPC.TAZ, CAZ, and CFPM. In the next month, the radii of PIPC, PIPC.TAZ, CAZ, and CFPM recovered to the level of the 16th month. Some antibiotics, such as AMK, showed a prolonged recovery until the 18th month and beyond. The radii of CPFX and LVFX fluctuated, and the changes were small. The radii of AZT and GM remained small during the data collection period. The causes of these changes, however, were unknown.

We used a model of "one-mode two-way data" in the analysis in this study [24]. We independently analyzed the relationships between antibiotics by one-month intervals and considered that there was no chronological influence on the individual analyses. That is, we assumed that the coordinate vectors did not lie in the same space. On the other hand, when comparing the location of antibiotics and the temporal variation of the radius for each antibiotic, it is more appropriate to use a two-mode three-way data analysis model. A method to collectively analyze the relationships between antibiotics and the survey time point has been proposed [4, 24]. These methods are worth adopting in our research in the future. In addition, Okada has proposed another type of method of asymmetric multidimensional scaling [26, 27]. Addressing these statistical techniques is left for further study.

6.4 *Implications for Progress of Epidemiological Study*

In April 2016, Japan's Cabinet decided to reduce the use of antibiotics to less than two-thirds of the amount used in 2016 by 2020 [32]. This decision may help slow the emergence of drug-resistant bacteria, but it cannot fundamentally solve the problem of AMR. Many previous studies emphasize that an early warning system for resistant bacteria should be established. We are expecting the proposal of new kinds of epidemiological systems by jointly using the data collection system of Chans and data science.

The process of determining the first or second-chosen antibiotics must follow strict guidelines. When sufficient therapeutic effects cannot be obtained with a single antibiotic, it is common to use two or more antibiotics in combination. Hence, it may be unrealistic yet to give priority to antimicrobial drugs by our research method with regards to the drugs recommended by the guidelines. However, the method presented in this study is a new attempt, and we think this method can be utilized to reconfirm the effectiveness of conventional combination antibiotic therapy, which has been proposed by validated procedures of guideline.

Few studies in the medical and pharmaceutical fields have adopted data mining for the purpose of preventive medicine, although it is a very popular technique in the field of social science. The major reason for the persistence of the classical methods, such as t-test, χ^2-test, and multiple logistic regression analysis, is because they have already been utilized for a long time and the evaluation process has been widely recognized throughout the world. This may be a matter of course in medical and pharmaceutical fields, where safety is the top priority, but we think there may be small hope to discover new findings on the background factors of medical problems simply by using the classical methods. We think new methods of data mining, such as the radius-distance model, should be more widely applied in analyzing medical big data, and they can contribute to the aspect of drug administration by medical staff.

7 Conclusion

We examined the practical utility of the method of the radius-distance model as focused on pharmaceutical data of the cross-resistance rate (*CRR*) of *Pseudomonas aeruginosa*. Through chronological monitoring of the configurations on a two-dimensional plane, we clarified that most drugs can be classified according to the same strain groups of the antibiotics. Furthermore, different antibiotics with a similar mechanism of action can also be distinguished by this method. The configurations and placement were rather stable. The radius-distance model has the potential for discovering new characteristics or specific features of data from a new perspective in the fields of medical and pharmaceutical research.

8 Appendix: Time-Series Change of Two-Dimensional Configurations Using the Data of *CRR* of *P. Aeruginosa*

The numbers (1)–(23) correspond to the age of month after the hospital relocation.

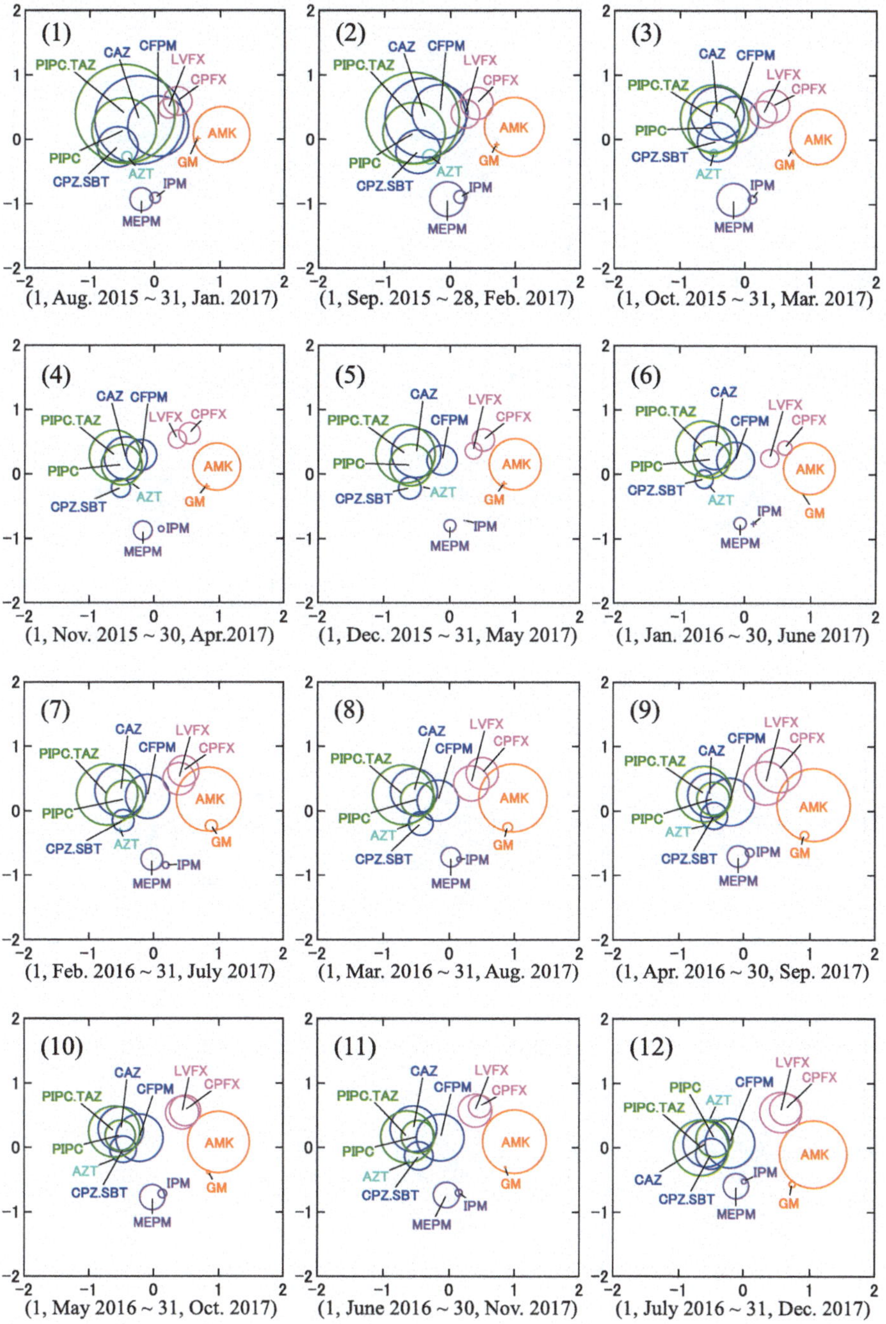
(1)
(1, Aug. 2015 ~ 31, Jan. 2017)
(2)
(1, Sep. 2015 ~ 28, Feb. 2017)
(3)
(1, Oct. 2015 ~ 31, Mar. 2017)
(4)
(1, Nov. 2015 ~ 30, Apr.2017)
(5)
(1, Dec. 2015 ~ 31, May 2017)
(6)
(1, Jan. 2016 ~ 30, June 2017)
(7)
(1, Feb. 2016 ~ 31, July 2017)
(8)
(1, Mar. 2016 ~ 31, Aug. 2017)
(9)
(1, Apr. 2016 ~ 30, Sep. 2017)
(10)
(1, May 2016 ~ 31, Oct. 2017)
(11)
(1, June 2016 ~ 30, Nov. 2017)
(12)
(1, July 2016 ~ 31, Dec. 2017)
CAZ
CFPM
LVFX
CPFX
PIPC.TAZ
PIPC
AMK
GM
AZT
CPZ.SBT
IPM
MEPM

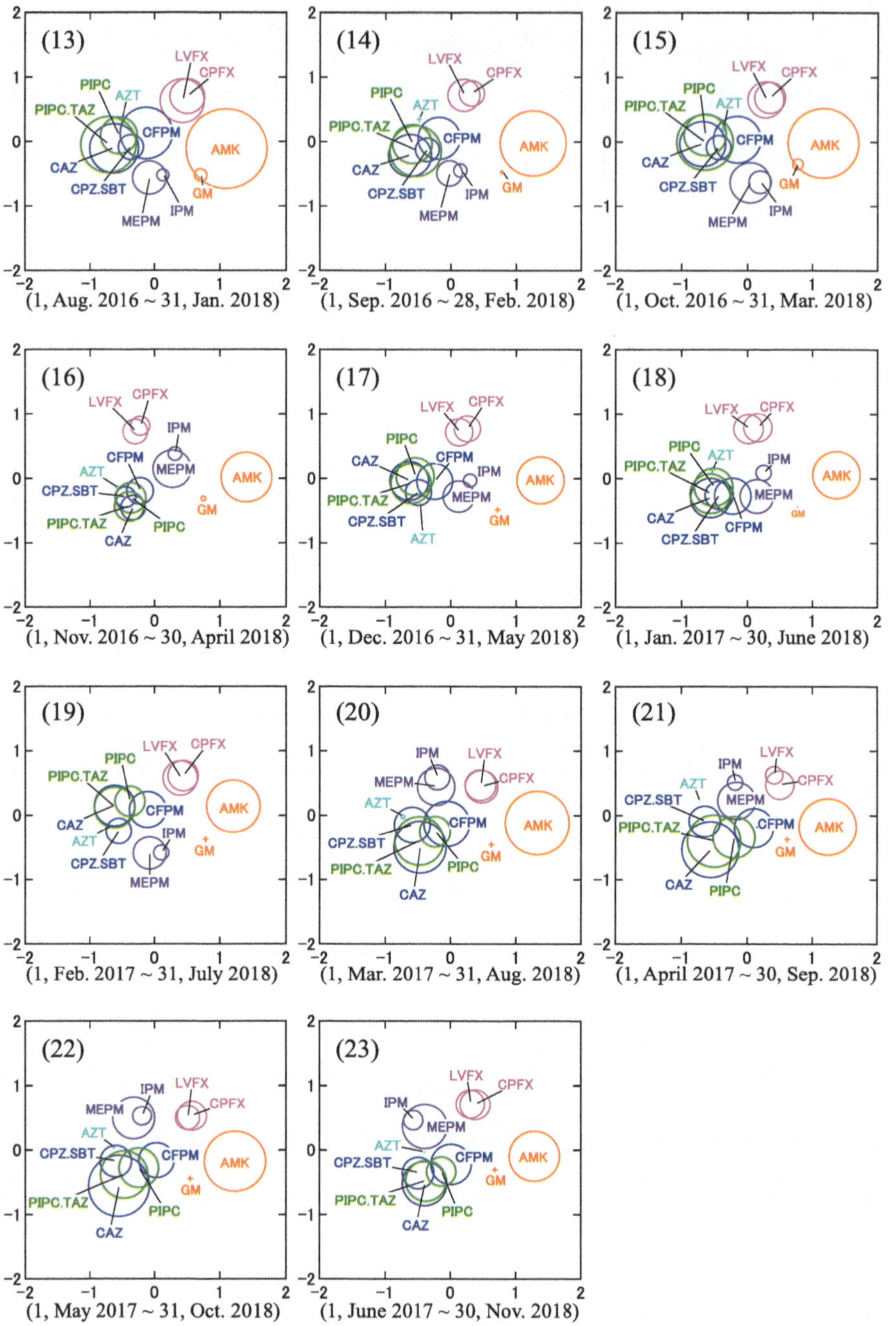

Acknowledgements This research was supported by the research funding of the Institute of Frontier Science and Technology, Okayama University of Science. This paper was proofread in English by Prof. Christopher Carman of the University of Occupational and Environmental Health. Akinori Okada, Prof. Emeritus of the Rikkyo University, gave us advice on an earlier version of this paper. We would like to express our deepest gratitude for their support.

References

1. Ahmad, S. I., Malak, H. A., & Abulreesh, H. H. (2021). Environmental antimicrobial resistance and its drivers: A potential threat to public health. *Journal of Global Antimicrobial Resistance, 27*, 101–111.
2. Aitken, C., & Jeffries, D. J. (2001). Nosocomial spread of viral disease. *Clinical Microbiology Reviews, 14*(3), 528–546.
3. Antimicrobial resistance: no action today, no cure tomorrow, Remarks at a high-level panel on World Health Day 2011, World Health Organization, 7 April 2011. https://www.who.int/director-general/speeches/detail/antimicrobial-resistance-no-action-today-no-cure-tomorrow
4. Bove, G., Okada, A., & Vicari, D. (2021). *Methods for the analysis of asymmetric proximity data*. Springer Nature.
5. Cell wall synthesis inhibitors: Penicillins, OSMOSIS from ELSEVIER. https://www.osmosis.org/learn/Cell_wall_synthesis_inhibitors:_Penicillins
6. "Cephalosporin" from Wikipedia, the free encyclopedia. https://en.wikipedia.org/wiki/Cephalosporin
7. Cuttelod, M., Senn, L., & Terletskiy V., et al. (2011). Molecular epidemiology of *Pseudomonas aeruginosa* in intensive care units over a 10-year period (1998–2007). *Clinical Microbiology and Infection, 17*(1), 57–62. https://doi.org/10.1111/j.1469-0691.2010.03164.x
8. Drugs. Com. (2022). Drug Interactions between amikacin and gentamicin, https://www.drugs.com/drug-interactions/amikacin-with-gentamicin-153-0-1168-0.html#:~:text=Using%20gentamicin%20together%20with%20amikacin,or%20have%20preexisting%20kidney%20disease
9. European Medicines Agency. (2018). Disabling and potentially permanent side effects lead to suspension or restrictions of quinolone and fluoroquinolone antibiotics, EMA 795349. https://www.ema.europa.eu/en/news/disabling-potentially-permanent-side-effects-lead-suspension-restrictions-quinolone-fluoroquinolone#:~:text=These%20serious%20side%20effects%20include,and%20altered%20taste%20and%20smell.
10. Fazeli, H., Akbari, R., Moghim, S., et al. (2012). *Pseudomonas aeruginosa* infections in patients, hospital means, and personnel's specimens. *Journal of Research in Medical Sciences: The Official Journal of Isfahan University of Medical Sciences, 17*(4), 332–337.
11. Godijk, N. G., Bootsma, M. C. J., & Bonten, M. J. M. (2022). Transmission routes of antibiotic resistant bacteria: A systematic review. *BMC Infectious Diseases, 22*, 482. https://doi.org/10.1186/s12879-022-07360-z
12. Guerrini, V., Rosa, M. D., & Pasquini, S., et al. (2013). New fluoroquinolones active against fluoroquinolones-resistant Mycobacterium tuberculosis strains. *Tuberculosis (Edinb), 93*(4), 405–411. https://doi.org/10.1016/j.tube.2013.02.017
13. Hatsuda, Y., Maki, S., Ishimaki, Y., et al. (2019). Development of a new correlation diagram to visualize the trends of antimicrobial cross-resistance. *Journal of Biomedical Soft Computing and Human Sciences, 24*(1), 39–48.
14. Hatsuda, Y., Ishizaka, T., Koizumi, N., et al. (2020). Monitoring antimicrobial cross-resistance with cross-resistance rate correlation diagrams: Changes in antibiotic susceptibility of *Pseudomonas aeruginosa* due to hospital relocation. *Journal of Clinical Pharmacy and Therapeutics, 45*, 13296. https://doi.org/10.1111/jcpt.13296
15. Hatsuda, Y., Maki, S., & Ishizaka, T., et al. (2021). Visualization of cross-resistance between antimicrobials by asymmetric multidimensional scaling. *Journal of Clinical Pharmacy and Therapeutics, 46*, 13564. https://doi.org/10.1111/jcpt.13564

16. Kapoor, G., Saigal, S., & Elongavan, A. (2017). Action and resistance mechanisms of antibiotics: A guide for clinicians. *Journal of Anaesthesiology, Clinical Pharmacology, 33*(3), 300–305. https://doi.org/10.4103/joacp.JOACP_349_15, https://www.ncbi.nlm.nih.gov/pmc/articles/PMC5672523/
17. Kovačević, T., Avram, S., & Milaković, D. et al (2016). Therapeutic monitoring of amikacin and gentamicin in critically and noncritically ill patients. *Journal of Basic and Clinical Pharmacy, 7*(3), 65–69. https://doi.org/10.4103/0976-0105.183260, https://www.ncbi.nlm.nih.gov/pmc/articles/PMC4910469/#:~:text=Results%3A,patients%20(P%20%3D%200.457
18. Moradali, M. F., Ghods S., & Rehm B. H. A. (2017). *Pseudomonas aeruginosa* lifestyle: a paradigm for adaptation, survival, and persistence. *Frontiers in Cellular and Infection Microbiology, 7*, 39. https://doi.org/10.3389/fcimb.2017.00039
19. Mukae, H., Kawanami, T., & Yatera, K., et al. (2014). Efficacy and safety of levofloxacin in patients with bacterial pneumonia evaluated according to the new Clinical Evaluation Methods for New Antimicrobial Agents to Treat Respiratory Infections (Second Version). *Journal of Infection and Chemotherapy, 20*(7), 417–422. https://doi.org/10.1016/j.jiac.2014.03.009
20. Nielsen, S. S., Bicout, D. J., & Calistri, P., et al. (2022). Assessment of listing and categorisation of animal diseases within the framework of the Animal Health Law (Regulation (EU) No 2016/429): antimicrobial-resistant *Pseudomonas aeruginosa* in dogs and cats. *EFSA Journal, 20*(5), 7310. https://doi.org/10.2903/j.efsa.2022.7310
21. Lila, G., Mulliqi, G., Raka, L., et al. (2018). Molecular epidemiology of *Pseudomonas aeruginosa* in university clinical center of Kosovo. *Infection and Drug Resistance, 11*, 2039–2046.
22. Okada, A., & Imaizumi, T. (1987). Nonmetric multidimensional scaling of asymmetric proximities. *Behaviormetrika, 14*(21), 81–96.
23. Okada, A., & Imaizumi, T. (1994). *Method of PC multidimensional scaling configuration* (pp. 10–29). Kyoritsu Shuppan Co., Ltd. (*in Japanese*). ISBN4–320–01472–3.
24. Okada, A., & Imaizumi, T. (1997). Asymmetric multidimensional scaling of two-mode three-way proximities. *Journal of Classification, 14*, 195–224.
25. Okada, A., & Imaizumi, T. (2003). *Developing a layout of a supermarket through asymmetric multidimensional scaling and cluster analysis of purchase data between data science and applied data analysis* (pp. 587–594). Springer.
26. Okada, A. (2011). Supekutoru bunnkai niyoru gaikokujinn ni taisuru shinnkinnkann no bunnseki [Analyzing social affinity for foreigners by spectral decomposition: Asymmetric multidimensional scaling of EASS 2008 Data]. *GSS Research Series, 11*, 119–128. (In Japanese).
27. Okada, A., & Tsurumi, H. (2012). Asymmetric multidimensional scaling of brand switching among margarine brands. *Behaviormetrika, 39*, 111–126.
28. Pham, T. M., Kretzschmar, M., Bertrand, X., & Bootsma, M. (2019). Tracking *Pseudomonas aeruginosa* transmissions due to environmental contamination after discharge in ICUs using mathematical models. *PLoS Computational Biology, 15*(8), e1006697. https://doi.org/10.1371/journal
29. Pham, T. D. M., Ziora, Z. M., & Blaskovich, M. A. T. (2019). Quinolone antibiotics. *Medicinal Chemistry Communication, 10*, 1719–1739. https://doi.org/10.1039/C9MD00120D, https://pubs.rsc.org/en/content/articlehtml/2019/md/c9md00120d#:~:text=The%20quinolone%20class%20of%20antibiotics,supercoiling%20required%20for%20DNA%20synthesis
30. Pham, T. M., Büchler, A. C., & Voor in 't holt, A. F., et al. (2022). Routes of transmission of VIM-positive *Pseudomonas aeruginosa* in the adult intensive care unit-analysis of 9 years of surveillance at a university hospital using a mathematical model. *Antimicrobial Resistance and Infection Control, 11*, 55. https://doi.org/10.1186/s13756-022-01095-x
31. Poole, K. (2005). Aminoglycoside resistance in *Pseudomonas aeruginosa. Antimicrobial Agents and Chemotherapy, 49*(2), 479–487. https://doi.org/10.1128/AAC.49.2.479-487.2005, https://www.ncbi.nlm.nih.gov/pmc/articles/PMC547279/
32. Public Awareness Council for Promotion of Antimicrobial Resistance (AMR) Countermeasures, International Infectious Disease Control Coordination Office, Japan Cabinet, October

18, 2016. (*in Japanese*). https://www.kantei.go.jp/jp/singi/kokusai_kansen/amr_taisaku/dai1/gijiroku.pdf, https://www.kantei.go.jp/jp/singi/kokusai_kansen/amr_taisaku/dai4/index.html
33. Queenan, A. M., & Bush, K. (2007). Carbapenemases: the versatile β-Lactamases. *Clinical Microbiology Reviews, 20*(3), 440–458. https://doi.org/10.1128/CMR.00001-07, https://www.ncbi.nlm.nih.gov/pmc/articles/PMC1932750/
34. Reuter, S., Sigge, A., Wiedeck, H., & Trautmann, M. (2002). Analysis of transmission pathways of *Pseudomonas aeruginosa* between patients and tap water outlets. *Critical Care Medicine, 30* (10), 2222–2228. https://doi.org/10.1097/00003246-200210000-00008
35. Safeyah, A., Maqbul, M. S., & Safeyah, M. (2018). Role of *Pseudomonas aeruginosa* in nosocomial infection. *International Journal of Health Sciences, 6*(1), 211–217.
36. Schmidt T. M. (Eds.). (2019). *Encyclopedia of microbiology* (4th ed., pp. 127–139). Elsevier. ISBN 978–0–12–811737–8
37. Spagnolo, A. M., Sartini, M., & Cristina, M. L. (2021). *Pseudomonas aeruginosa* in the healthcare facility setting. *Reviews in Medical Microbiology, 32*, 169–175.
38. The Society for Bacterial Drug Resistance, Gunma University Graduate School of Medicine, Japan. http://yakutai.dept.med.gunma-u.ac.jp/society/QandA.html
39. Website of "Shiny from R studio". https://shiny.rstudio.com
40. Willmann, M., Bezdan, D., & Zapata, L., et al. (2015). Analysis of a long-term outbreak of XDR *Pseudomonas aeruginosa*: a molecular epidemiological study. *Journal of Antimicrobial Chemotherapy, 70*(5), 1322–1330. https://doi.org/10.1093/jac/dku546

The Multivariate Power-Gamma Distribution Using Factor Analysis Models

Haruhiko Ogasawara

Abstract Moments and the moment-generating function of the univariate power-gamma distribution are obtained. This distribution was proposed elsewhere when a power of a variable follows the gamma distribution. The moments give those of the marginal distributions of the multivariate power-gamma distribution introduced in this paper. The multivariate power-gamma distribution is based on the multivariate gamma distribution when each variable follows the distribution of the sum of independent gammas with a similar pattern found in factor analysis. Then, power transformations of the variables give the multivariate power-gamma. The probability density function (pdf) of the distribution is derived using integral or series expressions. The one-factor models with some restrictions yield simplified pdfs.

Keywords Multivariate gamma · Power-gamma · Moments · Factor analysis · One-factor model · Series expression

1 Introduction

It is well-known that the gamma distribution has a variety of special cases, e.g., the exponential and chi-squares with positive integers or real values for the degrees of freedom (df) corresponding to the shape parameter ([12], Chap. 17). Transformations of gamma distributed variables, e.g., the inverse gamma are also used. Among them, the inverse chi-square ([4], Sect. 3.2.2), chi and inverse chi are relatively well-known as transformations of the chi-square. Especially, the inverse chi is recently focused on due to the convenience to yield the t-distribution by the product of a scaled inverse chi and an independent standard normal ([13]; [21], Chap. 9).

The above transformations are seen as special cases of power transformations of the gamma. The power-gamma distribution was introduced by Ogasawara [21], Chap. 9 as a special case of the Amoroso [2] distribution, whose probability density function (pdf, see [7], Eq. (1)) is

H. Ogasawara (✉)
Otaru University of Commerce, 3-5-21 Midori, Otaru 047-8501, Japan
e-mail: emt-hogasa@emt.otaru-uc.ac.jp

© The Author(s), under exclusive license to Springer Nature Singapore Pte Ltd. 2023
A. Okada et al. (eds.), *Facets of Behaviormetrics*, Behaviormetrics: Quantitative Approaches to Human Behavior 4, https://doi.org/10.1007/978-981-99-2240-6_13

$$g_{\text{Amoroso}}(x|a,\theta,\alpha,\beta) = \frac{|\beta/\theta|}{\Gamma(\alpha)}\left(\frac{x-a}{\theta}\right)^{\alpha\beta-1}\exp\left\{-\left(\frac{x-a}{\theta}\right)^{\beta}\right\}$$
$$(x \geq a \text{ if} \theta > 0;\ x \leq a \text{ if} \theta < 0;\ \ \beta \in \mathrm{R},\ \beta \neq 0;\ \ \alpha > 0;\ \ \theta \in \mathrm{R},\ \theta \neq 0).$$

Let Y be power-gamma distributed, which is denoted by $Y \sim \text{Power-}\Gamma(\alpha,\beta,\gamma)$. Then, the pdf of $Y = y$ is

$$\begin{aligned} g_{\text{Power-}\Gamma}(y|\alpha,\beta,\gamma) &= \frac{\beta^{\alpha} y^{\gamma(\alpha-1)}|\gamma|y^{\gamma-1}}{\Gamma(\alpha)}\exp(-\beta y^{\gamma}) \\ &= \frac{\beta^{\alpha}|\gamma|y^{\gamma\alpha-1}}{\Gamma(\alpha)}\exp(-\beta y^{\gamma}) \qquad (1) \\ &(0 < y < \infty,\ 0 < \alpha < \infty,\ 0 < \beta < \infty,\ -\infty < \gamma < \infty,\ \gamma \neq 0). \end{aligned}$$

The power-gamma is also seen as a reparametrization of Stacy's [27] generalized gamma, which takes the pdf using our notation:

$$\begin{aligned} g_{\text{Generalized-}\Gamma}(y|\alpha,\beta^{*},\gamma) &= \frac{\gamma y^{\alpha-1}}{\beta^{*\alpha}\Gamma(\alpha/\gamma)}\exp\{-(y/\beta^{*})^{\gamma}\} \qquad (2) \\ &(0 < y < \infty,\ 0 < \alpha < \infty,\ 0 < \beta^{*} < \infty,\ 0 < \gamma < \infty). \end{aligned}$$

It is seen that the power-gamma has an extension of the non-zero real power over the positive power of the generalized gamma. Note that β^{*} in (2) is a scale parameter while $\beta(= 1/\beta^{*})$ in (1) is the rate parameter when $\gamma = 1$ (the term "rate" originates from the occurrence rate of an event corresponding to the Poisson parameter or the inverted scale parameter of the exponential distribution).

It is found that special cases of the power-gamma are usual gamma ($\gamma = 1$), inverse gamma ($\gamma = -1$) and members of the power chi-square subfamily, e.g., chi-square ($\alpha = \nu/2,\ \beta = 1/2,\ \gamma = 1$), inverse chi-square ($\alpha = \nu/2,\ \beta = 1/2,\ \gamma = -1$), chi ($\alpha = \nu/2,\ \beta = 1/2,\ \gamma = 2$) and inverse chi ($\alpha = \nu/2,\ \beta = 1/2,\ \gamma = -2$) with ν being a typically integer-valued df. Other special cases are half-normal ($\alpha = 1/2,\ \beta = 1/(2\sigma^2),\ 0 < \sigma^2 < \infty,\ \gamma = 2$; a special case of the scaled chi), basic power half-normal ($\alpha = 1/2,\ \beta = 1/(2\sigma^2),\ 0 < \sigma^2 < \infty$), exponential ($\alpha = 1,\ \gamma = 1$), power exponential ($\alpha = 1$; for the Box–Cox power exponential, see [24, 28]) and a parametrization of the Weibull ($\alpha = \gamma > 0$). The basic power half-normal proposed above is seen as a reparametrization of a similar case given by Gómez and Bolfarine [10]. For Box–Cox symmetric or more generally elliptical distributions including power transformations, see [8, 18]. For an associated review, (see [12], Sect. 8.7).

Bivariate and multivariate extensions of the gamma distribution have also been given in various forms ([14], Chap. 48). Among the extensions satisfying marginal gammas, [6] gave a bivariate gamma with a form of the sums of a single common gamma and unique independent ones with equal-scale parameters. The common and unique independent gammas in this form correspond to the common and unique factors in the one-factor model of exploratory factor analysis (EFA; for FA, see, e.g., [11], Chap. 2, [5], pp. 226–232).

One of the properties of FA including latent variables is the inflation of the dimensionality of associated variables. In the Cherian bivariate gamma, the number of independent gammas is 3, which is given by the single common gamma and two unique gammas and is larger than the number of two manifest or observable gammas. In the one-factor model of EFA with p observable variables, we have a single common factor and p unique factors, yielding an inflation of the dimensionality by 1 as in the Cherian model. Principal component analysis (PCA) is used with similar purposes as FA. However, PCA is not a latent variable model since the number of components including minor ones is equal to that of observable ones yielding no inflation of dimensionality.

The Cherian bivariate gamma has been extended to the multivariate cases by Ramabhadran [23] when independent exponentially distributed variables are used, and by Prékopa and Szántai [22] for various patterns for sums of independent gammas corresponding to various patterns of factor loadings in FA (see also [14], Chap. 48, Sect. 2.2). Mathai and Moschopoulos ([16], Definition 1) gave a multivariate gamma of the one-factor type. The term "factorial" corresponding to the FA type was used by Royen [25, 26] in the formulations of the multivariate gammas including the FA-type structures of the inverses of covariance/correlation matrices, which are different from those of the current paper. Recently, Furman ([9], Theorem 3.1) showed a multivariate gamma with p fully or partially common gammas and a single unique gamma for p observable variables, where the pattern of the "ladder type" corresponding to that of the factor loading/pattern matrix in FA is employed.

In order to have, e.g., the pdf and moment-generating function (mgf) in the multivariate gammas of the FA type, some method of dimension reduction using single or multiple integrals for partialing out extra gamma(s) is required, which is called "trivariate reduction" or "variables-in-common method" in the bivariate case ([3], Chap. 7) and "multivariate reduction" for general cases ([9], Sect. 2).

The multivariate gammas using the same number of independent gammas as that for observable variables were given by Mathai and Moschopoulos ([17], Theorem 1.1), which is seen as a PCA-type model. When one of the two independent "unique" gammas is omitted in the Cherian model, the bivariate gamma of the PCA type is obtained, where the single common factor becomes equal to one of the two observable variables with no trivariate reduction required. One of the limitations of models of the PCA type is the unexchangeability of observable variables, which is easily found in the bivariate gamma of this type mentioned above. On the other hand, an advantage for the PCA type is that the vector of independent chi-squares is reconstructed from that of observable gammas with the inverse of a square matrix corresponding to the factor pattern matrix in FA.

The purpose of this paper is to introduce the multivariate power-gamma distribution with some properties using the multivariate gammas of the FA type. The remainder of this paper is organized as follows. In Sect. 2, moments of the univariate power-gamma are shown. Section 3 gives the multivariate power-gamma of the FA type with some properties. In Sect. 4, three special cases of the one-factor models are dealt with, where series or closed expressions of the pdfs are derived with numerical illustrations for bivariate cases.

2 Moments of the Power-Gamma Distribution

The pdf of the power-gamma defined in (1) gives the following properties.

Result 1 Let $Y \sim \text{Power-}\Gamma(\alpha, \beta, \gamma)$. Then, we have

$$\begin{aligned}
\mathrm{E}(Y^k) &= \int_0^\infty \frac{\beta^\alpha |\gamma| y^{\gamma\alpha-1} y^k}{\Gamma(\alpha)} \exp(-\beta y^\gamma)\mathrm{d}y \\
&= \int_0^\infty \frac{\beta^\alpha |\gamma| y^{\gamma\{\alpha+(k/\gamma)\}-1}}{\Gamma(\alpha)} \exp(-\beta y^\gamma)\mathrm{d}y \\
&= \frac{\Gamma\{\alpha+(k/\gamma)\}\beta^\alpha}{\Gamma(\alpha)\beta^{\alpha+(k/\gamma)}} \int_0^\infty \frac{\beta^{\alpha+(k/\gamma)}|\gamma| y^{\gamma\{\alpha+(k/\gamma)\}-1}}{\Gamma\{\alpha+(k/\gamma)\}} \exp(-\beta y^\gamma)\mathrm{d}y \\
&= \frac{\Gamma\{\alpha+(k/\gamma)\}}{\Gamma(\alpha)\beta^{k/\gamma}}, \\
&\quad 0 < \alpha + (k/\gamma),
\end{aligned}$$

where k is real-valued with $0 < \alpha + (k/\gamma)$.

$$\begin{aligned}
\mathrm{E}(Y) &= \frac{\Gamma\{\alpha+(1/\gamma)\}}{\Gamma(\alpha)\beta^{1/\gamma}}, \\
\mathrm{var}(Y) &= \frac{1}{\beta^{2/\gamma}}\left[\frac{\Gamma\{\alpha+(2/\gamma)\}}{\Gamma(\alpha)} - \frac{\Gamma^2\{\alpha+(1/\gamma)\}}{\Gamma^2(\alpha)}\right],
\end{aligned}$$

$$\begin{aligned}
\mathrm{sk}(Y) &= \left[\frac{\Gamma\{\alpha+(3/\gamma)\}}{\Gamma(\alpha)} - 3\frac{\Gamma\{\alpha+(2/\gamma)\}\Gamma\{\alpha+(1/\gamma)\}}{\Gamma^2(\alpha)} + 2\frac{\Gamma^3\{\alpha+(1/\gamma)\}}{\Gamma^3(\alpha)}\right] \\
&\quad \times \left[\frac{\Gamma\{\alpha+(2/\gamma)\}}{\Gamma(\alpha)} - \frac{\Gamma^2\{\alpha+(1/\gamma)\}}{\Gamma^2(\alpha)}\right]^{-3/2} \\
&= \left\{\Gamma\left(\alpha+\frac{3}{\gamma}\right)\Gamma^2(\alpha) - 3\Gamma\left(\alpha+\frac{2}{\gamma}\right)\Gamma\left(\alpha+\frac{1}{\gamma}\right)\Gamma(\alpha) + 2\Gamma^3\left(\alpha+\frac{1}{\gamma}\right)\right\} \\
&\quad \times \left\{\Gamma\left(\alpha+\frac{2}{\gamma}\right)\Gamma(\alpha) - \Gamma^2\left(\alpha+\frac{1}{\gamma}\right)\right\}^{-3/2},
\end{aligned}$$

$$\begin{aligned}
\mathrm{kt}(Y) &= \left[\frac{\Gamma\{\alpha+(4/\gamma)\}}{\Gamma(\alpha)} - 4\frac{\Gamma\{\alpha+(3/\gamma)\}\Gamma\{\alpha+(1/\gamma)\}}{\Gamma^2(\alpha)}\right.\\
&\quad \left.+6\frac{\Gamma\{\alpha+(2/\gamma)\}\Gamma^2\{\alpha+(1/\gamma)\}}{\Gamma^3(\alpha)} - 3\frac{\Gamma^4\{\alpha+(1/\gamma)\}}{\Gamma^4(\alpha)}\right]\\
&\quad \times\left[\frac{\Gamma\{\alpha+(2/\gamma)\}}{\Gamma(\alpha)} - \frac{\Gamma^2\{\alpha+(1/\gamma)\}}{\Gamma^2(\alpha)}\right]^{-2} - 3\\
&= \left[\frac{\Gamma\{\alpha+(4/\gamma)\}}{\Gamma(\alpha)} - 4\frac{\Gamma\{\alpha+(3/\gamma)\}\Gamma\{\alpha+(1/\gamma)\}}{\Gamma^2(\alpha)} - 3\frac{\Gamma^2\{\alpha+(2/\gamma)\}}{\Gamma^2(\alpha)}\right.\\
&\quad \left.+12\frac{\Gamma\{\alpha+(2/\gamma)\}\Gamma^2\{\alpha+(1/\gamma)\}}{\Gamma^3(\alpha)} - 6\frac{\Gamma^4\{\alpha+(1/\gamma)\}}{\Gamma^4(\alpha)}\right]\\
&\quad \times\left[\frac{\Gamma\{\alpha+(2/\gamma)\}}{\Gamma(\alpha)} - \frac{\Gamma^2\{\alpha+(1/\gamma)\}}{\Gamma^2(\alpha)}\right]^{-2},
\end{aligned}$$

where $\Gamma^j(\cdot) = \{\Gamma(\cdot)\}^j$, and $\mathrm{sk}(\cdot)$ and $\mathrm{kt}(\cdot)$ are the skewness and excess kurtosis of a random quantity in parentheses, respectively. The mgf is given by a series form:

$$\begin{aligned}
\mathrm{M}_Y(t) &= \mathrm{E}\{\exp(tY)\} = \int_0^\infty \frac{\beta^\alpha|\gamma|y^{\gamma\alpha-1}}{\Gamma(\alpha)}\exp(-\beta y^\gamma + ty)\,\mathrm{d}y\\
&= \int_0^\infty \frac{\beta^\alpha|\gamma|y^{\gamma\alpha-1}}{\Gamma(\alpha)}\exp(-\beta y^\gamma)\sum_{j=0}^\infty \frac{(ty)^j}{j!}\,\mathrm{d}y\\
&= \sum_{j=0}^\infty \frac{t^j\Gamma\{\alpha+(j/\gamma)\}}{\beta^{j/\gamma}\Gamma(\alpha)j!}\int_0^\infty \frac{\beta^{\{\alpha+(j/\gamma)\}}|\gamma|y^{\gamma\{\alpha+(j/\gamma)\}-1}}{\Gamma\{\alpha+(j/\gamma)\}}\exp(-\beta y^\gamma)\mathrm{d}y\\
&= \sum_{j=0}^\infty \frac{t^j\Gamma\{\alpha+(j/\gamma)\}}{\Gamma(\alpha)\beta^{j/\gamma}j!}\\
&= \sum_{j=0}^\infty \frac{\Gamma(\alpha+j)(t/\beta^{1/\gamma})^j}{\Gamma(\alpha)j!}\frac{\Gamma\{\alpha+(j/\gamma)\}}{\Gamma(\alpha+j)}\\
&= \sum_{j=0}^\infty \frac{(\alpha)_j(t/\beta^{1/\gamma})^j}{j!}\frac{\Gamma\{\alpha+(j/\gamma)\}}{\Gamma(\alpha+j)}\\
&= {}_1F_{0W}\big[\alpha;\,;t/\beta^{1/\gamma};\Gamma\{\alpha+(j/\gamma)\}/\Gamma(\alpha+j),\ j=0,1,\ldots\big]\quad(\gamma>0),
\end{aligned}$$

where $(\alpha)_j = \Gamma(\alpha+j)/\Gamma(\alpha) = \alpha(\alpha+1)\cdots(\alpha+j-1)$ $(j=1,\ 2,\ldots;\ \alpha>0)$ with $(\cdot)_0 = 1$ is the rising or ascending factorial using the Pochhammer notation, and ${}_1F_{0W}(c;\,;z;w_j,\ j=0,1,\ldots) = \sum_{j=0}^\infty (c)_j z^j w_j/j!$ is a weighted hypergeometric function or the weighted negative binomial expansion when the weight for the jth term is w_j, which was defined by Ogasawara ([20], Theorem 1) and reduces to a usual hypergeometric function or the negative binomial expansion ${}_1F_0(c;\,;z) = \sum_{j=0}^\infty (c)_j z^j/j! = (1-z)^{-c}$ (see, e.g., [1], Eq. (5)) when $w_j = 1$ $(j=0,1,\ldots)$. Note that the unweighted case is obtained when $\gamma = 1$, i.e., the case of the untransformed gamma yielding the known mgf $\{1 - (t/\beta)\}^{-\alpha}$.

The expression $\sum_{j=0}^{\infty} \frac{t^j \Gamma\{\alpha+(j/\gamma)\}}{\Gamma(\alpha)\beta^{j/\gamma} j!}$ in the above formula gives the second derivation of $\mathrm{E}(Y^k) = \frac{\Gamma\{\alpha+(k/\gamma)\}}{\Gamma(\alpha)\beta^{k/\gamma}}$ when k is a positive integer, as shown earlier. The mgf does not exist when $\gamma < 0$, which is known in the case of the inverse gamma, i.e., $\gamma = -1$.

When $\gamma = 1/m$ $(m = 1, 2, \ldots)$, the distribution of $Y \sim$ Power-$\Gamma(\alpha, \beta, \gamma)$ reduces to that of $Y = X^m$, where X is gamma distributed with the shape and rate parameters α and β, respectively, which is denoted by $X \sim \mathrm{Gamma}(\alpha, \beta)$. Then, Result 1 or a known property of the gamma gives
$\mathrm{E}(Y^k|\gamma = 1/m) = \frac{\Gamma\{\alpha+(k/\gamma)\}}{\Gamma(\alpha)\beta^{k/\gamma}} = \frac{\Gamma(\alpha+km)}{\Gamma(\alpha)\beta^{km}} = \frac{(\alpha)_{km}}{\beta^{km}}$ $(k = 1, 2, \ldots; m = 1, 2, \ldots)$,

which yields the following results.

Result 2 Let $Y \sim$ Power-$\Gamma(\alpha, \beta, \gamma = 1/m)$ $(m = 1, 2, \ldots)$. Then, we have

$$\mathrm{E}(Y|\gamma = 1/m) = \frac{(\alpha)_m}{\beta^m},$$

$$\mathrm{var}(Y|\gamma = 1/m) = \frac{1}{\beta^{2m}}\{(\alpha)_{2m} - (\alpha)_m^2\},$$

$$\mathrm{sk}(Y|\gamma = 1/m) = \frac{(\alpha)_{3m} - 3(\alpha)_{2m}(\alpha)_m + 2(\alpha)_m^3}{\{(\alpha)_{2m} - (\alpha)_m^2\}^{3/2}},$$

$$\begin{aligned}\mathrm{kt}(Y|\gamma = 1/m) &= \frac{(\alpha)_{4m} - 4(\alpha)_{3m}(\alpha)_m + 6(\alpha)_{2m}(\alpha)_m^2 - 3(\alpha)_m^4}{\{(\alpha)_{2m} - (\alpha)_m^2\}^2} - 3 \\ &= \frac{(\alpha)_{4m} - 4(\alpha)_{3m}(\alpha)_m - 3(\alpha)_{2m}^2 + 12(\alpha)_{2m}(\alpha)_m^2 - 6(\alpha)_m^4}{\{(\alpha)_{2m} - (\alpha)_m^2\}^2},\end{aligned}$$

where $(\cdot)_m^j = \{(\cdot)_m\}^j$.

When $m = 1$, we see that Result 2 with $(\alpha)_1 = \alpha$ and similar simplified expressions gives the well-documented results of the usual gamma:

$$\begin{aligned}&\mathrm{E}(Y|\gamma = 1) = \alpha/\beta, \ \mathrm{var}(Y|\gamma = 1) = \alpha/\beta^2, \\ &\mathrm{sk}(Y|\gamma = 1) = \{(\alpha)_3 - 3(\alpha)_2\alpha + 2\alpha^3\}/\alpha^{3/2} = 2/\alpha^{1/2}, \\ &\mathrm{kt}(Y|\gamma = 1) = \{(\alpha)_4 - 4(\alpha)_3\alpha + 6(\alpha)_2\alpha^2 - 3\alpha^4\}\alpha^{-2} - 3 = 6/\alpha,\end{aligned}$$

which are also obtained from the cumulant-generating function (cgf) giving the jth one:
$\kappa_j(Y|\gamma = 1) = -\mathrm{d}^j \alpha \ln\left(1 - \frac{t}{\beta}\right)/\mathrm{d}t^j|_{t=0} = (j-1)!\alpha/\beta^j$ $(j = 1, 2, \ldots)$.

Though Result 2 was presented when m is a positive integer, it is found that the expressions of Result 2 hold if an extended definition of $(\cdot)_m$ is employed as follows.

Result 3 Let $Y \sim$ Power-$\Gamma(\alpha, \beta, \gamma = 1/m)$ $(0 < m < \infty)$. Define $(\alpha)_m = \Gamma(\alpha + m)/\Gamma(\alpha)$ with a positive real number m. Then, Result 2 holds when the positive integer m is seen as a positive real number.

Further, consider the case of a negative real number γ or equivalently a negative real m. Recall that Result 1 gives
$\mathrm{E}(Y^k) = \frac{\Gamma\{\alpha+(k/\gamma)\}}{\Gamma(\alpha)\beta^{k/\gamma}} = \frac{\Gamma(\alpha+km)}{\Gamma(\alpha)\beta^{km}}, \quad 0 < \alpha + (k/\gamma) = \alpha + km,$

which holds as long as the condition $0 < \alpha+(k/\gamma) = \alpha+km$ is satisfied irrespective of $-\infty < k < 0$, $-\infty < \gamma = 1/m < 0$ or both.

The case of both negative k and m is trivial in that
$\mathrm{E}(Y^k) = \mathrm{E}\{\ (X^m)^k\} = \mathrm{E}(X^{km}) = \mathrm{E}(X^{|km|}) = \frac{\Gamma(\alpha+km)}{\Gamma(\alpha)\beta^{km}},$

which reduces to the case of both positive k and m, and always exists.

Result 4 Let $Y \sim$ Power-$\Gamma(\alpha, \beta, \gamma = 1/m)$ $(m \neq 0)$. When $m < 0$, define $(\alpha)_m = \Gamma(\alpha + m)/\Gamma(\alpha)$ in the same expression as positive m in Result 3. Then, we have
$\mathrm{E}(Y^k) = \frac{(\alpha)_{km}}{\beta^{km}}, \quad 0 < \alpha + (k/\gamma) = \alpha + km.$

The case with $m = \gamma = -1$ reduces to the inverse gamma. Then, Result 4 gives the moments when k is a positive integer as

$$\mathrm{E}(Y^k|m = \gamma = -1) = \frac{(\alpha)_{km}}{\beta^{km}}$$

$$= (\alpha)_{-k}\beta^k = \frac{\beta^k}{(\alpha-1)(\alpha-2)\cdots(\alpha-k)} \qquad (k < \alpha).$$

Recall that for positive integer k, $(\alpha)_k = \alpha(\alpha+1)\cdots(\alpha+k-1)$, while $(\alpha)_{-k}$ is the reciprocal of the descending factorial $(\alpha-1)(\alpha-2)\cdots(\alpha-k)$. When $\gamma = -1$, the role of β becomes a scale parameter, yielding β^k in the numerator of the above expression, which was located in the denominator when β was a rate parameter or the reciprocal of a scale parameter for $\gamma = 1$.

3 The Multivariate Power-Gamma Distribution of the FA Type

In this section, based on the multivariate gamma distribution of the FA type, the multivariate power-gamma of the FA type is defined.

Definition 1 Let $\mathbf{Y}^* = (Y_1^{\gamma_1}, ..., Y_p^{\gamma_p})^{\mathrm{T}} = \Lambda\mathbf{F} = (\mathbf{\Lambda}_0\ \mathbf{I}_p)\mathbf{F}$ $(-\infty < \gamma_i < \infty,\ \gamma_i \neq 0,\ i = 1, ..., p)$,

where $\mathbf{\Lambda}_0$ is a $p \times q$ $(1 \leq q \leq 2^p - p - 1)$ matrix consisting of 0 or 1; $\mathbf{F} = (F_1, ..., F_{p+q})^{\mathrm{T}}$ is a random vector whose $p + q$ elements are independently gamma distributed as $F_i \sim \mathrm{Gamma}(\alpha_i, \beta)$ $(i = 1, ..., p+q)$; α_i and β are the shape and rate parameters, respectively, with β common to the $p + q$ gammas; $\mathbf{I}_p$ is the

$p \times p$ identity matrix; and it is assumed that each column of $\mathbf{\Lambda}_0$ has at least two 1s. The form $\mathbf{Y}^* = \Lambda\mathbf{F} = \mathbf{\Lambda}(\mathbf{F}_0^{\mathrm{T}}, \mathbf{F}_1^{\mathrm{T}})^{\mathrm{T}}$ is seen as a FA model, where $\mathbf{F}_0 = (F_1, ..., F_q)^{\mathrm{T}}$ and $\mathbf{F}_1 = (F_{q+1}, ..., F_{p+q})^{\mathrm{T}}$ are the vectors of common and unique factors, respectively; and $\mathbf{\Lambda}$ is the factor loading/pattern matrix. Then, $\mathbf{Y} = (Y_1, ..., Y_p)^{\mathrm{T}}$ is said to have the multivariate power-gamma distribution of the "FA type", which is denoted by
$\mathbf{Y} \sim \text{Power-}\Gamma_p(\Lambda\boldsymbol{\alpha}, \beta, \boldsymbol{\gamma}) = \text{Power-}\Gamma_p\{\mathbf{\Lambda}(\alpha_1, ..., \alpha_{p+q})^{\mathrm{T}}, \beta, (\gamma_1, ..., \gamma_p)^{\mathrm{T}}\}$.

In Definition 1, $2^p - p - 1$ for the upper limit of q is due to the 0/1 pattern of $\mathbf{\Lambda}_0$ excluding the single column of all zeroes, and p columns of a single 1 corresponding to unique factors.

Lemma 1 *Define* $\boldsymbol{\lambda}_{\cdot j} = (\lambda_{1j}, ..., \lambda_{pj})^{\mathrm{T}}$ *as the jth column of* $\mathbf{\Lambda}_0 (j = 1, ..., q)$. *Then, the mgf of* $\mathbf{Y}^*$ *is*

$$\mathrm{M}_{\mathbf{Y}^*}(\mathbf{t}) = \left[\prod_{j=1}^{q} \frac{1}{\{1 - (\mathbf{t}^{\mathrm{T}}\boldsymbol{\lambda}_{\cdot j}/\beta)\}^{\alpha_j}}\right] \prod_{i=1}^{p} \frac{1}{\{1 - (t_i/\beta)\}^{\alpha_{q+i}}}$$
$$(0 \le \mathbf{t}^{\mathrm{T}}\boldsymbol{\lambda}_{\cdot j}/\beta < 1, \ j = 1, ..., q; \ 0 \le t_i/\beta < 1, \ i = 1, ..., p).$$

Proof The mgf of $\mathbf{Y}^*$ is given from that of $\mathbf{F}$:
$\mathrm{M}_{\mathbf{F}}(\mathbf{t}^*) = \left[\prod_{j=1}^{q} \frac{1}{\{1-(t_j^*/\beta)\}^{\alpha_j}}\right] \prod_{i=1}^{p} \frac{1}{\{1-(t_{q+i}^*/\beta)\}^{\alpha_{q+i}}}$.
Since $\mathbf{Y}^* = \Lambda\mathbf{F}$, the mgf of $\mathbf{Y}^*$ is

$$\mathrm{M}_{\mathbf{Y}^*}(\mathbf{t}) = \mathrm{E}\{\ \exp(\mathbf{t}^{\mathrm{T}}\mathbf{Y}^*)\} = \mathrm{E}\{\ \exp(\mathbf{t}^{\mathrm{T}}\mathbf{\Lambda}_0\mathbf{F}_0 + \mathbf{t}^{\mathrm{T}}\mathbf{F}_1)\}$$
$$= \left\{\prod_{j=1}^{q} \mathrm{E}(\mathbf{t}^{\mathrm{T}}\boldsymbol{\lambda}_{\cdot j}F_j)\right\} \prod_{i=1}^{p} \mathrm{E}(t_i F_{q+i})$$
$$= \left[\prod_{j=1}^{q} \frac{1}{\{1 - (\mathbf{t}^{\mathrm{T}}\boldsymbol{\lambda}_{\cdot j}/\beta)\}^{\alpha_j}}\right] \prod_{i=1}^{p} \frac{1}{\{1 - (t_i/\beta)\}^{\alpha_{q+i}}}$$
$$(0 \le \mathbf{t}^{\mathrm{T}}\boldsymbol{\lambda}_{\cdot j}/\beta < 1, \ j = 1, ..., q; \ 0 \le t_i/\beta < 1, \ i = 1, ..., p),$$

which gives the required result. Q.E.D.

A special case of Lemma 1 in the case of the one-factor model including location parameters was given by Mathai and Moschopoulos ([16], Eq. (2)).

Result 5 Lemma 1 gives the following.

(i) Let $\boldsymbol{\lambda}_{i\cdot}^{\mathrm{T}} = (\lambda_{i1}, ..., \lambda_{i,p+q})$ be the ith row of $\mathbf{\Lambda}$ $(i = 1, ..., p)$. The mgf of the marginal distribution of Y_i^* is given when $\mathbf{t} = (0, ..., 0, t_i, 0, ..., 0)^{\mathrm{T}}$ $(i = 1, ..., p)$ as

$$\mathrm{M}_{Y_i^*}(t_i) = \left[\prod_{j=1}^{q} \frac{1}{\{1 - (t_i 1\{\lambda_{ij} = 1\}/\beta)\}^{\alpha_j}}\right] \frac{1}{\{1 - t_i/\beta\}^{\alpha_{q+i}}}$$
$$= \frac{1}{\{1 - (t_i/\beta)\}^{\alpha_{q+i} + \sum_{j=1}^{q} 1\{\lambda_{ij}=1\}\alpha_j}} = \frac{1}{\{1 - (t_i/\beta)\}^{\boldsymbol{\lambda}_{i\cdot}^{\mathrm{T}}\boldsymbol{\alpha}}},$$

where $1\{\cdot\}$ is the indicator function. Then, it is found that the marginal distributions are gamma:

$Y_i^{y_i} = Y_i^* \sim \text{Gamma}(\boldsymbol{\lambda}_i^{\mathrm{T}}\boldsymbol{\alpha}, \beta) \equiv \text{Gamma}(\alpha_{(i)}, \beta)\ (i = 1, ..., p).$

The above result is expected since Y_i^* is the sum of independent gammas with the same rate parameters, whose number of variables and the sum of their shape parameters are $\boldsymbol{\lambda}_{i\cdot}^{\mathrm{T}}\mathbf{1}_{p+q}$ and $\boldsymbol{\lambda}_{i\cdot}^{\mathrm{T}}\boldsymbol{\alpha}$, respectively.

(ii) Let $\boldsymbol{\alpha}_0 = (\alpha_1, ..., \alpha_q)^{\mathrm{T}}$ and $\boldsymbol{\alpha}_1 = (\alpha_{q+1}, ..., \alpha_{p+q})^{\mathrm{T}}$. Then, the mean and variance of Y_i^*, and the covariance matrix of $\mathbf{Y}^*$ are

$$\mathrm{E}(Y_i^*) = \alpha_{(i)}/\beta,\ \ \mathrm{var}(Y_i^*) = \alpha_{(i)}/\beta^2\ \ (i = 1, ..., p),$$
$$\mathrm{cov}(\mathbf{Y}^*) = \boldsymbol{\Lambda}\mathrm{diag}(\boldsymbol{\alpha})\boldsymbol{\Lambda}^{\mathrm{T}}/\beta^2 = \{\boldsymbol{\Lambda}_0\mathrm{diag}(\boldsymbol{\alpha}_0)\boldsymbol{\Lambda}_0^{\mathrm{T}} + \mathrm{diag}(\boldsymbol{\alpha}_1)\}/\beta^2.$$

The last result of $\mathrm{cov}(\mathbf{Y}^*)$ shows a typical covariance structure of the orthogonal FA model, where $\mathrm{diag}(\boldsymbol{\alpha}_0)/\beta^2$ and $\mathrm{diag}(\boldsymbol{\alpha}_1)/\beta^2$ are the covariance matrices of q orthogonal common factors and p unique factors, respectively.

(iii) The kth order moments $(k = 1, 2, ...)$ of Y_i^* is given by $\mathrm{M}_{\mathbf{Y}^*}(\mathbf{t})$ obtained in Lemma 1 or more easily by $\mathrm{M}_{Y_i^*}(t_i)$ given in (i) as a mgf of a gamma:

$$\mathrm{E}(Y_i^{*k}) = \frac{\mathrm{d}^k}{\mathrm{d}t_i^k}\mathrm{M}_{Y_i^*}(t_i)|_{t_i=0} = \frac{\mathrm{d}^k}{\mathrm{d}t_i^k}\frac{1}{\{1-(t_i/\beta)\}^{\alpha_{(i)}}}|_{t_i=0}$$
$$= \alpha_{(i)}(\alpha_{(i)}+1)\cdots(\alpha_{(i)}+k-1)/\beta^k = \frac{(\alpha_{(i)})_k}{\beta^k} = \frac{\Gamma(\alpha_{(i)}+k)}{\beta^k\Gamma(\alpha_{(i)})}\quad (k = 1, 2, ...).$$

The real-valued kth moment $(k \neq 0,\ \alpha_{(i)} + k > 0)$ should be given from the definition:

$$\mathrm{E}(Y_i^{*k}) = \int_0^\infty \frac{\beta^{\alpha_{(i)}} y_i^{*k} y_i^{*(\alpha_{(i)}-1)}}{\Gamma(\alpha_{(i)})}\exp(-\beta y_i^*)\mathrm{d}y_i^*$$
$$= \frac{\beta^{\alpha_{(i)}}\Gamma(\alpha_{(i)}+k)}{\beta^{\alpha_{(i)}+k}\Gamma(\alpha_{(i)})}\int_0^\infty \frac{\beta^{\alpha_{(i)}+k} y_i^{*(\alpha_{(i)}+k-1)}}{\Gamma(\alpha_{(i)}+k)}\exp(-\beta y_i^*)\mathrm{d}y_i^*$$
$$= \frac{\Gamma(\alpha_{(i)}+k)}{\beta^k\Gamma(\alpha_{(i)})} = \frac{(\alpha_{(i)})_k}{\beta^k},$$

where the extended notation defined by $(\alpha_{(i)})_k = \Gamma(\alpha_{(i)}+k)/\Gamma(\alpha_{(i)})$ with possibly negative real value k introduced in Result 4 is used.

The cumulant-generating function for Y_i^* becomes

$\mathrm{K}_{Y_i^*}(t_i) = \ln \mathrm{M}_{Y_i^*}(t_i) = -\alpha_{(i)}\ln\{1-(t_i/\beta)\},$

which gives the kth cumulant:

$\frac{\mathrm{d}^k}{\mathrm{d}t_i^k}\mathrm{K}_{Y_i^*}(t_i)|_{t_i=0} = -\frac{\mathrm{d}^k}{\mathrm{d}t_i^k}\alpha_{(i)}\ln\{1-(t_i/\beta)\}|_{t_i=0} = \frac{\alpha_{(i)}(k-1)!}{\beta^k}\ (k = 1, 2, ...).$

The $(k_1, ..., k_r)$th product cumulant of $Y_i^*(i = 1, ..., r;\ r = 2, ..., p)$ is given by

$$\frac{\mathrm{d}^{k_1+\ldots+k_r}}{\mathrm{d}t_1^{k_1}\cdots\mathrm{d}t_r^{k_r}}\mathrm{K}_{\mathbf{Y}^*}(\mathbf{t})|_{\mathbf{t}=\mathbf{0}}$$

$$=-\frac{\mathrm{d}^{k_1+\ldots+k_r}}{\mathrm{d}t_1^{k_1}\cdots\mathrm{d}t_r^{k_r}}\left[\sum_{m=1}^{q}\alpha_m\ln\{1-(\mathbf{t}^{\mathrm{T}}\boldsymbol{\lambda}_{\cdot m}/\beta)\}+\sum_{m=1}^{p}\alpha_{q+m}\ln\{1-(t_m/\beta)\}\right]|_{\mathbf{t}=\mathbf{0}}$$

$$=\sum_{m=1}^{q}\alpha_m\frac{1\{\lambda_{1m}\times\cdots\times\lambda_{rm}=1\}(k_1+\ldots+k_r-1)!}{\beta^{k_1+\ldots+k_r}}$$

$(k_i = 1, 2, \ldots)$,

where the results for the first r variables are given without loss of generality excluding the case of $r = 1$. As expected, the last result shows that the cross product cumulants depend only on those of the common factors (compare the above results with the corresponding ones of [17], Sect. 3 and [9], Sect. 3.

The distribution of $\mathbf{Y}^* = \Lambda\mathbf{F}$ is a special case of Power-$\Gamma_p(\Lambda\boldsymbol{\alpha}, \beta, \boldsymbol{\gamma})$ when $\boldsymbol{\gamma} = \mathbf{1}_p$, where $\mathbf{1}_p$ is the $p \times 1$ vector of 1s. Then, $\mathbf{Y}^*$ is said to have the multivariate gamma distribution of the FA type:

$\mathbf{Y}^* \sim$ Power-$\Gamma_p(\Lambda\boldsymbol{\alpha}, \beta, \mathbf{1}_p) \equiv \Gamma_p(\Lambda\boldsymbol{\alpha}, \beta)$.

The multivariate power chi-square distribution is defined as a special case of the multivariate power-gamma as

$$\text{Power-}\Gamma_p(\Lambda\boldsymbol{\alpha}, \beta, \boldsymbol{\gamma}) = \text{Power-}\Gamma_p(\Lambda\mathbf{n}/2,\ 1/2, \boldsymbol{\gamma}) \equiv \text{Power-}\chi_p^2(\Lambda\mathbf{n}, \boldsymbol{\gamma})$$

with $\mathbf{n} = (n_1, \ldots, n_{p+q})^{\mathrm{T}}$, which is obtained when

$$F_i \sim \text{gamma}(\alpha_i, \beta) = \text{gamma}(n_i/2, 1/2)(i = 1, \ldots, p+q),$$

i.e., F_i is chi-square distributed with n_i degrees of freedom (df), denoted by $F_i \sim \chi^2(n_i)$. Note that $Y_i^* = \boldsymbol{\lambda}_i^{\mathrm{T}}\mathbf{F} \sim \chi^2(\boldsymbol{\lambda}_i^{\mathrm{T}}\mathbf{n})$, which gives the definition of the multivariate chi-square of the FA type:

$\mathbf{Y}^* \sim$ Power-$\chi_p^2(\Lambda\mathbf{n}, \mathbf{1}_p) = \chi_p^2(\Lambda\mathbf{n})$.

Typically, n_is are positive integers though positive real values can also be used.

Recall the definitions $\boldsymbol{\alpha}_0 = (\alpha_1, \ldots, \alpha_q)^{\mathrm{T}}$ and $\boldsymbol{\alpha}_1 = (\alpha_{q+1}, \ldots, \alpha_{p+q})^{\mathrm{T}}$ with the corresponding notations $\mathbf{F}_0$ and $\mathbf{F}_1$. Recall also that $\boldsymbol{\lambda}_{\cdot j}$ is the jth column of $\boldsymbol{\Lambda}_0$ ($j = 1, \ldots, q$). Let $\alpha_0 = \mathbf{1}_q^{\mathrm{T}}\boldsymbol{\alpha}_0$, $\alpha_+ = \mathbf{1}_p^{\mathrm{T}}\boldsymbol{\alpha}_1$, $\boldsymbol{\lambda}_{0i\cdot} = (\lambda_{i1}, \ldots, \lambda_{iq})^{\mathrm{T}}$ ($i = 1, \ldots, p$), and $y_+^* = \mathbf{1}_p^{\mathrm{T}}\mathbf{y}^*$. Then, we have the following result.

Lemma 2 *The pdf of* $\mathbf{Y}^* \sim \Gamma_p(\Lambda\alpha, \beta)$ *of the FA type at* $\mathbf{Y}^* = \mathbf{y}^*$ *is*

$$
\begin{aligned}
&g_{\Gamma,p}(\mathbf{y}^*|\Lambda\alpha,\beta)\\
&=\frac{\beta^{\alpha_+ + \alpha_0}\exp(-\beta\mathbf{1}_p^{\mathrm{T}}\mathbf{y}^*)}{\prod_{i=1}^{p+q}\Gamma(\alpha_i)}\\
&\times\sum_{j=0}^{\infty}\int_{\mathbf{y}^*>\boldsymbol{\Lambda}_0\mathbf{f}_0}\left\{\prod\nolimits_{i=1}^{p}(y_i^*-\boldsymbol{\lambda}_{0i}^{\mathrm{T}}\mathbf{f}_0)^{\alpha_{q+i}-1}\right\}\prod\nolimits_{i=1}^{q}\frac{\{\beta(\mathbf{1}_p^{\mathrm{T}}\boldsymbol{\lambda}_{\cdot i}-1)\}^j f_i^{\alpha_i+j-1}}{j!}\mathrm{d}\mathbf{f}_0.
\end{aligned}
$$

Proof The FA model gives
$Y_i^* = \boldsymbol{\lambda}_i^{\mathrm{T}}\mathbf{F} = \boldsymbol{\lambda}_{0i}^{\mathrm{T}}\mathbf{F}_0 + F_{q+i}\ (i = 1, \ldots, p)$.

Consider the joint distribution of $\mathbf{F}$ and use the variable transformation from $\mathbf{F}_1$ to $\mathbf{Y}^* = \Lambda\mathbf{F} = \boldsymbol{\Lambda}_0\mathbf{F}_0 + \mathbf{F}_1$ with unchanged $\mathbf{F}_0$ and the unit Jacobian. Let $g_\Gamma(f_i|\alpha_i,\beta)$ be the pdf of $F_i \sim \mathrm{Gamma}(\alpha_i,\beta)$ at $F_i = f_i\ (i = 1, \ldots, p+q)$. Noting that $\mathbf{F}_0$ and the elements of $\mathbf{Y}^* - \boldsymbol{\Lambda}_0\mathbf{F}_0 = \mathbf{F}_1$ are independently distributed, the pdf of $\mathbf{Y}^*$ is obtained by integrating out $\mathbf{F}_0$:

$$
\begin{aligned}
&g_{\Gamma,p}(\mathbf{y}^*|\Lambda\boldsymbol{\alpha},\beta)\\
&=\int_{\mathbf{y}^*>\boldsymbol{\Lambda}_0\mathbf{f}_0}\left\{\prod\nolimits_{i=1}^{p}g_\Gamma(y_i^*-\boldsymbol{\lambda}_{0i}^{\mathrm{T}}\mathbf{f}_0|\alpha_{q+i},\beta)\right\}\prod\nolimits_{i=1}^{q}g_\Gamma(f_i|\alpha_i,\beta)\,\mathrm{d}\mathbf{f}_0\\
&=\frac{\beta^{\alpha_+ + \alpha_0}\exp(-\beta\mathbf{1}_p^{\mathrm{T}}\mathbf{y}^*)}{\prod_{i=1}^{p+q}\Gamma(\alpha_i)}\\
&\times\int_{\mathbf{y}^*>\boldsymbol{\Lambda}_0\mathbf{f}_0}\left\{\prod\nolimits_{i=1}^{p}(y_i^*-\boldsymbol{\lambda}_{0i}^{\mathrm{T}}\mathbf{f}_0)^{\alpha_{q+i}-1}\right\}\left(\prod\nolimits_{i=1}^{q}f_i^{\alpha_i-1}\right)\exp\{\beta(\mathbf{1}_p^{\mathrm{T}}\boldsymbol{\Lambda}_0\mathbf{f}_0-\mathbf{1}_q^{\mathrm{T}}\mathbf{f}_0)\}\mathrm{d}\mathbf{f}_0,
\end{aligned}
$$

where $\int_{\mathbf{y}^*>\boldsymbol{\Lambda}_0\mathbf{f}_0}(\cdot)\mathrm{d}\mathbf{f}_0$ indicates the multiple integral with respect to the elements of $\mathbf{f}_0 = (f_1, \ldots, f_q)^{\mathrm{T}}$ over the region $\mathbf{y}^* > \boldsymbol{\Lambda}_0\mathbf{f}_0$, i.e., $\bigcap_{i=1}^{q}\{y_i^* > \boldsymbol{\lambda}_{0i\cdot}^{\mathrm{T}}\mathbf{f}_0\}$. Using

$$
\begin{aligned}
\exp\{\beta(\mathbf{1}_p^{\mathrm{T}}\boldsymbol{\Lambda}_0\mathbf{f}_0-\mathbf{1}_q^{\mathrm{T}}\mathbf{f}_0)\}&=\exp\left\{\sum\nolimits_{i=1}^{q}\beta(\mathbf{1}_p^{\mathrm{T}}\boldsymbol{\lambda}_{\cdot i}-1)f_i\right\}\\
&=\prod\nolimits_{i=1}^{q}\sum\nolimits_{j=0}^{\infty}\{\beta(\mathbf{1}_p^{\mathrm{T}}\boldsymbol{\lambda}_{\cdot i}-1)f_i\}^j/j!,
\end{aligned}
$$

the previous result becomes

$$
\begin{aligned}
&\int_{\mathbf{y}^*>\boldsymbol{\Lambda}_0\mathbf{f}_0}\left\{\prod\nolimits_{i=1}^{p}g_\Gamma(y_i^*-\boldsymbol{\lambda}_{0i}^{\mathrm{T}}\mathbf{f}_0|\alpha_{q+i},\beta)\right\}\prod\nolimits_{i=1}^{q}g_\Gamma(f_i|\alpha_i,\beta)\,\mathrm{d}\mathbf{f}_0\\
&=\frac{\beta^{\alpha_+ + \alpha_0}\exp(-\beta\mathbf{1}_p^{\mathrm{T}}\mathbf{y}^*)}{\prod_{i=1}^{p+q}\Gamma(\alpha_i)}\\
&\times\int_{\mathbf{y}^*>\boldsymbol{\Lambda}_0\mathbf{f}_0}\left\{\prod\nolimits_{i=1}^{p}(y_i^*-\boldsymbol{\lambda}_{0i}^{\mathrm{T}}\mathbf{f}_0)^{\alpha_{q+i}-1}\right\}\prod\nolimits_{i=1}^{q}f_i^{\alpha_i-1}\sum_{j=0}^{\infty}\frac{\{\beta(\mathbf{1}_p^{\mathrm{T}}\boldsymbol{\lambda}_{\cdot i}-1)f_i\}^j}{j!}\mathrm{d}\mathbf{f}_0
\end{aligned}
$$

$$= \frac{\beta^{\alpha_+ + \alpha_0} \exp(-\beta \mathbf{1}_p^{\mathrm{T}} \mathbf{y}^*)}{\prod_{i=1}^{p+q} \Gamma(\alpha_i)}$$

$$\times \sum_{j=0}^{\infty} \int_{\mathbf{y}^* > \boldsymbol{\Lambda}_0 \mathbf{f}_0} \left\{ \prod_{i=1}^{p} (y_i^* - \boldsymbol{\lambda}_{0i}^{\mathrm{T}} \mathbf{f}_0)^{\alpha_{q+i}-1} \right\} \prod_{i=1}^{q} \frac{\{\beta(\mathbf{1}_p^{\mathrm{T}} \boldsymbol{\lambda}_{\cdot i} - 1)\}^j f_i^{\alpha_i + j - 1}}{j!} \mathrm{d}\mathbf{f}_0.$$

Q.E.D.

Among the FA-type models, consider the one-factor model with $q = 1$, $\boldsymbol{\Lambda}_0 = \mathbf{1}_p$, $\Lambda \mathbf{F} = (\boldsymbol{\Lambda}_0 \; \mathbf{I}_p)\mathbf{F} = \mathbf{1}_p F_0 + \mathbf{F}_1$ and $\Lambda \alpha = \mathbf{1}_p \alpha_0 + \boldsymbol{\alpha}_1$,

where $\mathbf{F} = (F_0, \mathbf{F}_1^{\mathrm{T}})^{\mathrm{T}} = (F_0, F_1, ..., F_p)^{\mathrm{T}}$ and $\boldsymbol{\alpha} = (\alpha_0, \boldsymbol{\alpha}_1^{\mathrm{T}})^{\mathrm{T}} = (\alpha_0, \alpha_1, ..., \alpha_p)^{\mathrm{T}}$ with $\alpha_0, \alpha_1, ..., \alpha_p$ redefined.

Lemma 3 *The pdf of* $\mathbf{Y}^* \sim \Gamma_p(\mathbf{1}_p \alpha_0 + \boldsymbol{\alpha}_1, \beta)$ *of the one-factor type at* $\mathbf{Y}^* = \mathbf{y}^*$ *is*

$$g_{\Gamma,p}(\mathbf{y}^* | \mathbf{1}_p \alpha_0 + \boldsymbol{\alpha}_1, \beta)$$

$$= \frac{\beta^{\alpha_+ + \alpha_0} \exp(-\beta y_+^*)}{\prod_{i=0}^{p} \Gamma(\alpha_i)} \sum_{j=0}^{\infty} \int_0^{\min\{\mathbf{y}^*\}} \left\{ \prod_{i=1}^{p} (y_i^* - f_0)^{\alpha_i - 1} \right\} \frac{\{\beta(p-1)\}^j f_0^{\alpha_0 + j - 1}}{j!} \mathrm{d} f_0.$$

Proof The one-factor model gives

$Y_i^* = \boldsymbol{\lambda}_i^{\mathrm{T}} \mathbf{F} = F_0 + F_i \;\; (i = 1, ..., p)$.

Consider the joint distribution of $\mathbf{F}$ and use the variable transformation from $\mathbf{F}_1$ to $\mathbf{Y}^* = \Lambda \mathbf{F} = \mathbf{1}_p F_0 + \mathbf{F}_1$ with unchanged F_0 and the unit Jacobian. Let $g_{\Gamma}(f_0 | \alpha_0, \beta)$ be the pdf of $F_0 \sim \text{gamma}(\alpha_0, \beta)$ at $F_0 = f_0$. Noting that F_0 and the elements of $\mathbf{Y}^* - \mathbf{1}_p F_0 = \mathbf{F}_1$ are independently distributed, the pdf of $\mathbf{Y}^*$ is obtained by integrating out F_0:

$$g_{\Gamma,p}(\mathbf{y}^* | \mathbf{1}_p \alpha_0 + \boldsymbol{\alpha}_1, \beta)$$

$$= \int_0^{\min\{y_1^*, ..., y_p^*\}} \left\{ \prod_{i=1}^{p} g_{\Gamma}(y_i^* - f_0 | \alpha_i, \beta) \right\} g_{\Gamma}(f_0 | \alpha_0, \beta) \, \mathrm{d} f_0$$

$$= \frac{\beta^{\mathbf{1}_p^{\mathrm{T}} \boldsymbol{\alpha}_1 + \alpha_0} \exp(-\beta \mathbf{1}_p^{\mathrm{T}} \mathbf{y}^*)}{\prod_{i=0}^{p} \Gamma(\alpha_i)} \int_0^{\min\{\mathbf{y}^*\}} \left\{ \prod_{i=1}^{p} (y_i^* - f_0)^{\alpha_i - 1} \right\} f_0^{\alpha_0 - 1} \exp\{\beta(p-1) f_0\} \, \mathrm{d} f_0$$

$$= \frac{\beta^{\alpha_+ + \alpha_0} \exp(-\beta y_+^*)}{\prod_{i=0}^{p} \Gamma(\alpha_i)} \int_0^{\min\{\mathbf{y}^*\}} \left\{ \prod_{i=1}^{p} (y_i^* - f_0)^{\alpha_i - 1} \right\} f_0^{\alpha_0 - 1} \sum_{j=0}^{\infty} \frac{\{\beta(p-1) f_0\}^j}{j!} \mathrm{d} f_0$$

$$= \frac{\beta^{\alpha_+ + \alpha_0} \exp(-\beta y_+^*)}{\prod_{i=0}^{p} \Gamma(\alpha_i)} \sum_{j=0}^{\infty} \int_0^{\min\{\mathbf{y}^*\}} \left\{ \prod_{i=1}^{p} (y_i^* - f_0)^{\alpha_i - 1} \right\} \frac{\{\beta(p-1)\}^j f_0^{\alpha_0 + j - 1}}{j!} \mathrm{d} f_0.$$

Q.E.D.

Mathai and Moschopoulos ([16], Sect. 5) gave an expression for a model essentially equivalent to that in Lemma 3 with no integral though they use $(p + 1)$-fold infinite series.

Result 6 The pdf of $Y_+^* \equiv \mathbf{1}_p^{\mathrm{T}}\mathbf{Y}^*$ with $\mathbf{Y}^* \sim \Gamma_p(\mathbf{1}_p\alpha_0 + \boldsymbol{\alpha}_1, \beta)$ of the one-factor type at $Y_+^* = y_+^* = \mathbf{1}_p^{\mathrm{T}}\mathbf{y}^*$ when $\mathbf{Y}^* = \mathbf{y}^*$ *is*

$$
\begin{aligned}
& g_{Y_+^*}(y_+^*|\mathbf{1}_p\alpha_0 + \boldsymbol{\alpha}_1, \beta) \\
&= \frac{\beta^{\alpha_+ + \alpha_0}\exp(-\beta y_+^*)}{\Gamma(\alpha_+)\Gamma(\alpha_0)} \int_0^{\min\{\mathbf{y}^*\}} (y_+^* - pf_0)^{\alpha_+ - 1} f_0^{\alpha_0 - 1} \exp\{\beta(p-1)f_0\}\,\mathrm{d}f_0 \\
&= \frac{\beta^{\alpha_+ + \alpha_0}\exp(-\beta y_+^*) y_+^{*(\alpha_+ + \alpha_0 - 1)}}{\Gamma(\alpha_+)\Gamma(\alpha_0)p^{\alpha_0}} \sum_{j=0}^{\infty} \frac{\{\beta(p-1)y_+^*\}^j}{p^j j!} \mathrm{B}(p\min\{\mathbf{y}^*\}/y_+^*|\alpha_0 + j, \alpha_+),
\end{aligned}
$$

where $\mathrm{B}(c|a, b) = \int_0^c t^{a-1}(1-t)^{b-1}\mathrm{d}t$ $(0 \le c \le 1)$ is the incomplete beta function.

Proof Noting that $\mathbf{1}_p^{\mathrm{T}}\mathbf{Y}^* - pf_0 \sim \mathrm{Gamma}(\mathbf{1}_p^{\mathrm{T}}\boldsymbol{\alpha}_1, \beta)$, as in Lemma 4, we have

$$
\begin{aligned}
& \int_0^{\min\{\mathbf{y}^*\}} g_\Gamma(\mathbf{1}_p^{\mathrm{T}}\mathbf{y}^* - pf_0|\mathbf{1}_p^{\mathrm{T}}\boldsymbol{\alpha}_1, \beta) g_\Gamma(f_0|\alpha_0, \beta)\,\mathrm{d}f_0 \\
&= \frac{\beta^{\mathbf{1}_p^{\mathrm{T}}\boldsymbol{\alpha}_1 + \alpha_0}\exp(-\beta\mathbf{1}_p^{\mathrm{T}}\mathbf{y}^*)}{\Gamma(\mathbf{1}_p^{\mathrm{T}}\boldsymbol{\alpha}_1)\Gamma(\alpha_0)} \int_0^{\min\{\mathbf{y}^*\}} (\mathbf{1}_p^{\mathrm{T}}\mathbf{y}^* - pf_0)^{\mathbf{1}_p^{\mathrm{T}}\boldsymbol{\alpha}_1 - 1} f_0^{\alpha_0 - 1} \exp\{\beta(p-1)f_0\}\,\mathrm{d}f_0 \\
&= \frac{\beta^{\alpha_+ + \alpha_0}\exp(-\beta y_+^*)}{\Gamma(\alpha_+)\Gamma(\alpha_0)} \int_0^{\min\{\mathbf{y}^*\}} (y_+^* - pf_0)^{\alpha_+ - 1} f_0^{\alpha_0 - 1} \exp\{\beta(p-1)f_0\}\,\mathrm{d}f_0.
\end{aligned}
$$

Define $e_0 = pf_0/y_+^*$ with $\mathrm{d}f_0/\mathrm{d}e_0 = y_+^*/p$. Then, the above integral becomes

$$
\begin{aligned}
& \int_0^{\min\{\mathbf{y}^*\}} (y_+^* - pf_0)^{\alpha_+ - 1} f_0^{\alpha_0 - 1} \exp\{\beta(p-1)f_0\}\,\mathrm{d}f_0 \\
&= \int_0^{p\min\{\mathbf{y}^*\}/y_+^*} (y_+^* - y_+^* e_0)^{\alpha_+ - 1} (y_+^* e_0/p)^{\alpha_0 - 1} (y_+^*/p) \exp\{\beta(p-1)y_+^* e_0/p\}\,\mathrm{d}e_0 \\
&= \frac{y_+^{*(\alpha_+ + \alpha_0 - 1)}}{p^{\alpha_0}} \int_0^{p\min\{\mathbf{y}^*\}/y_+^*} (1 - e_0)^{\alpha_+ - 1} e_0^{\alpha_0 - 1} \sum_{j=0}^{\infty} \frac{\{\beta(p-1)y_+^* e_0\}^j}{p^j j!}\,\mathrm{d}e_0 \\
&= \frac{y_+^{*(\alpha_+ + \alpha_0 - 1)}}{p^{\alpha_0}} \sum_{j=0}^{\infty} \frac{\{\beta(p-1)y_+^*\}^j}{p^j j!} \int_0^{p\min\{\mathbf{y}^*\}/y_+^*} e_0^{\alpha_0 + j - 1}(1 - e_0)^{\alpha_+ - 1}\,\mathrm{d}e_0 \\
&= \frac{y_+^{*(\alpha_+ + \alpha_0 - 1)}}{p^{\alpha_0}} \sum_{j=0}^{\infty} \frac{\{\beta(p-1)y_+^*\}^j}{p^j j!} \mathrm{B}(p\min\{\mathbf{y}^*\}/y_+^*|\alpha_0 + j, \alpha_+),
\end{aligned}
$$

which gives the required result. Q.E.D.

In Result 6, note that $\mathbf{1}_p^{\mathrm{T}}\mathbf{Y}^* - pf_0 = f_1 + \ldots + f_p$ is gamma distributed, whereas $\mathbf{1}_p^{\mathrm{T}}\mathbf{Y}^* = f_1 + \ldots + f_p + pf_0$ is not gamma since the latter is a linear combination of independent gammas with equal scales rather than their sum unless $p = 1$.

Theorem 1 *The pdf of multivariate power-gamma distributed* $\mathbf{Y} = (Y_1, \ldots, Y_p)^{\mathrm{T}}$ *at* $\mathbf{Y} = \mathbf{y} = (y_1, \ldots, y_p)^{\mathrm{T}}$ *with* $\mathbf{y}^* = (y_1^{\gamma_1}, \ldots, y_p^{\gamma_p})^{\mathrm{T}}$ *when* $\mathbf{Y}^* = (Y_1^{\gamma_1}, \ldots, Y_p^{\gamma_p})^{\mathrm{T}} \sim \Gamma_p(\boldsymbol{\Lambda\alpha}, \beta)$ *of the FA type is*

$$
\begin{aligned}
& g_{\text{Power-}\Gamma,p}(\mathbf{y}|\Lambda\boldsymbol{\alpha},\beta,\boldsymbol{\gamma}) \\
& = \frac{\beta^{\alpha_+ + \alpha_0}\exp(-\beta\mathbf{1}_p^{\mathrm{T}}\mathbf{y}^*)\prod_{i=1}^p |\gamma_i| y_i^{\gamma_i - 1}}{\prod_{i=1}^{p+q}\Gamma(\alpha_i)} \\
& \times \sum_{j=0}^{\infty}\int_{\mathbf{y}^*>\boldsymbol{\Lambda}_0\mathbf{f}_0}\left\{\prod_{i=1}^p (y_i^* - \boldsymbol{\lambda}_{0i}^{\mathrm{T}}\mathbf{f}_0)^{\alpha_{q+i}-1}\right\}\prod_{i=1}^q \frac{\{\beta(\mathbf{1}_p^{\mathrm{T}}\boldsymbol{\lambda}_{\cdot i} - 1)\}^j f_i^{\alpha_i + j - 1}}{j!} \mathrm{d}\mathbf{f}_0.
\end{aligned}
$$

Proof Noting that the Jacobian of the variable transformation from $\mathbf{Y}^*$ to $\mathbf{Y}$ is $\prod_{i=1}^p |\mathrm{d}y_i^*/\mathrm{d}y_i| = \prod_{i=1}^p |\gamma_i| y_i^{\gamma_i-1}$, Lemma 2 gives the required pdf. Q.E.D.

Corollary 1 *The pdf of multivariate power-gamma distributed* $\mathbf{Y} = (Y_1, \ldots, Y_p)^{\mathrm{T}}$ *at* $\mathbf{Y} = \mathbf{y} = (y_1, \ldots, y_p)^{\mathrm{T}}$ *with* $\mathbf{y}^* = (y_1^{\gamma_1}, \ldots, y_p^{\gamma_p})^{\mathrm{T}}$ *when* $\mathbf{Y}^* = (Y_1^{\gamma_1}, \ldots, Y_p^{\gamma_p})^{\mathrm{T}} \sim \Gamma_p(\mathbf{1}_p\alpha_0 + \boldsymbol{\alpha}_1, \beta)$ *of the one-factor type is*

$$
\begin{aligned}
& g_{\text{Power-}\Gamma,p}(\mathbf{y}|\mathbf{1}_p\alpha_0 + \boldsymbol{\alpha}_1, \beta, \boldsymbol{\gamma}) \\
& = \frac{\beta^{\alpha_+ + \alpha_0}\exp(-\beta y_+^*)\prod_{i=1}^p |\gamma_i| y_i^{\gamma_i-1}}{\prod_{i=0}^p \Gamma(\alpha_i)} \sum_{j=0}^{\infty}\int_0^{\min\{\mathbf{y}^*\}}\left\{\prod_{i=1}^p (y_i^* - f_0)^{\alpha_i - 1}\right\}\frac{\{\beta(p-1)\}^j f_0^{\alpha_0 + j - 1}}{j!}\mathrm{d}f_0.
\end{aligned}
$$

Proof Using the same Jacobian as in Theorem 1, Lemma 3 gives the required pdf. Q.E.D.

Recall that $\boldsymbol{\lambda}_{i\cdot}^{\mathrm{T}}$ is the ith row of $\boldsymbol{\Lambda}$ in the FA-type model with the definition of $\alpha_{(i)} = \boldsymbol{\lambda}_{i\cdot}^{\mathrm{T}}\boldsymbol{\alpha}$ $(i = 1, .., p)$.

Theorem 2 *The kth order moment with k being possibly non-integer and/or negative value for the marginal distribution of power-gamma distributed* Y_i *when* $Y_i^* = Y_i^{\gamma_i} \sim \mathrm{Gamma}(\alpha_{(i)}, \beta)$ $(i = 1, \ldots, p)$ *in the FA-type model is*

$\mathrm{E}(Y_i^k) = (\alpha_{(i)})_{k/\gamma_i}/\beta^{k/\gamma_i}, \ \ 0 < \alpha_{(i)} + (k/\gamma_i)\ (\gamma_i \neq 0, \ i = 1, \ldots, p).$

Proof Note that $Y_i \sim \text{Power-}\Gamma(\alpha_{(i)}, \beta, \gamma_i)$ $(\gamma_i \neq 0)$. Using the extended notation $(\alpha_{(i)})_{1/\gamma_i} = \Gamma\{\alpha_{(i)} + (1/\gamma_i)\}/\Gamma(\alpha_{(i)})$ $(0 < \alpha_{(i)} + (1/\gamma_i))$ for possibly non-integer and/or negative γ_i *and k*, Result 4 gives the required result. Q.E.D.

Lemma 4 *The product moment of multivariate power-gamma distributed* $\mathbf{Y} = (Y_1, \ldots, Y_p)^{\mathrm{T}}$ *at* $\mathbf{Y} = \mathbf{y} = (y_1, \ldots, y_p)^{\mathrm{T}}$ *with* $\mathbf{y}^* = (y_1^{\gamma_1}, \ldots, y_p^{\gamma_p})^{\mathrm{T}}$ *when* $\mathbf{Y}^* = (Y_1^{\gamma_1}, \ldots, Y_p^{\gamma_p})^{\mathrm{T}} \sim \Gamma_p(\Lambda\boldsymbol{\alpha}, \beta)$ *of the FA type is*

$\mathrm{E}\left(\prod_{i=1}^p Y_i^{k_i}\right) = \mathrm{E}\left(\prod_{i=1}^p Y_i^{*k_i/\gamma_i}\right).$

Proof By definition,

$$
\begin{aligned}
\mathrm{E}\left(\prod_{i=1}^p Y_i^{k_i}\right) & = \int_{\mathbf{0}}^{\infty}\frac{\beta^{\alpha_+ + \alpha_0}\exp(-\beta\mathbf{1}_p^{\mathrm{T}}\mathbf{y}^*)\prod_{i=1}^p |\gamma_i| y_i^{\gamma_i + k_i - 1}}{\prod_{i=1}^{p+q}\Gamma(\alpha_i)} \\
& \times \sum_{j=0}^{\infty}\int_{\mathbf{y}^*>\boldsymbol{\Lambda}_0\mathbf{f}_0}\left\{\prod_{i=1}^p (y_i^* - \boldsymbol{\lambda}_{0i}^{\mathrm{T}}\mathbf{f}_0)^{\alpha_{q+i}-1}\right\}\prod_{i=1}^q \frac{\{\beta(\mathbf{1}_p^{\mathrm{T}}\boldsymbol{\lambda}_{\cdot i} - 1)\}^j f_i^{\alpha_i + j - 1}}{j!}\mathrm{d}\mathbf{f}_0\mathrm{d}\mathbf{y}.
\end{aligned}
$$

Employ the variable transformation from $\mathbf{y}$ to $\mathbf{y}^*$ with the Jacobian $\prod_{i=1}^{p} |\,\mathrm{d}y_i/\mathrm{d}y_i^*| = \prod_{i=1}^{p} |1/\gamma_i| y_i^{*\{(1/\gamma_i)-1\}}$. Then, the above result becomes

$$\mathrm{E}\left(\prod\nolimits_{i=1}^{p} Y_i^{k_i}\right) = \int_{\mathbf{0}}^{\infty} \frac{\beta^{\alpha_+ + \alpha_0} \exp(-\beta \mathbf{1}_p^{\mathrm{T}} \mathbf{y}^*) \prod_{i=1}^{p} y_i^{*\{\gamma_i + k_i - 1\}/\gamma_i} y_i^{*\{(1/\gamma_i)-1\}}}{\prod_{i=1}^{p+q} \Gamma(\alpha_i)}$$

$$\times \sum_{j=0}^{\infty} \int_{\mathbf{y}^* > \mathbf{\Lambda}_0 \mathbf{f}_0} \left\{\prod\nolimits_{i=1}^{p} (y_i^* - \boldsymbol{\lambda}_{0i}^{\mathrm{T}} \mathbf{f}_0)^{\alpha_{q+i}-1}\right\} \prod\nolimits_{i=1}^{q} \frac{\{\beta(\mathbf{1}_p^{\mathrm{T}} \boldsymbol{\lambda}_{\cdot i} - 1)\}^j f_i^{\alpha_i + j - 1}}{j!} \mathrm{d}\mathbf{f}_0 \mathrm{d}\mathbf{y}^*$$

$$= \int_{\mathbf{0}}^{\infty} \frac{\beta^{\alpha_+ + \alpha_0} \exp(-\beta \mathbf{1}_p^{\mathrm{T}} \mathbf{y}^*) \prod_{i=1}^{p} y_i^{*k_i/\gamma_i}}{\prod_{i=1}^{p+q} \Gamma(\alpha_i)}$$

$$\times \sum_{j=0}^{\infty} \int_{\mathbf{y}^* > \mathbf{\Lambda}_0 \mathbf{f}_0} \left\{\prod\nolimits_{i=1}^{p} (y_i^* - \boldsymbol{\lambda}_{0i}^{\mathrm{T}} \mathbf{f}_0)^{\alpha_{q+i}-1}\right\} \prod\nolimits_{i=1}^{q} \frac{\{\beta(\mathbf{1}_p^{\mathrm{T}} \boldsymbol{\lambda}_{\cdot i} - 1)\}^j f_i^{\alpha_i + j - 1}}{j!} \mathrm{d}\mathbf{f}_0 \mathrm{d}\mathbf{y}^*$$

$$= \mathrm{E}\left(\prod\nolimits_{i=1}^{p} Y_i^{*k_i/\gamma_i}\right).$$

Q.E.D.

The result of Lemma 4 is expected since the variable transformation used in the proof restores the original gammas.

Theorem 3 *The* $(k_1, ..., k_p)$th *product moment and cumulant of* $Y_1, ..., Y_p$ *are equal to the corresponding* $(k_1/\gamma_1, ..., k_p/\gamma_p)$*th ones of* $Y_1^*, ..., Y_p^*$. *When* $k_i/\gamma_i \equiv k_i^* (i = 1, ..., p)$ *are positive integers, the cumulants are given by Result 5 (iii) when* k_i *is replaced by* $k_i^* (i = 1, ..., p)$.

Theorem 3 is not trivial in that k_i and γ_i can be real-valued and/or negative as long as the condition is satisfied.

4 Special Cases of the One-Factor Model and Numerical Illustrations

In this section, three special cases of the one-factor model are shown with their simplified properties. The first special case has a restriction of integer-valued shape parameters for unique factors, i.e., $F_i (i = 1, ..., p)$ with an unconstrained single common factor F_0.

Lemma 5 *The pdf of* $\mathbf{Y}^* \sim \Gamma_p(\mathbf{1}_p\alpha_0 + \boldsymbol{\alpha}_1, \beta)$ *of the one-factor model at* $\mathbf{Y}^* = \mathbf{y}^*$, *when the elements of* $\boldsymbol{\alpha}_1 = (\alpha_1, ..., \alpha_p)^{\mathrm{T}}$ *are all positive integers, is*

$$
\begin{aligned}
& g_{\Gamma,p}(\mathbf{y}^*|\mathbf{1}_p\alpha_0+\boldsymbol{\alpha}_1,\beta) \\
&= \frac{\beta^{\alpha_+ +\alpha_0}\exp(-\beta y_+^*)}{\Gamma(\alpha_0)\prod_{i=1}^p(\alpha_i-1)!}\sum\nolimits_{h_1=0}^{\alpha_1-1}\cdots\sum\nolimits_{h_p=0}^{\alpha_p-1}\left\{\prod\nolimits_{i=1}^p\binom{\alpha_i-1}{h_i}y_i^{*(\alpha_i-1-h_i)}(-1)^{h_i}\right\} \\
&\quad\times\sum_{j=0}^{\infty}\frac{\{\beta(p-1)\}^j(\min\{\mathbf{y}^*\})^{\alpha_0+j+h_1+\ldots+h_p}}{j!(\alpha_0+j+h_1+\ldots+h_p)}.
\end{aligned}
$$

Proof When $\alpha_1, ..., \alpha_p$ are positive integers, the pdf derived in Lemma 3 for the one-factor model can be expanded as

$$
\begin{aligned}
& g_{\Gamma,p}(\mathbf{y}^*|\mathbf{1}_p\alpha_0+\boldsymbol{\alpha}_1,\beta) \\
&= \frac{\beta^{\alpha_+ +\alpha_0}\exp(-\beta y_+^*)}{\prod_{i=0}^p\Gamma(\alpha_i)}\sum_{j=0}^{\infty}\int_0^{\min\{\mathbf{y}^*\}}\left\{\prod\nolimits_{i=1}^p(y_i^*-f_0)^{\alpha_i-1}\right\}\frac{\{\beta(p-1)\}^j f_0^{\alpha_0+j-1}}{j!}\mathrm{d}f_0 \\
&= \frac{\beta^{\alpha_+ +\alpha_0}\exp(-\beta y_+^*)}{\prod_{i=0}^p\Gamma(\alpha_i)} \\
&\quad\times\sum_{j=0}^{\infty}\int_0^{\min\{\mathbf{y}^*\}}\left\{\prod\nolimits_{i=1}^p\sum\nolimits_{h_i=0}^{\alpha_i-1}\binom{\alpha_i-1}{h_i}y_i^{*(\alpha_i-1-h_i)}(-f_0)^{h_i}\right\}\frac{\{\beta(p-1)\}^j f_0^{\alpha_0+j-1}}{j!}\mathrm{d}f_0 \\
&= \frac{\beta^{\alpha_+ +\alpha_0}\exp(-\beta y_+^*)}{\prod_{i=0}^p\Gamma(\alpha_i)}\sum\nolimits_{h_1=0}^{\alpha_1-1}\cdots\sum\nolimits_{h_p=0}^{\alpha_p-1}\left\{\prod\nolimits_{i=1}^p\binom{\alpha_i-1}{h_i}y_i^{*(\alpha_i-1-h_i)}(-1)^{h_i}\right\} \\
&\quad\times\sum_{j=0}^{\infty}\int_0^{\min\{\mathbf{y}^*\}}\frac{\{\beta(p-1)\}^j f_0^{\alpha_0+j+h_1+\ldots+h_p-1}}{j!}\mathrm{d}f_0 \\
&= \frac{\beta^{\alpha_+ +\alpha_0}\exp(-\beta y_+^*)}{\prod_{i=0}^p\Gamma(\alpha_i)}\sum\nolimits_{h_1=0}^{\alpha_1-1}\cdots\sum\nolimits_{h_p=0}^{\alpha_p-1}\left\{\prod\nolimits_{i=1}^p\binom{\alpha_i-1}{h_i}y_i^{*(\alpha_i-1-h_i)}(-1)^{h_i}\right\} \\
&\quad\times\sum_{j=0}^{\infty}\frac{\{\beta(p-1)\}^j(\min\{\mathbf{y}^*\})^{\alpha_0+j+h_1+\ldots+h_p}}{j!(\alpha_0+j+h_1+\ldots+h_p)}.
\end{aligned}
$$

Q.E.D.

The second special case is given when $\alpha_i = 1$ ($i = 1, ..., p$). That is, all the unique factors are exponentially distributed.

Lemma 6 *The pdf of* $\mathbf{Y}^* \sim \Gamma_p(\mathbf{1}_p\alpha_0+\boldsymbol{\alpha}_1,\beta)$ *of the one-factor model at* $\mathbf{Y}^* = \mathbf{y}^*$, *when the elements of* $\boldsymbol{\alpha}_1 = (\alpha_1, ..., \alpha_p)^{\mathrm{T}}$ *are all one, i.e.,* $\boldsymbol{\alpha}_1 = \mathbf{1}_p$, *is*

$$
\begin{aligned}
& g_{\Gamma,p}\{\mathbf{y}^*|\mathbf{1}_p(\alpha_0+1),\beta\} \\
&= \frac{\beta^{p+\alpha_0}\exp(-\beta y_+^*)}{\Gamma(\alpha_0)}\int_0^{\min\{\mathbf{y}^*\}}f_0^{\alpha_0-1}\exp\{\beta(p-1)f_0\}\,\mathrm{d}f_0 \\
&= \frac{\beta^{p+\alpha_0}\exp(-\beta y_+^*)}{\Gamma(\alpha_0)}\sum_{j=0}^{\infty}\frac{\{\beta(p-1)\}^j(\min\{\mathbf{y}^*\})^{\alpha_0+j}}{j!(\alpha_0+j)}.
\end{aligned}
$$

Proof When the additional condition of $\boldsymbol{\alpha}_1 = \mathbf{1}_p$ in Lemma 5 is imposed, the required result follows. Alternatively, the result is directly obtained from the above integral expression. Q.E.D.

As mentioned earlier, [23] dealt with the case of $\alpha_0 = 1$ with $\beta = 1$ as well as $\boldsymbol{\alpha}_1 = \mathbf{1}_p$, which indicates that all the $p + 1$ common/unique factors are exponentially distributed. Then, consider a scaled Ramabhadran's case with $0 < \beta < \infty$.

Lemma 7 *The pdf of* $\mathbf{Y}^* \sim \Gamma_p(\mathbf{1}_p\alpha_0 + \boldsymbol{\alpha}_1, \beta)$ *of the one-factor model at* $\mathbf{Y}^* = \mathbf{y}^*$, *when the elements of* $\boldsymbol{\alpha} = (\alpha_0, \boldsymbol{\alpha}_1^{\mathrm{T}})^{\mathrm{T}} = (\alpha_0, \alpha_1, ..., \alpha_p)^{\mathrm{T}}$ *are all one, is*
$g_{\Gamma,p}(\mathbf{y}^*|2\mathbf{1}_p, \beta) = \beta^p \frac{\exp\{\beta(p-1)\min\{\mathbf{y}^*\}\}-1}{\exp(\beta y_+^*)(p-1)}$.

Proof When the condition of α_0 is added in Lemma 6, we have

$$\begin{aligned}
& g_{\Gamma,p}(\mathbf{y}^*|\mathbf{1}_p(\alpha_0+1), \beta) = g_{\Gamma,p}(\mathbf{y}^*|2\mathbf{1}_p, \beta) \\
& = \beta^{p+1}\exp(-\beta y_+^*)\sum_{j=0}^{\infty}\frac{\{\beta(p-1)\}^j(\min\{\mathbf{y}^*\})^{j+1}}{(j+1)!} \\
& = \beta^{p+1}\exp(-\beta y_+^*)\frac{1}{\beta(p-1)}\sum_{j=0}^{\infty}\frac{\{\beta(p-1)\}^{j+1}(\min\{\mathbf{y}^*\})^{j+1}}{(j+1)!} \\
& = \beta^p\frac{\exp\{\beta(p-1)\min\{\mathbf{y}^*\}\}-1}{\exp(\beta y_+^*)(p-1)},
\end{aligned}$$

which is also given by the expression
$\beta^{p+1}\exp(-\beta y_+^*)\int_0^{\min\{\mathbf{y}^*\}}\exp\{\beta(p-1)f_0\}\,\mathrm{d}f_0$.
Q.E.D.

Lemmas 5, 6 and 7 yield the following results.

Theorem 4 *(i) The pdf of multivariate power-gamma distributed* $\mathbf{Y} = (Y_1, ..., Y_p)^{\mathrm{T}}$ *at* $\mathbf{Y} = \mathbf{y} = (y_1, ..., y_p)^{\mathrm{T}}$ *with* $\mathbf{y}^* = (y_1^{\gamma_1}, ..., y_p^{\gamma_p})^{\mathrm{T}}$ *when* $\mathbf{Y}^* = (Y_1^{\gamma_1}, ..., Y_p^{\gamma_p})^{\mathrm{T}} \sim \Gamma_p(\mathbf{1}_p\alpha_0 + \boldsymbol{\alpha}_1, \beta)$ *of the one-factor model with the elements of* $\boldsymbol{\alpha}_1 = (\alpha_1, ..., \alpha_p)^{\mathrm{T}}$ *being all positive integers is*

$$\begin{aligned}
& g_{\text{Power-}\Gamma,p}(\mathbf{y}|\mathbf{1}_p\alpha_0 + \boldsymbol{\alpha}_1, \beta, \boldsymbol{\gamma}) \\
& = \frac{\beta^{\alpha_+ + \alpha_0}\exp(-\beta y_+^*)\prod_{i=1}^p|\gamma_i|y_i^{\gamma_i-1}}{\Gamma(\alpha_0)\prod_{i=1}^p(\alpha_i-1)!}\sum\nolimits_{h_1=0}^{\alpha_1-1}\cdots\sum\nolimits_{h_p=0}^{\alpha_p-1}\left\{\prod\nolimits_{i=1}^p\binom{\alpha_i-1}{h_i}y_i^{*(\alpha_i-1-h_i)}(-1)^{h_i}\right\} \\
& \times\sum_{j=0}^{\infty}\frac{\{\beta(p-1)\}^j(\min\{\mathbf{y}^*\})^{\alpha_0+j+h_1+\ldots+h_p}}{j!(\alpha_0+j+h_1+\ldots+h_p)}.
\end{aligned}$$

(ii) In the above case, if the additional condition of $\boldsymbol{\alpha}_1 = \mathbf{1}_p$ *is employed, we have*

$$
\begin{aligned}
&g_{\text{Power-}\Gamma,p}(\mathbf{y}|\mathbf{1}_p(\alpha_0+1),\beta,\boldsymbol{\gamma}) \\
&= \frac{\beta^{p+\alpha_0}\exp(-\beta y_+^*)\prod_{i=1}^p |\gamma_i| y_i^{\gamma_i-1}}{\Gamma(\alpha_0)} \int_0^{\min\{\mathbf{y}^*\}} f_0^{\alpha_0-1}\exp\{\beta(p-1)f_0\}\mathrm{d}f_0 \\
&= \frac{\beta^{p+\alpha_0}\exp(-\beta y_+^*)\prod_{i=1}^p |\gamma_i| y_i^{\gamma_i-1}}{\Gamma(\alpha_0)} \sum_{j=0}^{\infty} \frac{\{\beta(p-1)\}^j(\min\{\mathbf{y}^*\})^{\alpha_0+j}}{j!(\alpha_0+j)}.
\end{aligned}
$$

(iii) When all the elements of $\boldsymbol{\alpha} = (\alpha_0, \alpha_1, ..., \alpha_p)^{\mathrm{T}}$ *are one, we obtain*

$$
g_{\text{Power-}\Gamma,p}(\mathbf{y}|2\mathbf{1}_p,\beta,\boldsymbol{\gamma}) = \beta^p \frac{[\exp\{\beta(p-1)\min\{\mathbf{y}^*\}\}-1]\prod_{i=1}^p |\gamma_i| y_i^{\gamma_i-1}}{\exp(\beta y_+^*)(p-1)}.
$$

Before presenting numerical illustrations, we provide the following properties for clarity.

Result 7 Consider $\mathbf{Y}^* = (Y_1^*, ..., Y_p^*)^{\mathrm{T}} \sim \Gamma_p(\mathbf{1}_p\alpha_0 + \boldsymbol{\alpha}_1, \beta)$, i.e., variables before power transformation for the one-factor model.

$$
\begin{aligned}
\text{(i) } 0 < \alpha_0/\beta^2 &= \mathrm{cov}(Y_i^*, Y_j^*) \\
&< \min\{\mathrm{var}(Y_i^*), \mathrm{var}(Y_j^*)\} = (\alpha_0 + \min\{\alpha_i, \alpha_j\})/\beta^2 \ (i \neq j).
\end{aligned}
$$

(ii) When α_0 goes to $+\,0$, F_0 approaches the one-point distribution of zero in probability with $\mathbf{Y}^*$ approaching p independent gammas. When $\boldsymbol{\alpha}_1 \to \mathbf{0}$, $Y_i^*(i = 1, ..., p)$ approach the identical gamma $\Gamma(\alpha_0, \beta)$.

(iii) $\mathrm{cov}(1/Y_i^*, Y_i^*) < 0$ when the covariance exists. This property is confirmed as

$$
\begin{aligned}
&\mathrm{cov}(1/Y_i^*, Y_i^*) = 1 - \mathrm{E}(1/Y_i^*)\mathrm{E}(Y_i^*) \\
&= 1 - \frac{\Gamma(\alpha_0+\alpha_i-1)\beta}{\Gamma(\alpha_0+\alpha_i)} \frac{\Gamma(\alpha_0+\alpha_i+1)}{\Gamma(\alpha_0+\alpha_i)\beta} = 1 - \frac{\alpha_0+\alpha_i}{\alpha_0+\alpha_i-1} = \frac{-1}{\alpha_0+\alpha_i-1} < 0 \\
&(\alpha_0+\alpha_i > 1).
\end{aligned}
$$

Result 7(i) of positive covariances smaller than the associated variances is a limitation of the model in applications as noted by [22], which can be relaxed by considering the sum of gammas with different scale parameters as in the usual FA model or linear combinations of equal-scale gammas. However, these relaxations generally yield non-gamma distributed marginals with complicated pdfs (see [15, 19]). The obvious properties of Result 7(ii) show that the one-factor model is situated between those of a single gamma and independent p gammas. The negative covariance of Result 7(iii) is expected since $1/Y_i^*$ is a decreasing function in the support.

Numerical illustrations of the pdfs of the bivariate power-gammas when the one-factor models of Lemmas 6 and 7 hold are shown in Figs. 1 and 2, respectively. In Fig. 1 with a single common gamma, $\alpha_0 = 0.2$ is used with two unique independent exponentials, which gives $\text{cor}(Y_1^*, Y_2^*) = \alpha_0/(\alpha_0 + 1) = 1/6$, while in Fig. 2 with three independent common/unique exponentials, we always have $\text{cor}(Y_1^*, Y_2^*) = 1/2$. Three pairs of powers $\boldsymbol{\gamma} = (-2, -1)^T, (-2, 2)^T, (2, 2)^T$ are used for transformation in the figures. The first case of $\boldsymbol{\gamma} = (-2, -1)^T$ gives two decreasing transformations of the original gammas while the second case of $\boldsymbol{\gamma} = (-2, 2)^T$ yields a decreasing and an increasing transform as in Result 7(iii). The rate parameter $\beta = 0.8$ is used for all the cases. The pdfs are shown by a mesh with 30^2 points of $Y_i = 0\ (0.1)\ 3\ (i = 1, 2)$ using finite approximations of the series expression in Lemma 6. As suggested by Result 7(iii), in Fig. 1, the relatively less correlated Y_i^*s by construction are found to be conveyed to the corresponding transformed ones. In Fig. 2, increased correlations are shown in the transformed variables.

Table 1 gives the approximate moments for the power-gamma distributed variables, i.e., Y_is shown in Figs. 1 and 2. The approximations are obtained by the mesh points with the corresponding pdfs. The largest correlation coefficient 0.48 is given by Ex. (2.3) and the smallest −0.40 by Ex. (2.2).

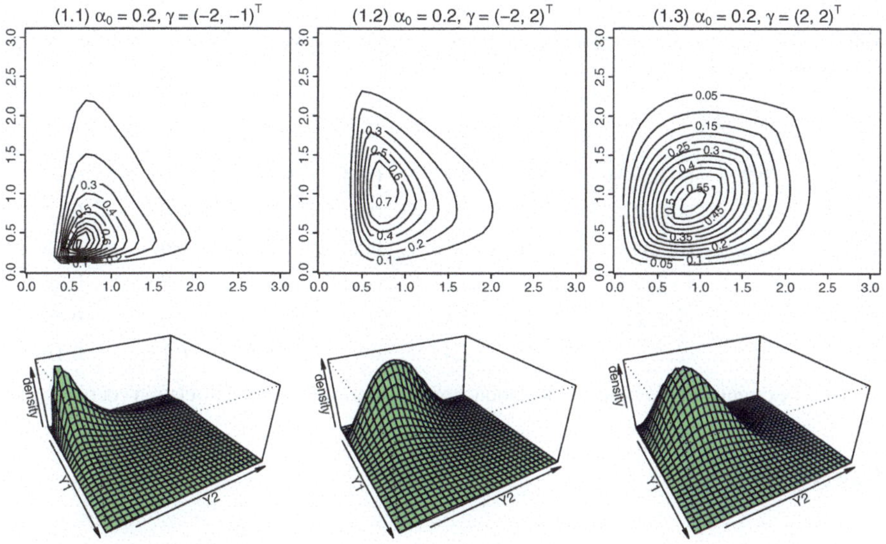

Fig. 1 Density plots of the bivariate power-gamma distributions with a common gamma and unique exponentials (Y_1 = the horizontal axis, Y_2 = the vertical axis for upper plots, $\beta = 0.8$)

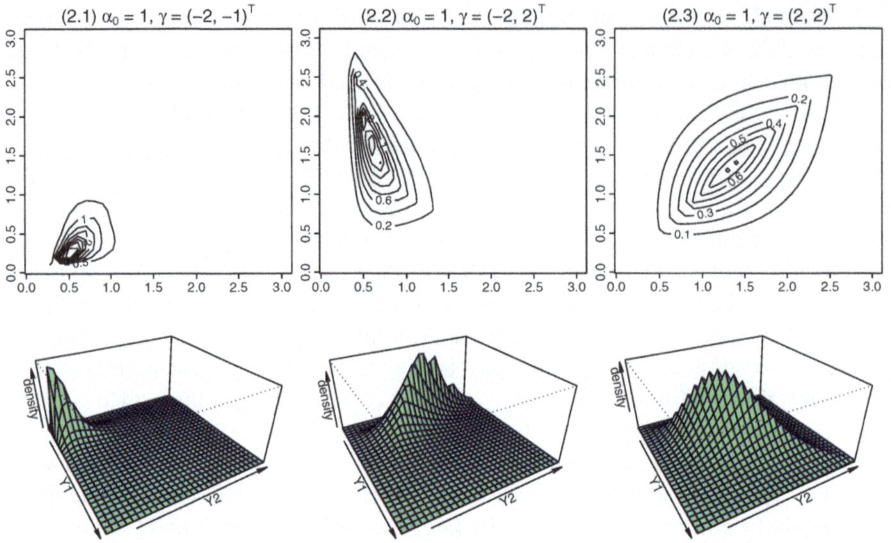

Fig. 2 Density plots of the bivariate power-gamma distributions with common and unique exponentials (Y_1 = the horizontal axis, Y_2 = the vertical axis for upper plots, $\beta = 0.8$)

Table 1 The approximate moments for the bivariate power-gamma distributions for the examples used in Figs. 1 and 2

Ex. No.	α_0	γ^T	mean		SD		cov	cor
			Y_1	Y_2	Y_1	Y_2		
(1.1)	0.2	(−2, −1)	1.06	0.95	0.53	0.68	0.04	0.10
(1.2)	0.2	(−2, 2)	1.07	1.11	0.53	0.52	−0.04	−0.13
(1.3)	0.2	(2, 2)	1.10	1.10	0.52	0.52	0.04	0.16
(2.1)	1	(−2, −1)	0.77	0.64	0.34	0.51	0.06	0.35
(2.2)	1	(−2, 2)	0.78	1.48	0.36	0.52	−0.07	−0.40
(2.3)	1	(2, 2)	1.47	1.47	0.52	0.52	0.13	0.48

Note Ex. = example, SD = standard deviation, cov = covariance, cor = correlation coefficient

References

1. Abadir, K. M. (1999). An introduction to hypergeometric functions for economists. *Econometric Reviews, 18*, 287–330.
2. Amoroso, L. (1925). Ricerche intorno alla curva dei redditi. *Annali di Mathematica Pura ed Applicata, Serie IV, Tomo, II*, 123–159.
3. Balakrishnan, N., & Lai, C. D. (2009). *Continuous bivariate distribution* (2nd ed.). Springer.
4. Bernard, J. M., & Smith, A. F. M. (1994). *Bayesian theory*. Wiley.
5. Bollen, K. A. (1989). *Structural equations with latent variables*. Wiley.
6. Cherian, K. C. (1941). A bi-variate correlated gamma-type distribution function. *The Journal of the Indian Mathematical Society, 5*, 133–144.

7. Crooks, G. E. (2015) *The Amoroso distribution.* arXiv:1005.3274v2 [math.ST]
8. Ferrari, S. L. P., & Fumes, G. (2017). Box-Cox symmetric distributions and applications to nutritional data. *AStA Advances in Statistical Analysis, 101*, 321–344.
9. Furman, E. (2008). On a multivariate gamma distribution.*Statistics and Probability Letters, 78*, 2353-2360. See also Su, J., & Furman, E. (2012). Erratum to "On a multivariate gamma distribution by E. Furman" [Statist. Probab. Lett. 78 (2008) 2353–2360]. *Statistics and Probability Letters, 82*, 1040–1041.
10. Gómez, Y. M., & Bolfarine, H. (2015). Likelihood-based inference for the power half-normal distribution. *Journal of Statistical Theory and Applications, 14*, 383–398.
11. Harman, H. H. (1976). *Modern factor analysis* (3rd ed.). University of Chicago Press.
12. Johnson, N. L., Kotz, S., & Balakrishnan, N. (1994). *Continuous univariate distributions* (2nd ed., Vol. 1). Wiley.
13. Kollo, T., Käärik, M., & Selart, A. (2021). Multivariate skew *t*-distribution: Asymptotics for parameter estimators and extension to skew *t*-copula. *Symmetry, 13*, 1059. https://doi.org/10.3390/sym13061059
14. Kotz, S., Balakrishnan, N., & Johnson, N. L. (2000). *Continuous multivariate distributions* (2nd ed.). Vol. 1: Models and applications. Wiley.
15. Mathai, A. M. (1982). Storage capacity of a dam with gamma type inputs. *Annals of the Institute of Statistical Mathematics, 34*(A), 591–597.
16. Mathai, A. M., & Moschopoulos, P. G. (1991). On a multivariate gamma. *Journal of Multivariate Analysis, 39*, 135–153.
17. Mathai, A. M., & Moschopoulos, P. G. (1992). A form of multivariate gamma distribution. *Annals of the Institute of Statistical Mathematics, 44*, 97–106.
18. Morán-Vásquez, R. A., & Ferrari, S. L. P. (2019). Box–Cox elliptical distributions with application.*Metrika, 82*, 547–571.
19. Moschopoulos, P. G. (1985). The distribution of the sum of independent gamma random variables. *Annals of the Institute of Statistical Mathematics, 37*(A), 541–544.
20. Ogasawara, H. (2022). The density of the sample correlations under elliptical symmetry with or without the truncated variance-ratio. *Journal of Multivariate Analysis* (online published). https://doi.org/10.1016/j.jmva.2022.105152. Preprint at https://doi.org/10.13140/RG.2.2.35609.08807
21. Ogasawara, H. (2022). *Expository moments for pseudo distributions.* Springer Nature. https://doi.org/10.1007/978-981-19-3525-1/.
22. Prékopa, A., & Szántai, T. (1978). A new multivariate gamma distribution and its fitting to empirical stream flow data. *Water Resources Research, 14*, 19–29.
23. Ramabhadran, V. K. (1951). A multivariate gamma-type distribution. *Sankhyā, 11*, 45–46.
24. Rigby, R. A., & Stasinopoulos, D. M. (2004). Smooth centile curves for skew and kurtotic data modelled using the Box-Cox power exponential distribution. *Statistics in Medicine, 23*, 3053–3076.
25. Royen, T. (1991). Multivariate gamma distributions with one-factorial accompanying correlation matrices and applications to the distribution of the multivariate range. *Metrika, 38*, 299–315.
26. Royen, T. (2007). Integral representations and approximations for multivariate gamma distributions. *Annals of the Institute of Statistical Mathematics, 59*, 499–513.
27. Stacy, E. W. (1962). A generalization of the gamma distribution. *The Annals of Mathematical Statistics, 33*, 1187–1192.
28. Voudouris, V., Gilchrist, R., Rigby, R., Sedgwick, J., & Stasinopoulos, D. (2012). Modelling skewness and kurtosis with the BCPE density in GAMLSS. *Journal of Applied Statistics, 39*, 1279–1293.

Looking Back on Japan in the Heisei Era from the Perspective of Repeated Survey Data: Focusing on a Sense of "Giri Ninjō (Obligation-Human Feeling)" and "Kurashi-Kata" (Attitudes Towards Life) in the Surveys on the Japanese National Character

Yoosung Park and Tadahiko Maeda

Abstract This paper looks at the 30-year period of Japan in the Heisei era (1989–2019), based mainly on the results of different waves of the Surveys on the Japanese National Character. The paper outlines changes and constants in certain aspects of Japanese attitudes, behaviours and values during the Heisei period. Japan in the Heisei era was a time of reform and change in all aspects of society, from the bursting of the economic bubble to the global recession, including the collapse of Lehman Brothers and the Great East Japan Earthquake. This may have led to some changes in the traditional Japanese view of human relationships and attitudes. This paper therefore seeks to clarify the values and attitudes of the Japanese people based on data from the 9th (1998) to 14th (2018) waves of the Surveys on the Japanese National Character, focusing on the survey's long-term, continuous items, "Giri ninjō" and "kurashi-kata" (attitudes towards life). The specific findings were as follows: (1) Looking at the responses to the "Giri ninjō" item, it was confirmed that "Giri ninjō" relationships were still favoured in the Heisei period, but there was a slight change in the relationship among age groups. (2) Furthermore, it was also confirmed that "kurashi-kata" (attitudes towards life), including the fulfilment of one's own comfort, does not lead to an attitude that values the image of "Giri ninjō" relationships such as righteous and humane human relationships.

Keywords Survey on the Japanese national character · "Giri ninjō" · "kurashi-kata" (attitudes towards Life) · Repeated cross-sectional surveys

Y. Park (✉) · T. Maeda
The Institute of Statistical Mathematics, 10-3 Midori-Cho, Tachikawa City, Tokyo 190-8562, Japan
e-mail: parkys@ism.ac.jp

T. Maeda
e-mail: maeda@ism.ac.jp

© The Author(s), under exclusive license to Springer Nature Singapore Pte Ltd. 2023

A. Okada et al. (eds.), *Facets of Behaviormetrics*, Behaviormetrics: Quantitative Approaches to Human Behavior 4, https://doi.org/10.1007/978-981-99-2240-6_14

1 Introduction

This paper looks at the 30-year period of Japan in the Heisei era (1989–2019), based mainly on the results of multiple waves of the Surveys on the Japanese National Character. The paper outlines changes and consistencies in certain aspects of Japanese attitudes, behaviours and values during this Heisei period. The purpose of the "Surveys on the Japanese National Character" is to conduct statistical surveys on people's attitudes, opinions and sentiments in everyday situations and to quantitatively clarify the characteristics of the "Japanese way of thinking and viewing things" [6]. The Institute of Statistical Mathematics (ISM) has conducted a nationwide survey every 5 years since 1953. The 14th nationwide survey was conducted from October to December 2018, the 65th anniversary of the start of the survey. It has been carried out as repeated cross-sectional surveys, using basically the same sample design and survey mode (face-to-face interviewing) and the same questionnaire.

In characterizing the Heisei period, [31] describes these 30 years as a "period of failure" and argues that this is partly because the fundamental transformation of the social system, namely, globalization and the information age, coincided with a period of decline in the population structure. On the other hand, according to the [4], there was a widespread perception during the Heisei period that people's lifestyles and consumption motives were geared towards neither consumption nor savings and that their outlook for the future of the world was neither good nor bad, even though they held deeply "middle-class" self-identities. The perception that the world would neither improve nor worsen in the future was widespread. However, these are only macroeconomic approaches that neglect consideration of actual events.

It is expected that the particular values and ways of thinking of people during the Heisei period will emerge only by comparing them with those of people during the Showa period, which was characterized by rapid development and growth after World War II aimed at transitioning from a traditional society to a modern society. This kind of comparison over time can be made most effectively through long-term continuous research, but the changes in Japanese values and attitudes over time have not been sufficiently examined in light of the various concepts that describe the process of social transformation, such as industrialization, urbanization, individualization and rationalization, which have been discussed as characteristics of the transition from a premodern society to a modern society. Therefore, it is also effective to examine the recent survey from the viewpoint of the linkage between the question items to capture the awareness of people in the Heisei period and the question items to capture traditional values, as well as from the viewpoint of the linkage between the items to capture traditional values.

Based on this awareness of the issue, [20], for example, based on data from the 14th (2018) nationwide Surveys on the Japanese National Character, argues that the evaluation of intent of out-migration from urban areas to rural areas question "Do you want to make a U-turn or I-turn?" tends to differ among age groups, with the younger and middle-aged groups being more likely to migrate to rural areas than older groups. Furthermore, Park finds that (a) a sense of "Giri ninjō has a positive effect on the intent to engage in out-migration from urban areas to rural areas and (b) this tendency is more pronounced in the young- and middle-aged age groups. This implies that the penetration of the mentality of "Giri ninjō", which is an ambiguous but deep-rooted preference in interpersonal relationships that Japanese people have long felt, is one factor influencing the psychological mechanism that promotes migration and settlement in rural areas. In other words, the results suggest that young- and middle-aged Japanese are breaking away from individualization and urbanization, which are considered to be characteristics of modernization.

The Survey on the Japanese National Character Committee, led by Chikio Hayashi, has long inquired about "Giri ninjō", which is often depicted in art and literature as a characteristic of Japanese human relationships, and has developed a scale to measure this concept over the years ISM National Character Committee (1961), [6], etc. Hayashi and his colleagues have also submitted this concept translated from Japanese to English. For example, [8], at a relatively early stage, loosely defined this concept as "duty and affection", contrasting righteousness and humanity, and later they (1992, p. 241) described it as "duty and humanity". Hayashi and Kuroda [10] also explained this concept as " Obligation-Human feeling". However,there is no settled translation for these concepts, and this paper will adopt the [10]'s "Giri ninjō (Obligation-Human feeling)" as a provisional translation.

This scale has been used for many years since the third survey (1963). Hayashi and Sakuraba [13] stated that "Giri ninjō" is one of the fundamental characteristics of modern Japanese that are accepted by today's youth without any sense of discomfort.

The definition of "Giri ninjō" has not been clearly defined, but examples exist. According to the Gakken 4-character phrase dictionary,[1] it is "a social relationship or consideration for others that cannot be avoided because of past circumstances or ties". In other words, the concept of "Giri ninjō" does not indicate a clear and limited relationship with a rational mindset and norms, such as a mutually beneficial relationship formed between an employer and employee, but is rather defined as a relationship that tends to be difficult to avoid because of past experiences or ties and that is more emotional than rational, such as consideration for others. Hayashi [14] states that "Giri ninjō" does not indicate merely a parallel between righteousness and humanity but suggests a conflict between righteousness and humanity, and people are moved by this conflict, which is a theme in kabuki and many other stories.

On the other hand, social capital, a concept that describes relationships with others, is "the characteristics of social mechanisms such as trust, norms and networks that can enhance the efficiency of society by stimulating cooperative behavior among people" [22]. Inaba [15] further summarizes the concept in the context of Japanese society as "trust, norms of reciprocity and mutuality such as 'the good you do for others is good you do yourself', 'helping and being helped', and ties among people and groups". One thing to note here is that [15] and many social capital theorists focus on relationships with others under a clear norm of reciprocity.

---

[1] Online web dictionary, referenced on 6 November 2022. https://dictionary.goo.ne.jp/word/%E7%BE%A9%E7%90%86%E4%BA%BA%E6%83%85_%28%E5%9B%9B%E5%AD%97%E7%86%9F%E8%AA%9E%29/

In contrast, the concept of "Giri ninjō" proposed by Hayashi and his colleagues represents human relationships beyond clear norms of reciprocity. Therefore, their view of this concept is that it includes altruistic attitudes towards others in a broader sense that go beyond norms of reciprocity and so-called humanistic feelings.

Hayashi [6] confirms, based on data from the "Surveys on the Japanese National Character" from 1963 to 1998, that the degree of "Giri ninjō" is highly stable across time, although there are some variations. The same tendency was also found in an analysis using the data from the 12th Survey on the Japanese National Character (2008) after the latter half of the 2000s [14], and [14] pointed out that the degree of "Giri ninjō" remained high across all age groups. Hayashi [14] points out that the degree of "Giri ninjō" is still high, regardless of age group. In other words, as Hayashi and his colleagues pointed out, "Giri ninjō" is by no means an outdated concept; it is one of the expressions of Japanese human relationships that persist even among contemporary Japanese. However, the degree of "Giri ninjō" over the whole Heisei period has not yet been examined.

This paper, focusing on the Heisei period, is based on data from the 9th (1993) to 14th (2018) waves of the Surveys on the Japanese National Character, specifically on two long-term, continuous items, the "Giri ninjō" scale used since the 3rd (1963) survey and the "kurashi-kata" ("attitude towards life", item number #2.4) item, to clarify the values and attitudes of the Japanese people during the Heisei era so as to develop an idea of human relationships unique to the Japanese people.

2 Research Methods

2.1 *Overview of the "Surveys on the Japanese National Character": Sample and Data*

The Research Committee on the Study of the Japanese National Character of the Institute of Statistical Mathematics (ISM) has conducted the "Surveys on the Japanese National Character" once every 5 years since the first survey in 1953. Every survey was conducted by face-to-face interviews with a sample drawn from the population by probability sampling. Specifically, individuals selected by stratified multistage random sampling included Japanese men and women aged 20 years or older residing throughout Japan. The age range of the sample has been set between 20 and 84 years old since the 13th survey (2013), and it was between 20 and 80 years old for the 11th (2003) and 12th (2008) surveys.

In the "Surveys on the Japanese National Character", two types of questionnaires (K-type and M-type), each with different questions, are assigned equally among the sample at each survey site. In this study, we use the K-type questionnaire items, including the relevant items of "Giri ninjō" and "kurashi-kata" (attitude towards life), which are representative, long-term and continuous questions.

The sample size (intended sample) of the K-type questionnaire for the 2018 survey was 3209, with 1584 finally collected, for a response rate of 49.4%. For details on the recent administration of the "Surveys on the Japanese National Character", including the sample design, survey methodology and size of the K-type and M-type samples in the 13th survey, see [18, 28].

2.2 Measures

The questions in the "Surveys on the Japanese National Character" are mainly related to people's values, attitudes and consciousness and are numbered and organized by question domain (for details, see [18]) and the "List of Survey Items" by the Research Committee on the Study of the Japanese National Character (2021). The terms in < > are the terms used in this paper and may differ from the standard item terms in the National Character Surveys, which are indicated by quotation marks (") after the number.

The analysis uses seven items related to "Giri ninjō" (#4.4, #5.1, #5.1b, #5.1c1, #5.1c2, #5.6, #5.1d) and "kurashi-kata" (attitudes towards life) (#2.4), which are described below. The categories related to these items were treated in the same way as in the past, given that these items are long-term, continuous items.

In addition, the following basic attributes of respondents were used: #1.1 "gender", #1.2 "age", #1.3 "educational background", #1.5 "urban vs. rural" and #1.6 "regions". The age groups were divided into six levels, ranging from their 20 s to their 70 s and above. Educational background was divided into three levels: "low education (junior high school graduate or lower)", "middle education (high school graduate)", and "high education (college graduate or higher)". In addition to these variables, we also used #1.5 "urban vs. rural" to reflect the size of the city where respondents reside. For example, the degree of solidarity in relationships within a community may differ depending on the city size. Therefore, we need to consider the possibility that city size may affect responses. In this study, the seven-level classification of the original #1.5 "urban vs. rural" question was recoded into three categories: "town and villages/city with a population of up to 49,999", "city with a population of 50,000 to 499,999" and "city with a population of more than 500,000 or 6 major cities". Note that the survey does not include any item that asks respondents their income level, and therefore it is not possible to conduct an analysis using objective variables related to the economic status of individuals or households. SPSS Statistics 21.0 J for Windows was used for the analysis.

In this study, we confirm the relationship between the responses to the "Giri ninjō" and "kurashi-kata" (attitudes towards life) questions and the respondents' attributes. Then, we use ordinary least squares (OLS) regression to clarify the psychological mechanism by which attitudinal factors such as "kurashi-kata" (attitudes towards life) affect "Giri ninjō".

3 Results and Discussion

3.1 Preferred Human Relationships as Revealed by the Distribution of Answers to Questions on "Giri Ninjō" Scale

The "Giri ninjō" scale measure what is related to humane "Giri ninjō" relationships, a significant characteristic of Japan [6]. The scale is based on a combination of answers to several questions. These questions do not simply contrast between the rational mindset of the "norm of reciprocity" and emotionality found in consideration of others ("humane feelings") but attempt to capture a person's "preference for warmhearted human relations" by combining the two [11].

The questions asked to capture the sense of "Giri ninjō" are shown in Tables 1 and 2 which shows the distribution of responses to each of the seven questions in each survey round, mainly the Heisei period. To calculate the "Giri ninjō" scale, we scored the number of "positive" responses to the seven questions shown in Table 2, three of which are defined by answers to the specific category of the respective item, and two of which are formed by cross-classifying answers to two items (Research Committee of National Character and The Institute of Statistical Mathematics 1961). Specifically, for #4.4 "rumour about teacher", #5.6 "preferred type of boss", #5.1d "important values", each response is scored as one, while the responses to #5.1 "benefactor on his or her deathbed" and #5.1b "father on his deathbed" and to #5.1c1 "employment examination: relative" and #5.1c2 "employment examination: child of a benefactor" are combined and scored as one [6]. Thus, the scale has a maximum of 5 points.

We decided to follow [6] scoring principles. Specifically, for <Scale item 1>, #4.4, "rumour about teacher", the two options are "deny it" and "tell the truth"; the former is scored as one for a positive ("Giri ninjō"-oriented) response. For <Scale item 2>, responses to #5.1 and #5.1b are combined. For #5.1, "benefactor on his or her deathbed", there are two options: "leave everything and go back home", and "however worried he might be about the benefactor, he should go to the meeting". For #5.1b, "father on his deathbed", the response options are "leave everything and go back home" and "however worried he might be about his father, he should go to the meeting". For these items, simultaneously choosing to go back home for #5.1 and to go to the meeting for #5.1b is scored as one for a positive response. For <Scale item 3>, #5.1c1 "employment examination: relative", the response choices are "the one with the highest grade" and "your relative", and for #5.1c2 "employment examination: child of a benefactor", the response choices are "the one with the highest grade" and "the son of your benefactor". A positive response is the combination of "the one with the highest grade" for the former and "the son of your benefactor" for the latter. Making different choices in two settings such as these is a typical example of what was stated earlier about measuring relationship preferences in terms of the combination of a rational mindset and emotionality. For <Scale Item 4>#5.6 "preferred type of boss", choosing the response choice "a person who sometimes

Table 1 Questions and answers in the "Giri ninjō" scale

Question
<"Giri ninjō" 1 >#4.4 Rumor about teacher Suppose that a child comes home and says that he has heard a rumor that his teacher had done something to get himself into trouble, and suppose that the parent knows this to be true. Do you think it is better for the parent to tell the child the truth, or to deny it? **1** Deny it **2** Tell the truth
<"Giri ninjō" 2 >#5.1 Benefactor on death-bed #5.1b Father on death-bed (Hand card) Imagine this situation. Mr.A was orphaned at an early age and was brought up by Mr.B, a kind neighbor. Mr.B gave him a good education, sent him to a university, and now Mr.A has become the president of a company. One day he gets a telegram saying that Mr.B who brought him up, is seriously ill and asking if he would come at once. This telegram arrives as he is leaving to attend an important meeting which will decide whether his firm is to go bankrupt or to survive. Which of the following do you think he should do? **1** Leave everything and go back home **2** However worried he might be about Mr.B, he should go to the meeting (Hand card) The last question supposed that Mr.B had taken him in as an orphan in his youth and brought him up. Suppose that it was his real father who was seriously ill. Which would have been your answer then? **1** Leave everything and go back home **2** However worried he might be about his father, he should go to the meeting
<"Giri ninjō" 3 >#5.1c1 Employment examination: relative #5.1c2 Employment examination: a child of benefactor (Hand card) Suppose that you are the president of a company. The company decides to employ one person, and then carries out an employment examination. The supervisor in charge reports to you saying, "Your relative who took the examination got the second highest grade. But I believe that either your relative or the candidate who got the highest grade would be satisfactory. What shall we do?" In such a case, which person **1** One with the highest grade **2** Your relative (Hand card) In the last question we supposed that the one getting the second highest grade was your relative. Suppose that the one who got the second highest grade was the son of parents to whom you felt **1** One with the highest grade **2** Son of your benefactor
<"Giri ninjō" 4 >#5.6 Type of boss preferred (Hand card) Suppose you are working in a firm. Which of the following department chiefs would you prefer to work under? **1** A person who always sticks to the work rules and never demands any unreasonable work, but who, on the other hand, never does anything for you personally in matters not **2** A person who sometimes demands extra work in spite of rules against it, but who, on the other hand, looks after you personally in matters not connected with work
<"Giri ninjō" 5 >#5.1d Important values (Hand card) If you were asked to choose the two most important items listed on this card, which two would you choose? **1** Respect for parents **2** Repaying people who have helped you in the past **3** Respect for the rights of the individual **4** Respect for the freedom of the individual

Note 1. [Hand card] after question numbers indicates that a card (list) of options was shown to the respondents for their response.

Note 2. Each question shows only the major options, with the common options of 'Other' and 'D.K.' (don't know) omitted.

Table 2 Distribution of answers to "Giri ninjō" questions (%)

		IX (1993) (%)	X (1998) (%)	XI (2003) (%)	XII (2008) (%)	XIII (2013) (%)	XIV (2018) (%)
4.4 Rumor about teacher	Deny it	24	20	15	21	21	20
	Tell the truth	59	64	64	63	60	60
#5.1 Benefactor on death-bed	Leave everything and go back home	49	46	41	43	39	39
	However worried he might be about Mr. B, he should go to the meeting	42	45	48	50	53	53
#5.1b Father on death-bed	Leave everything and go back home	48	44	40	46	44	46
	However worried he might be about his father, he should go to the meeting	44	47	51	48	49	47
#5.1c1 Employment examination: relative	One with the highest grade	67	70	73	79	78	79
	Your relative	24	22	18	16	18	14
#5.1c2 Employment examination: a child of benefactor	One with the highest grade	45	49	54	58	56	61
	Son of your benefactor	44	42	35	36	38	31
#5.6 Type of boss preferred	A person who always sticks to the work rules and never demands any unreasonable work, but who, on the other hand, never does anything for you personally in matters not connected with	12	16	18	15	17	22
	A person who sometimes demands extra work in spite of rules against it, but who, on the otherhand, looks after you personally in matters not connected with work	82	80	77	81	81	74

(continued)

Table 2 (continued)

		IX (1993) (%)	X (1998) (%)	XI (2003) (%)	XII (2008) (%)	XIII (2013) (%)	XIV (2018) (%)
#5.1d Important values	Respect for parents	69	70	73	76	75	69
	Repaying people who have helped you in the past	43	43	47	57	58	56
	Respect for the rights of the individual	38	40	37	27	32	35
	Respect for the freedom of the individual	42	43	37	36	33	35

demands extra work despite there being rules against it but who, on the other hand, looks after you personally in matters not connected with work" rather than "a person who always sticks to the work rules and never demands any unreasonable work, but who, on the other hand, never does anything for you personally in matters not connected with work" is treated as a positive response. Finally, for <Scale Item 5>, #5.1d "important morals", there are four opinions: "respect for parents", "repaying people who have helped you in the past", "respect for the rights of the individual" and "respect for the freedom of the individual". Respondents are asked to choose two of the four. Choosing both "respect for parents" and "repaying people who have helped you in the past" is treated as a positive response. The score given to each positive response is one, with no particular weight. The total value of these responses is used to evaluate the respondents' "Giri ninjō" orientation. Thus, the scale values range from 0 to 5, with higher values indicating a more "Giri ninjō" orientation.

Since it would be difficult to examine all changes across the 12 rounds in the "Giri ninj" scale starting in 1963, we focus on the Heisei period ranging from 1993 to 2018 (see Fig. 1).

The early years of the Heisei era (1993–1998) were a period of economic recession and political instability that accompanied a "bubble burst" [6]. As a result, a general sense of stagnation prevailed in society [6].

The periods of the survey that reveal similar distributions are (1998, 2003), (2008, 2013) and (2018), with 2008 and 2013 being close to 1988. The values that decreased from 1993 to 1998 in the early Heisei period rose in the following three surveys starting in 2003. In other words, it was expected that the values on the "Giri ninjō" scale in the Showa period would not be high and that the values on the "Giri ninjō" scale in the Heisei period would continue to decrease. However, the 2008 and 2013 surveys show a return to the high values of the 1988 level.

Table 3 shows the "Giri ninjō" scale scores by age group. Although the "Giri ninjō" scale scores may seem consistent, the criteria for categorizing the scores into "high" or "low" have not been theoretically determined. According to many previous studies by Hayashi and his colleagues [6, 12], the distribution of "Giri ninjō" scale values in

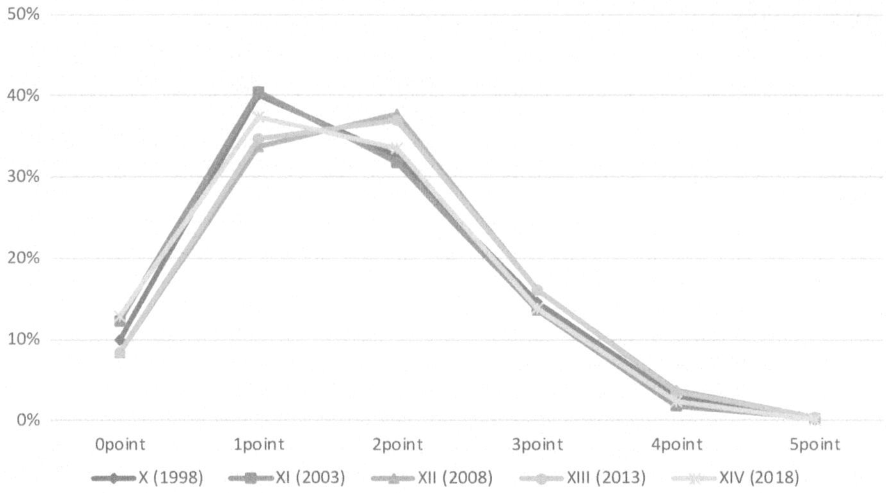

Fig. 1 Variation in the values of the "Giri ninjō" scale during the Heisei era

the waves of the Surveys on the Japanese National Character to date shows that there are peaks at scale values 1 and 2, with 0 being only a small portion of respondents. Since scale values of 1 and 2 are common among Japanese, a scale value of 3 or higher means that the respondents are fairly "Giri ninjō"-oriented according to the scale [11]. Thus, in a practical sense, the comparison of response patterns of typically low "Giri ninjō", at a scale value of 0, and that of high "Giri ninjō", at a scale value of three or higher, should be compared.

In the early stage of the project, the Surveys on the Japanese National Character focused on contrasts between "traditional" and the "modern" values during the postwar Showa period [6]. Chikio Hayashi and other members of the survey team investigated and interpreted changes in traditional Japanese attitudes during the postwar Showa period using the "Giri ninjō" scale and other multiquestion instruments that measure traditional attitudes (Hayashi 1973). Moreover, it has been pointed out that the path of the idea of traditional ways of thinking have changed in terms of content over time [14]. To trace the points made in previous studies and to examine the tendency of responses to the "Giri ninjō" scale during the Heisei period, focusing on differences among age groups, Table 3 shows the distribution of scale values by age group in 10-year increments. Table 3 also includes a column that shows the percentage of "Giri ninjō" scale values of 3 or higher (3 + points) by age group.

Looking at the change in the percentage of those with a scale value of 3 or higher, one can see that the percentage of those aged 60 and above with a scale value of 3 or higher was high in the late 1980s, when the scale values became higher overall, and the number of Japanese with a "Giri ninjō" attitude was expected to decrease as the proportion of older age groups that had shown high scale values until the 1998 survey decreased. However, as [14] also showed, the "Giri ninjō" scale values were again high for all age groups in the 2008 survey, indicating a revival trend. This can

Table 3 Distribution of scores on the "Giri ninjō" scale by age group and survey period (%)

		0point	1point	2point	3point	4point	5point	total	0point(%)	3point+(%)
X (1998)	20-29	18%	42%	28%	10%	1%	0%	173	18%	11%
	30-39	11%	41%	37%	10%	2%	0%	207	11%	12%
	40-49	10%	42%	32%	13%	3%	0%	265	10%	16%
	50-59	9%	42%	33%	13%	3%	0%	267	9%	16%
	60-69	6%	38%	30%	20%	5%	0%	255	6%	26%
	70+	7%	36%	34%	20%	2%	0%	172	7%	23%
XI (2003)	20-29	17%	42%	31%	9%	0%	0%	121	17%	9%
	30-39	13%	40%	33%	13%	0%	0%	209	13%	13%
	40-49	14%	41%	30%	14%	2%	0%	195	14%	16%
	50-59	12%	45%	29%	12%	1%	1%	268	12%	14%
	60-69	12%	38%	34%	12%	3%	0%	253	12%	15%
	70+	5%	35%	33%	22%	5%	0%	146	5%	27%
XII (2008)	20-29	10%	38%	34%	15%	4%	0%	186	10%	18%
	30-39	10%	35%	37%	16%	2%	0%	290	10%	18%
	40-49	9%	31%	39%	16%	5%	0%	283	9%	20%
	50-59	9%	33%	40%	13%	4%	1%	359	9%	18%
	60-69	6%	34%	36%	21%	3%	1%	381	6%	24%
	70+	7%	32%	40%	15%	5%	1%	230	7%	21%
XIII (2013)	20-29	11%	47%	29%	9%	4%	0%	150	11%	13%
	30-39	10%	36%	33%	18%	4%	0%	261	10%	22%
	40-49	12%	32%	38%	16%	2%	1%	250	12%	18%
	50-59	8%	34%	41%	12%	4%	1%	299	8%	17%
	60-69	8%	33%	37%	19%	3%	0%	316	8%	22%
	70+	4%	33%	40%	19%	4%	0%	315	4%	24%
XIV (2018)	20-29	20%	39%	29%	11%	1%	0%	131	20%	12%
	30-39	18%	36%	29%	14%	3%	0%	224	18%	17%
	40-49	13%	35%	32%	17%	3%	0%	288	13%	20%
	50-59	14%	39%	35%	11%	2%	0%	267	14%	13%
	60-69	9%	35%	41%	13%	2%	0%	310	9%	15%
	70+	9%	40%	33%	15%	3%	0%	364	9%	18%

also be seen from the overall response percentages to individual questions in Table 2, which confirms that "positive" responses are generally higher than "negative" ones.

In the subsequent 2013 survey, the "Giri ninjō" scale values were higher among the middle-aged group in their 30 and 40 s and the older age group in their 60 s and above than among other age groups, continuing the revival trend. Furthermore, in the 2018 survey, the "Giri ninjō" scale values were higher among middle-aged people in their 40 s than among other age groups. However, the higher scale values in the 60 and older age group, which were also noticeable in the Showa and early Heisei periods and in the 2008 and 2013 surveys, were not observed in the 2018 survey and decreased by 11 points from those in the 1998 survey (1998 survey: 26% for those aged 60–69 years; 2018 survey: 15% for those aged 60–69).

An interesting point that emerges in the mid-to-late Heisei period is the high rate of the response "going back home" to the questions about one's benefactor or father being on his deathbed. In the Showa period, "going to meetings" when one's parents were on their death beds increased, in accordance with the original meaning of the "Giri ninjō" way of thinking (e.g. [11], but in the Heisei period, the meaning of the "Giri ninjō" way of thinking seem to have changed.

A detailed examination of this table suggests that the distribution of scores must have undergone a subtle change during the Heisei period. The evidence for this claim is that while the tendency of the proportion of respondents scoring 0 being higher among younger age groups than older age groups remained almost the same throughout the period, the difference between the age groups of respondents who scored three or higher, i.e. those who favoured "Giri ninjō" answers, which was clear in the early Heisei period, seems to have gradually levelled off in the later survey periods. The fact that differences among respondents of different age groups who prefer "traditional" and "modern" answers became clear in recent years might indicate the scale has lost ability to function as a contrasting axis of "traditional" versus "modern" values.

3.2 Changes in Lifestyle Goals as Revealed by the Distribution of Answers to the "Kurashi-Kata" (Attitudes Towards Life) Question

A question about "kurashi-kata" (#2.4 "attitudes towards life"), referencing "preferred way of life" or "life goals", has been included in the survey since its inception in 1953.[2] This item was originally taken from a prewar "Sotei" survey (a kind of adult skills survey administered to the male population of a certain age) question; this representative question has appeared in multiple waves of the National Character

[2] The Japanese term for this item (#2.4) is "Kurashi-kata", a direct translation of which is "(preferred) way of living", rather than "attitudes towards life". However, the latter is the formal English term for this item in this project and is thus adopted here.

Survey and has been used as a straightforward indicator of changes in life goals in the postwar period [24].

The specific survey item is as follows:

(*Hand on a card*) There are all sorts of attitudes towards life. Which one of the following statements would you say comes closest to your way of life?

1 Work hard and get rich.
2 Study earnestly and make a name for yourself.
3 Don't think about money or fame; just live a life that suits your own taste.
4 Live each day as it comes, cheerfully and without worrying.
5 Resist all evils in the world and live a pure and just life.
6 Never think of yourself; give everything in service of society.
7 Other (Specify).
8 Don't know.

Furthermore, Tables 3 and 4 show changes in the distribution of responses in just the Heisei period. The relative stability in the percentages during that period is also evident in the table. Looking at the responses to this question from 1953 to 1998, as shown in Fig. 2., one can see that the response percentage of those answering "work hard and get rich", remained relatively stable at approximately 15%, indicating that a certain number of people (approximately 1/6) in each survey period set such goals for their lives. On the other hand, the proportion of those selecting "study earnestly and make a name for yourself" and "never think of yourself; give everything in the service of society" did not change noticeably. Although their percentage values were small from the beginning of the survey, they show an overall decreasing trend, seemingly bottoming out in recent years [6].

The most notable decrease in Fig. 2 is observed for "resist all evils in the world and live a pure and just life". To be more precise, this was the most common response (29%) in 1953, but it declined to 11% over the last 20 years of the study period. On the other hand, especially in the Showa period, two response options, "do not think about money or fame; just live a life that suits your taste" and "live each day as it comes, cheerfully and without worrying", increased. The former became the most common opinion as early as 1958. These two options remained the most common during the Heisei period, and this indicates that Japanese people in the Heisei period placed importance on enjoying the things around them and in their daily lives. Furthermore, this tendency does not seem to be greatly affected by differences in age group or gender (see Tables 5 and 6). We also note that [24] points that "most of these changes occurred by 1973, with little change in the following 30 years" remains valid today, more than 15 more years later.

We believe that one of the reasons for the trend that Japanese have continued to lose interest in economic success in life may be that the confidence in Japan's economic growth that many Japanese felt in the early years of the Heisei era diminished and that from the middle to the end of the Heisei era, people became more aware that Japan's relative prestige was becoming stagnant. These beliefs are shared by [31], who argues that the Heisei period was a "period of failure".

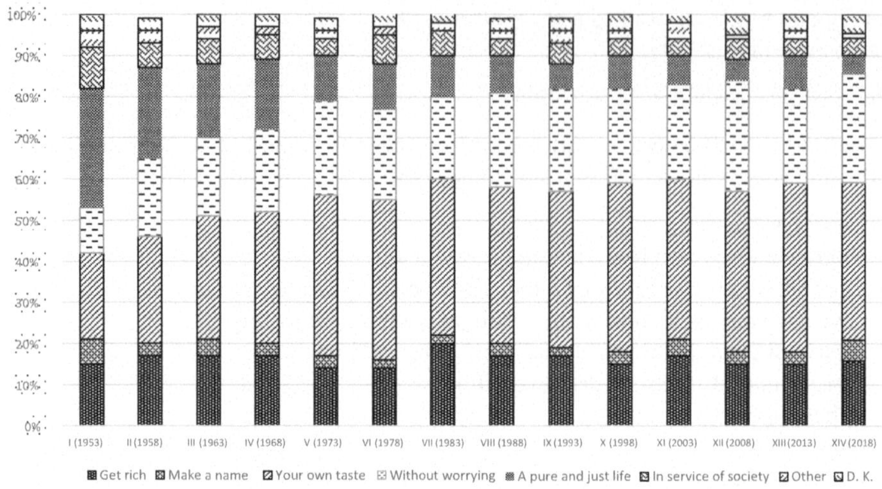

Fig. 2 Variation in the distribution of answers to the "kurashi-kata" (attitudes towards life) question from the 1st to the 14th survey

Table 4 Distribution of answers to the "kurashi-kata" (attitudes towards life) question in the Heisei era (%)

	Get rich (%)	Make a name (%)	Your own taste (%)	Without worrying (%)	A pure and just life (%)	In service of society (%)	Other (%)	D. K. (%)	Total
IX (1993)	17	2	38	25	6	5	3	3	1883
X (1998)	15	3	41	23	8	4	2	4	1339
XI (2003)	17	4	39	23	7	4	4	4	1192
XII (2008)	15	3	39	27	5	5	1	5	1729
XIII (2013)	15	3	41	23	8	4	2	4	1591
XIV (2018)	16	5	38	27	4	4	1	5	1584

On the other hand, "Do not think about money or fame; just live a life that suits your taste" and "Live each day as it comes, cheerfully and without worrying" are consistently supported by the respondents, indicating that one of their goals is to live peacefully without pessimism, not in pursuit of economic wealth or competition for it. In other words, a preference for "living life according to your taste" and "living life cheerfully and without worrying" is a unique characteristic that may

Table 5 Distribution of answers to the "kurashi-kata" (attitudes towards life) question by age group and survey period (%)

		Get rich (%)	Make a name (%)	Your own taste (%)	Without worrying (%)	A pure and just life (%)	In service of society (%)	Other (%)	D. K. (%)	Total
IX (1993)	20-29	17	4	50	20	4	2	1	2	583
	30-39	17	3	45	23	4	3	2	4	639
	40-49	19	2	43	21	5	4	2	4	901
	50-59	17	3	38	23	8	5	2	4	730
	60-69	14	2	30	36	9	5	1	2	620
	70+	14	3	23	42	7	7	1	3	265
X (1998)	20-29	10	2	57	21	5	2	1	3	173
	30-39	14	3	44	25	6	3	1	2	207
	40-49	10	3	56	18	7	2	2	3	265
	50-59	20	2	37	23	8	5	1	3	267
	60-69	16	4	31	27	8	6	4	4	255
	70+	19	3	23	27	12	6	2	8	172
XI (2003)	20-29	15	1	48	25	7	1	3	1	121
	30-39	14	3	44	25	4	2	4	3	209
	40-49	12	5	44	24	3	3	6	4	195
	50-59	20	3	39	23	5	3	3	4	268
	60-69	19	6	35	19	8	8	4	2	253
	70+	18	5	26	25	14	3	2	6	146
XII (2008)	20-29	20	4	38	26	5	4	1	2	186
	30-39	14	7	40	28	3	4	1	3	290
	40-49	13	3	50	19	4	5	2	4	283

(continued)

Table 5 (continued)

		Get rich (%)	Make a name (%)	Your own taste (%)	Without worrying (%)	A pure and just life (%)	In service of society (%)	Other (%)	D. K. (%)	Total
	50-59	13	1	39	33	5	4	1	4	359
	60-69	16	2	38	25	6	6	1	6	381
	70+	15	3	26	28	10	9	1	8	230
XIII (2013)	20-29	17	7	42	25	7	2	0	1	150
	30-39	20	6	36	25	5	4	1	3	261
	40-49	13	5	41	28	3	4	2	4	250
	50-59	19	4	34	29	4	3	2	5	299
	60-69	16	2	41	25	4	4	2	6	316
	70+	21	4	27	26	9	7	1	5	315
XIV (2018)	20-29	21	5	46	24	3	1	0	1	131
	30-39	21	5	38	26	3	3	0	4	224
	40-49	17	5	40	26	4	3	1	5	288
	50-59	13	4	41	24	3	6	2	7	267
	60-69	13	5	40	29	3	5	2	4	310
	70+	15	6	31	27	8	6	2	5	364

Table 6 Distribution of answers to the "kurashi-kata" (attitudes towards life) question by gender and survey period (%)

		Get rich (%)	Make a name (%)	Your own taste (%)	Without worrying (%)	A pure and just life (%)	In service of society (%)	Other (%)	D. K. (%)	Total
IX (1993)	Men	19	3	40	21	7	5	2	3	1724
	Women	15	2	39	29	5	3	2	4	2014
X (1998)	Men	16	4	42	19	9	5	2	3	615
	Women	14	2	41	27	7	3	2	4	724
XI (2003)	Men	18	4	40	17	9	5	3	4	551
	Women	15	4	38	28	5	2	5	3	641
XII (2008)	Men	18	4	41	19	6	6	1	5	815
	Women	12	3	37	33	5	5	2	4	914
XIII (2013)	Men	20	4	38	22	6	6	1	3	737
	Women	17	4	35	31	5	3	1	5	854
XIV (2018)	Men	16	6	41	22	4	6	1	4	765
	Women	16	4	36	31	5	3	1	5	819

have provided self-justification for low economic aspirations or compulsory pride and self-assurance in a society facing low economic growth in the Heisei era. This phenomenon may have the same roots as the discussion on the image of modern Japanese people who cherish simple happiness (e.g. [3, 17], and the World Youth Value Survey conducted by the Cabinet [2] from 1972 to 2003).

3.3 Relation Between "Giri Ninjō" and "Kurashi-Kata" (Attitudes Towards Life) Among Modern Japanese

Here, we examine the relationship between the level of "Giri ninjō" and "kurashi-kata" (attitudes towards life) among the Japanese using the data from the 14th Survey, the most recent survey. Section 3.2 shows that based on the trends in the responses to the "kurashi-kata" (attitudes towards life) question, Japanese in the Heisei period placed importance on familiar happiness and pleasures, preferring to "live a life that suits my taste without thinking about money or fame" and "live each day as it comes, cheerfully and without worrying".

Many authors, including [30], suggest that increased individualism in Japan has dampened friendships, resulting in a decline in happiness. Based on these points, we can infer that depending on the differences in life goals indicated in the "kurashi-kata" (attitudes towards life) question, it is possible that respondents' image of valuable human relationships may differ.

Table 7 shows the "Giri ninjō" scale scores for each "kurashi-kata" (attitude towards life) respondent category. In addition, Table 8 shows the cross tabulation between the two items when the responses to the "kurashi-kata" (attitude towards life) question are merged (when "do not think about money or fame; just live a life that suits your taste" is merged with "live each day as it comes", which we focused on in Sect. 3.2, and when "work hard and get rich", "study earnestly and make a name for yourself", "never think of yourself; give everything in service of society" and "resist all evils in the world and live a pure and just life" are merged. We interpret the former combined category as reflecting people's perceived importance of achieving a comfortable life, while the latter may reflect the perceived importance of economic achievements and social justice.

As a result, it is observed that the more important a person believes it is to work hard and succeed, to relate to others, and to contribute to society, the higher his or her "Giri ninjō", while the more one is self-paced and seeks to achieve a comfortable life, the lower one's "Giri ninjō", with values ranging from 0 to 2 points. A statistically significant association ($\chi 2(1) = 3.861$, $p = 0.049$) between the "kurashi-kata" (attitudes towards life) question and the "Giri ninjō" scale is observed from this table.

This tendency is more clearly shown in the results of the multiple regression analysis with the degree of "Giri ninjō" as the dependent variable, as shown in Table 9. Furthermore, as shown in Model 2 of the table, where the combined category of "do not think about money or fame; live a life that suits your taste" and "live each day

Table 7 Distribution of scores on the "Giri ninjō" scale by category of answer to the "kurashi-kata" (attitudes towards life) question (%)

	0point (%)	1point (%)	2point (%)	3point (%)	4point (%)	5point (%)	Total
Get rich	13	28	41	15	3	0	251
Make a name	14	35	25	23	4	0	80
Your own taste	14	39	32	13	1	0	606
Without worrying	14	39	32	13	2	0	420
A pure and just life	6	46	29	12	7	0	69
In service of society	3	39	45	9	4	0	67

Table8 Comparison of the "Giri ninjō" scale values between two groups based on their answers to the "kurashi-kata (attitudes towards life)" question

	0point-2point (%)	3point+ (%)	Total
Get rich/Make a name/A pure and just life/In service of society	81	19	467
Your own taste/Without worrying	85	15	1026

as it comes, cheerfully and without worrying" is added as an independent variable to the baseline models (Model 1), this category is negatively associated with the degree of "Giri ninjō" orientedness.

In the past, [7] has interpreted the choice of these two answers as a manifestation of the Japanese "zest for life", and this attitude was previously commonly possessed by many Japanese (Gekkan NIRA Vol. 4, No. 9, National Institute for Research Advancement, 1982, " Measuring the Heart" Japanese National Character 2004 pp. 103–122). However, based on the results of the 14th Surveys on the Japanese National Character, it is not clear how these attitudes relate to typical traditional Japanese values such as the sense of "Giri ninjō". Our view is that it is not easy to measure or define a typical Japanese "way of life", Japanese "joie de vivre" or "zest for living" using this item.

We additionally note the effect of different attributes on the degree of "Giri ninjō". The results in Table 9 show that women have a stronger "Giri ninjō" mentality than men. It is also confirmed that the middle-aged respondents in their 40 s (using their 20 s as the reference category) and the older age group in their 60 s have a stronger "Giri ninjō" attitude. With regard to educational attainment, there appears to be no effect in model 1(i.e. baseline model); however, people who get higher educational attainment have a weaker "Giri ninjō" attitude than people who get lower educational

Table 9 Results of the regression analysis with the "Giri ninjō" scale as a dependent variable

	Model1					Model2				
Explanatory variable	β *beta*	standardized β	P value	(95% confidence interval)	VIF	β *beta*	standardized β	P value	(95% confidence interval)	VIF
<Gender> (#1.1)										
Women(Reference category)				(ref.)					(ref.)	
Men	−0.121	−0.063	0.013	(−0.261, −0.026)	1.007	−0.150	−0.078	0.003	(−0.248, −0.052)	1.012
<Age range> (#1.2)										
20–29(Reference category)				(ref.)					(ref.)	
30–39	0.117	0.042	0.275	(−0.093, 0.327)	2.375	0.128	0.047	0.234	(−0.083, 0.339)	2.320
40–49	0.262	0.105	0.011	(0.061, 0.463)	2.689	0.261	0.105	0.011	(0.059, 0.463)	2.603
50–59	0.115	0.045	0.271	(−0.090, 0.319)	2.610	0.124	0.048	0.239	(−0.083, 0.331)	2.485
60–69	0.221	0.091	0.032	(0.019, 0.423)	2.837	0.230	0.095	0.026	(0.027, 0.433)	2.741
70+	0.201	0.088	0.054	(−0.003, 0.405)	3.269	0.182	0.079	0.084	(−0.024, 0.389)	3.174
<Educational background> (#1.3)										
Low (junior high school graduate or below) (Reference category)				(ref.)					(ref.)	
Medium (high school graduate)	0.009	0.005	0.911	(−0.151, 0.169)	2.783	−0.013	−0.007	0.875	(−0.178, 0.151)	2.807

(continued)

Table 9 (continued)

	Model1					Model2				
Explanatory variable	β *beta*	standardized β	P value	(95% confidence interval)	VIF	β *beta*	standardized β	P value	(95% confidence interval)	VIF
High (college graduate or above)	−0.165	−0.085	0.055	(−0.333, 0.004)	3.086	−0.184	−0.095	0.038	(−0.357, −0.010)	3.138
<Urban scale> (#1.5)										
Towns and villages to cities under 50,000 (Reference category)				(ref.)					(ref.)	
Cities of 50,000 up to under 500,000	−0.041	−0.019	0.582	(−0.186, 0.105)	1.892	−0.026	−0.012	0.733	(−0.177, 0.125)	1.914
Cities of 500,000 up to the six major urban areas	0.008	0.004	0.910	(−0.123, 0.138)	1.886	0.028	0.014	0.685	(−0.107, 0.163)	1.905
#2.4 Attitudes toward life										
Get rich/Make a name/A pure and just life/In service of society (Reference category)									(ref.)	
Your own taste/Without worrying						−0.184	−0.089	0.001	(−0.289, −0.078)	1.011

(continued)

Table 9 (continued)

	Model1					Model2				
Explanatory variable	β *beta*	standardized β	P value	(95% confidence interval)	VIF	β *beta*	standardized β	P value	(95% confidence interval)	VIF
	ANOVA			P<0.05		ANOVA			P<0.05	
	R^2			0.021		R^2			0.030	
	Adjusted R^2			0.014		Adjusted R^2			0.023	
	n			1564		n			1473	

attainment (as the reference category) in model 2. There also appears to be no effect regarding the urbanicity of the area of residence.

4 Conclusion

Hayashi [6], Chap. 6, and Sakamoto [24] point out that in the mid-1970s, right after the oil shock of 1973, a "return to tradition" occurred in Japanese views and thinking, in which the declining trend of traditional views turned into an increase or stagnation. This is the second characteristic of the consciousness trend among Japanese people through the 1970s. According to [24], the "traditional regression phenomenon" is not simply a return to the past. In other words, this "retro-traditional phenomenon" is not a simple regression in the same space but rather a spiralling retrogressive change that, if projected onto a space where the old questions could be mapped, would appear as if it were a simple regression. This seems to echo what Ulrich Beck, Anthony Giddens and Scott Lash have pointed out—that the values of society are changing recursively [1]. They point out that society is basically changing along the path of "premodern society (traditional society)/simple modernity/reflexive modernity". If we follow their argument and consider the above results, it is conceivable that the changes in Japanese society unique to the Heisei period may have prompted reflection among modern contemporary Japanese individuals, which in turn may have influenced the change in values during the Heisei period.

As noted in Sect. 3.2, the differences among age groups on the "Giri ninjō" scale have narrowed, superficially suggesting that the "traditional versus modern" axis of values has become ambiguous; as the OLS in Sect. 3.3 shows, the level of "Giri ninjō" is not affected by educational attainment or urbanicity. The results are different from one-way differences in age groups. This may be an example of a recursive regression that differs slightly from a simple regression to tradition. For example, since 1985, the lifetime unmarried rate has increased for both men and women (National Institute of Population and Social Security Research, 2021), and according to the National Survey of Living Standards (Ministry of Health and [16], the percentage of single-person households was from 18.2% in 1986 to 28.8% in 2019. This societal change means that compared to the typical family structure in the Showa period, family structures have become more diverse. Such individualistic changes, considered one of the characteristics of modernization, suggest that Japanese people's values gradually transformed throughout the Showa and Heisei periods.

Sakamoto [24] pointed out that one of the key trends in Japanese people's attitudes in the postwar period, as revealed by the trend in responses to the "kurashi-kata" (attitudes towards life) question obtained from the 50-year-long national character survey, was the emergence of values that prioritize private life. Additionally, in this survey, since the 1970s, the number of respondents selecting "family is the most important" for the item "the most important thing in life" (#2.7) has increased dramatically. In other words, the importance placed on having a less strenuous lifestyle and a desire to find pleasure and happiness in the little things consistently prevailed among the Japanese during the Heisei period of societal instability and low economic growth.

Furthermore, the negative relationship between the following one's own pace and seeking to achieve a comfortable life, as indicated by the question on "kurashi-kata", and the preference for "Giri ninjō"-style relationships is also suggestive. According to Chikio Hayashi, who has helped manage this survey since its inception, Japanese attitudes around 1988, the early years of the Heisei Era, were more "traditional-modern" (old versus new ways of thinking) and began to waver around 1978 (e.g. [6], Chap. 6); this tendency continued during the early Heisei era. Although it is not clear what best describes the "ways of thinking" of the Japanese in [6] opinion, the "Giri ninjō" and "kurashi-kata" (attitudes towards life) items capture some aspects of Japanese mentality in contemporary Japan in the late Heisei period. The fact that modern Japanese do not necessarily desire social prestige or economic success but value living life at their own pace and that this attitude is in the opposite direction of traditional human relationship preferences represents an aspect of contemporary Japanese society.

Since the "Giri ninjō" scale is considered a typical instrument for measuring "Japanese attitudes", previous studies have generally considered it by itself. In contrast, this study clarifies that there is a linkage between the "Giri-ninjō" scale and "kurashi-kata" (attitudes towards life). This study confirms the effectiveness of the "Giri ninjō" scale as an explanatory variable for the characteristics of the Japanese "kurashi-kata" (attitude towards life). This finding may contribute to the expansion of future research on the Japanese national character. [32], p. 18) pointed out that longitudinal surveys assume that the same item will be used for years to maintain comparability. He also indicated that if the survey series lasts more than a few decades, items may need to be changed or replaced appropriately for long-term comparability. This discussion also applies to the "Giri-ninjō" scale and further discussion is needed on how to modify and utilize these items in the future.

The present study has several limitations. One is that we did not have enough "objective" explanatory variables, for example, the socioeconomic status of respondents such as occupation and income, in our regression analysis. We did not incorporate topical background variables into the model, which would have been useful for interpreting the results. Therefore, it is necessary to consider expanding the model to include the economic growth rate and other objective variables while considering generational and historical factors.

Acknowledgements During the review process for this paper, we received valuable comments and advice from our anonymous reviewers, for which we are most grateful. We would also like to thank Ryozo Yoshino, Professor in the Faculty of Culture and Information Studies at Doshisha University, for his advice. We would like to express our deepest gratitude to him. Mr. Aguru Ishibashi of Senshu University also helped us prepare the manuscript, for which we thank him.

Declarations Conflicts of Interest The authors declare no potential conflicts of interest with respect to the research, authorship and/or publication of this article.

Funding This research was supported by grants from JSPS Kakenhi (Grant Number JP 18H04101).

Ethics approval Ethical approval was granted for this project. All procedures used in the 13th and 14th surveys were approved by the Institutional Review Board of the Institute of Statistical Mathematics (Approval Numbers ISM08-001, ISM13-003 and ISM18-003, respectively). Informed consent was obtained orally from the interviewees by the interviewer on site. No such approval was confirmed for the 8th–11th survey carried out, since such an internal review system was not established in Japan at that time.

References

1. Beck, U., Giddens, A., & Lash, S. (1994). *Reflexive modernization: Politics, tradition and aesthetics in the modern social order*. Polity Press
2. Cabinet Office. (2009). 'Sekai seisyonen kachikan chosa' the Eighth World Youth Survey. https://www8.cao.go.jp/youth/kenkyu/worldyouth8/html/mokuji.html
3. Furuichi, N. (2011). Zetsubo no kuni no kofuku na wakamonotachi (The happy youth of a desperate country: The disconnect between Japan's Malaise and its millennials). Kodansya.
4. Hakuhodo Institute of Life and Living. (2019). Seikatsusya no heisei 30 nenshi: Deta de yomu kachikan no henka (Heisei 30 year history of living person: changing values as seen in the data). Nihon Keizai Shimbun Shuppansha.
5. Hayashi, C. (1973). *Hikaku Nihonjin ron(Comparative Japanese theory)*. Tyukou-shinsyo.
6. Hayashi, C. (2001). *Nihonjin no kokuminsei kenkyu (Research on the Japanese national character)*. Nansosha.
7. Hayashi, C. (2004). *Kokoro wo hakaru : Nihonjin no kokuminsei (Japanese national characters: Measuring the heart*. Bensey Publishing.
8. Hayashi, C., & Suzuki, T. (1984). Changes in belief systems, quality of life issues and social conditions over 25 years in post-war Japan. *Annals of the Institute of Statistical Mathematics, 36*, 135–161.
9. Hayashi, C., Suzuki, T., & Sasaki, M. (1992). *Data Analysis for Comparative Social Research: International Perspectives*. Elsevier Science Publishers.
10. Hayashi, C., & Kuroda, Y. (1993). *Japanese Culture in Comparative Perspective*. Praeger.
11. Hayashi, C., & Hayashi, F. (1995). Kokuminsei no kokusaihikaku (Comparative study of national character). *Proceedings of the Institute of Statistical Mathematics, 43*(1), 27–80.
12. Hayashi, C., Yoshino, R., Suzuki, T., Hayashi, F., Kamano, S., Miyake, I., Murakami, M., & Sasaki, M. (1998). *Kokuminsei nanaka-koku hikaku (cross-national comparison of seven nations)*. Idemitsu-Syoten.
13. Hayashi, C., & Sakuraba, M. (2002). *Suji ga akasu nihonjin no senzairyoku (Japanese Potential Revealed by the Numbers)*. Kodansya.
14. Hayashi, F. (2012). Kangaekata no sujimichi (Ways of thinking). *Bullet Data Analysis Japan Classification Society, 2*(1), 1–16.
15. Inaba, Y. (2011). *Sosyaru kyapitaru nyumon (Primer on social capital)*. Chuokoronshinsya.
16. Ministry of Health, Labour and Welfare. (2020). Kokumin seikatsu kiso chosa no gaikyo (Overview of the 2019 comprehensive survey of living conditions). https://www.mhlw.go.jp/toukei/saikin/hw/k-tyosa/k-tyosa19/dl/14.pdf
17. Murata, H., & Masaki, M. (2013). Chukosei wa naze kohuku nanoka (Why are junior high and high school students "happy"?). In: The NHK monthly report on broadcast research (Vol 1, 2, pp. 34–43).
18. Nakamura, T., Tsuchiya, T., & Maeda, T. (2015). Kokuminsei no kenkyu: Dai-13-ji zenkoku chosa 2013-nen zenkoku chosa (Research on the national character: The 13th (2013) national survey). In *Tokei suri kenkyusho chosa kenkyo repoto (The institute of statistical mathematics research report)* (p. 116).
19. Nira, G. (2004). "Measuring the heart" Japanese national character. *Natl Inst Res Adv, 4*(9), 103–122.

20. Park, Y. (2021). Nani ga chiho he no ijuiko wo kimerunoka? (What factors determine the intent to out-migrate from urban areas to rural areas?). *Proceedings of the Institute of Statistical Mathematics, 69*(3), 315–337.
21. Park, Y., & Maeda, T. (2015). Dare ga doryoku wa mukuwareru to kanjite iruka (Who feels that effort is rewarded? Analysis about a feeling of effort from survey on the Japanese national character). *Proceedings of the Institute of Statistical Mathematics, 63*(2), 261–276.
22. Putnam, R. D. (1993). What makes democracy work? *National Civic Review, 82*(2), 101–107.
23. Research NIoPaSS. (2021). Population statistics data book. https://www.ipss.go.jp/syoushika/tohkei/Popular/Popular2021.asp?chap=0
24. Sakamoto, Y. (2005). Nihonjin no kokuminsei 50 nen no kiseki (A statistical research on the Japanese national character: General trends in attitude over the past fifty years). *Proceedings of the Institute of Statistical Mathematics, 53*(1), 3–33.
25. Sakamoto, Y. (2010). Tokeiteki nihonjin kenkyuzakkan:Aru kokuminsei chosakei no 36nen no omoide (Memories of a statistical study on the japanese national character: Looking back on 36 years of being a researcher at the institute of statistical mathematics). *Proceedings of the Institute of Statistical Mathematics, 58*(1), 61–82.
26. The Research Committee of the Japanese National Character of the Institute of Statistical Mathematics. (1961). Nihonjin no kokuminsei (Japanese national character). Shiseido.
27. The Research Committee on the Study of the Japanese National Character of the Institute of Statistical Mathematics. (1969). Nihonjin no kokuminsei (Japanese national character). Shiseido.
28. The Research Committee on the Study of the Japanese National Character of the Institute of Statistical Mathematics. (2015). Kokuminsei no kenkyu: Dai-13-ji zenkoku chosa 2013-nen zenkoku chosa (Research on the national character: The 13th (2013) national survey). In *Tokei suri kenkyusho chosa kenkyu repoto (The institute of statistical mathematics research report)*
29. The Research Committee on the Study of the Japanese National Character of the Institute of Statistical Mathematics. (2021). Kokuminsei no kenkyu: Dai-14-ji zenkoku chosa 2018-nen zenkoku chosa (Research on the national character: The 14th (2018) national survey). https://www.ism.ac.jp/survey/KSResults/ResultSummary.pdf
30. Uchida, Y., & Ogihara, Y. (2012). Bunkateki kohukukan: Bunkashinrigakuteki chiken to syorai e no tenbo (Cultural construal of happiness: Cultural psychological perspectives and future direction of happiness research). *Japanese Psychological Research, 55*(1), 26–42.
31. Yoshimi, S. (2019). *Heisei jidai (Heisei period).* Iwanami Syoten.
32. Yoshino, R. (2021). *Cultural manifod analysis on national character: methodology of cross-national and longitudinal survey.* Springer.

Closed Form and Lord-Wingersky Algorithm for the Total Score Distribution of a Test with a General Scoring Scheme: The Distribution of the Sum of Rolling Heterogeneous Dice

Kojiro Shojima

Abstract This study assumed a general test with a scoring scheme that was adopted in many academic tests, psychological and medical scales, and social and health survey questionnaires, in which the number of categories and their weights for each item can vary within a test, and the weight of each category can be any positive or negative rational number. We first showed that the distribution of the weighted sum score (total score) for the test was identical to the distribution of the sum of rolling heterogeneous dice. After introducing the characteristic function of the distribution, we derived a closed form of the probability mass function (PMF) of the distribution. We also provided a Lord-Wingersky (LW) algorithm that can produce the PMF of the distribution and compared the two methods in terms of accuracy and efficiency. Since the closed-form explicitly parameterizes the item weights, it is useful for modeling and predicting the behavior of the distribution or examining the properties of the distribution theoretically, whereas the LW algorithm is useful when the distribution is needed for routine computation because it is very fast and has no error.

1 Introduction

Using item response theory (IRT) [12, 18, 51], many (binary or true/false) achievement tests have been scaled, and many psychological and medical scales and social and health survey questionnaires have been standardized by the polytomous IRT model, which includes the graded response model [43], partial credit model [15, 30, 33], and nominal response model [1]. IRT assumes that items in a test, scale, or questionnaire, which we will collectively refer to as "test" hereafter, measure a single latent trait, denoted as $\theta (\in \mathbb{R})$, estimates a valid continuous value of θ from the responses of each respondent to the items, and feeds the value back to the respondent. A group of models that do not assume unidimensionality, called the multidimensional

K. Shojima (✉)
National Center for University, Meguro-Ku, Komaba 153-0041, Japan
e-mail: shojima@rd.dnc.ac.jp

© The Author(s), under exclusive license to Springer Nature Singapore Pte Ltd. 2023
A. Okada et al. (eds.), *Facets of Behaviormetrics*, Behaviormetrics: Quantitative Approaches to Human Behavior 4, https://doi.org/10.1007/978-981-99-2240-6_15

IRT model [42, 55], feeds several latent traits back into each respondent. In any case, a computer is required to compute the traits. However, if a student feeling low visits a school infirmary and takes a depression screening test, and if the screening test is not computerized, the latent trait value for the test result is usually not calculated using a computer; however, the total score of the test is calculated on the spot. In addition, IRT is rarely used to scale term and admission examinations in schools because the sample size is too small to apply IRT or the exams do not satisfy the assumption of local independence, which requires that item responses be stochastically independent when latent traits are given. Therefore, the practice of evaluating a respondent's traits using the total score will continue for a while in the fields of pedagogy, psychology, sociology, behavioral sciences, medical sciences, and health sciences.

The behavior of the total score of a test scaled under a dichotomous IRT model has already been thoroughly investigated by [10, 16, 29, 31]. In particular, the distribution of number-right scores conditioned on θ was referred to as the compound binomial (CB) distribution [27, 28], and [29] briefly described a recursive method for obtaining the probability mass function (PMF) of the CB. This recursive method is now referred to as the Lord-Wingersky (LW) algorithm.

The LW algorithm can quickly and accurately obtain the conditional score distribution on θ of a test and has a wide range of applications in various situations. For example, Chen and Thissen (2000) proposed a method for estimating item parameters using the algorithm, and [39] developed model fit indices based on the algorithm. In addition, [32] generalized the original LW algorithm to be feasible when each item has a weight, and [25] presented a different generalized version of the algorithm that is applicable even when the weight of each item is not an integer (but the weights are equal across items). Furthermore, [17, 48, 52] independently extended the original algorithm to work even when the IRT model applied is polytomous. Orland et al. [38] used this extended algorithm in total score-based test linking. In addition, [46] demonstrated that the algorithm is feasible when the polytomous items are on a Likert-type scale and the weight of each ordinal (graded) category is not an integer. Moreover, [5, 23] developed an algorithm estimating each respondent's multidimensional latent traits in multidimensional IRT models [2, 6, 34, 42].

This study assumes a test that is more general than those considered in previous studies, in which the number of categories of each item may be distinct, the scoring scheme of each item may not be uniform, and the weight of each item category can be a positive or negative rational number. This study presents an LW algorithm for calculating the PMF of the total score distribution for such a test with a general scoring scheme. However, this recursive algorithm does not necessarily examine the mathematical and theoretical properties of the test distribution. Therefore, before presenting the LW algorithm, we derive the theoretical properties of the distribution, such as the characteristic function and tail bounds, and show the closed form of the PMF of the distribution.

2 Method

2.1 Scoring Scheme

Let us assume that a test consists of J items and that the number of categories for item j is K_j. The categories of item can be categorized as nominal, ordinal, or partially ordinal. We also assume that the weight (score) for category k $(= 1, 2, \cdots, K_j)$ of item j is expressed as w_{jk}, and it can take any rational number ($w_{jk} \in \mathbb{Q}$). In other words,

$$w_{jk} = \frac{a}{b} \quad (a \in \mathbb{Z} \text{ and } b \in \mathbb{N} \text{ s.t. } GCD(a, b) = 1) \tag{1}$$

It should be noted that w_{jk} is defined as an irreducible fraction (the greatest common devisor (GCD) of the enumerator and denominator of the weight is one), and the negative weight is attributed to the numerator $a(\in \mathbb{Z})$ if $w_{jk} < 0$. Additionally, w_{jk} is an integer when $b = 1$. Let

$$\boldsymbol{w}_j = \left[w_{j1}\ w_{j2}\ \cdots w_{jKj}\right]' \quad (K_j \times 1)$$

denote a vector that collects all the weights for the K_j categories of item j and implies the scoring (weighting) scheme of the item. In this study, the scoring scheme $\boldsymbol{w}_j$ can be flexibly set as follows:

Example A: $\boldsymbol{w}_j = [0\ 1\ 2\ 3\ 4]'$
Example B: $\boldsymbol{w}_j = [0.0\ 0.8\ 1.6\ 2.4\ 3.2]'$
Example C: $\boldsymbol{w}_j = [0\ 0\ 0\ 2]'$
Example D: $\boldsymbol{w}_j = [0.5\ 1\ 0\ 2]'$
Example E: $\boldsymbol{w}_j = [0\ \ -1\ \ 0\ \ 0.25\ \ 2\ \ 0]'$

Example A illustrates the usual scoring scheme for a Likert-type item with five ordinal categories, as in [17, 48, 52]. Example B represents a scoring scheme used by [46] in which five ordinal categories in a Likert-type item have equispaced weights with increments of 0.8. Example C is a multiple-choice item with four nominal categories, where category 4 is the correct choice and two points are given to the respondents who choose this category. Example D illustrates a four-choice item where category 4 is the correct choice, as in Example C; however, the respondents selecting choices 1 or 2 are given partial scores of 0.5 and 1, respectively. In everyday school tests, choices other than the originally intended correct choices are sometimes later given a partial score because they are found to be not completely wrong. Finally, in Example E, the correct choice is category 5 in this six-choice item, and a partial point of 0.25 is given to category 4, however, one point is deducted when category 2 is selected. Point-deduction scoring is sometimes used to mark essay questions and some performance tests, such as oral presentations, interview examinations, and group discussions, where the presented description or performance is evaluated based on rubrics, and points are added and sometimes subtracted based on various perspectives. Such a scoring scheme is also common in gymnastics and figure skating, where points are added and subtracted for each individual move, following the code of points. Examples of the application of IRT to performance tests include [9, 22, 24, 50].

In general, the number of categories and scoring schemes for each item are not uniform, but distinct within a test. In addition, different item types such as nominal (e.g., multiple-choice items) and ordinal category items (e.g., Likert-type items) may be mixed as in [37, 47]. In this study, we assumed a test with high generality, each of which varied in the number of categories and scoring schemes, which covered various types of practical tests, scales, and questionnaires, and considered an LW algorithm and closed-form to obtain the PMF of the total score distribution for such a test.

2.2 Probability Space of Unfair Die

First, we consider the probability space of item j, $\{\Omega_j, \mathcal{F}_j, P_j\}$. For item j, the respondent with trait θ selects one category from $\{1, \cdots, K_j\}$ exclusively. That is, the sample space of item j, Ω_j, is expressed as follows:

$$\Omega_j = \{1, \cdots, K_j\}.$$

Accordingly, the event space can be automatically determined as $\mathcal{F}_j = 2^{\Omega_j}$, which is a set of subsets of Ω_j. In addition, using the IRT model, the probability measure on the sample space $\{\Omega_j, \mathcal{F}_j\}$, P_j, of the discrete random variable for item j, C_j, being observed as $k \in \mathcal{F}_j$ can be modeled as follows:

$$P_j(C_j = k|\theta) \to p_{jk}(\theta),$$

where $p_{jk}(\theta)$, hereafter is simply denoted as $p_{\theta jk}$, represents the probability of selecting category k in item j by the respondent with trait θ, which is called the category response function (CRF) in IRT. This study allows several different polytomous IRT models to be employed in a single test, as in [47], although they should be calibrated (equated) on the same θ scale. The above equation can be rewritten as follows:

$$C_j|\theta \sim Cat(\boldsymbol{p}_{\theta j}).$$

Hereafter, $C_j|\theta$ is denoted as $C_{\theta j}$ for readability, and $Cat(\boldsymbol{p})$ represents a categorical distribution with probability vector $\boldsymbol{p}$. Moreover,

$$\boldsymbol{p}_{\theta j} = \left[p_{\theta j1} \cdots p_{\theta jK_j}\right] \quad (\mathbf{1}'_{K_j}\boldsymbol{p}_{\theta j} = 1)$$

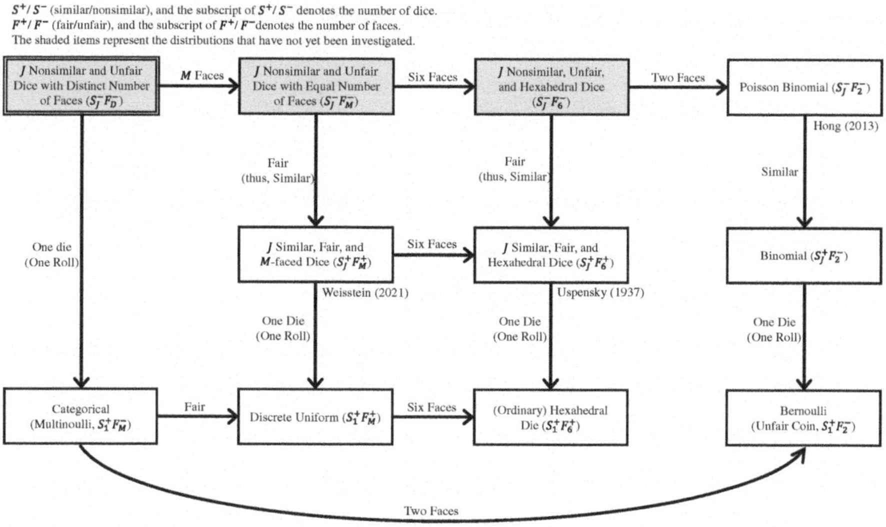

Fig. 1 Relations of discrete distributions

is a vector that collects all CRFs of item j and represents the PMF of the discrete random variable $C_{\theta j}$.

Figure 1 shows the relationship between the relevant discrete distributions in this study. The categorical distribution (bottom left) is sometimes referred to as a multinoulli distribution [35] because it is an extension of the Bernoulli distribution (bottom right) to being takable trichotomous or more values. In addition, a categorical distribution in which the probability of observing each category is equal ($\boldsymbol{p} = \mathbf{1}_M/M$) is identical to a discrete uniform distribution, which is further equivalent to the usual regular (fair) hexahedral die when the number of categories is six. Thus, $Cat(\boldsymbol{p}_{\theta j})$ is an unfair K_j-faced die.

Moreover, because category k is given a weight of w_{jk}, the score for item j of the respondent with trait θ, $U_{\theta j}(= U_j|\theta)$, is expressed as follows:

$$U_{\theta j} = w_{jC_{\theta j}}.$$

Accordingly, the characteristic function (CF) of the discrete random variable $U_{\theta j}$ is represented as follows:

$$\phi_{U_{\theta j}}(t) = E\left[e^{itU_{\theta j}}\right] = \sum_{k=1}^{K_j} p_{\theta jk} e^{itw_{jk}} \quad (i = \sqrt{-1}). \tag{2}$$

The CF of a discrete distribution is, by definition, the inverse discrete Fourier transform (inverse DFT) of the PMF distribution, and it holds all information regarding the moments of the distribution as well as the moment-generating function (MGF).

The mth moment of $U_{\theta j}$ about the origin is derived from CF as follows:

$$E\left[U_{\theta j}^{m}\right]=i^{-m}\left[\frac{d^{m}}{dt^{m}}\phi_{U_{\theta j}}(t)\right]_{t=0}=\sum_{k=1}^{K_j}w_{jk}^{m}p_{\theta jk}.$$

Therefore, the mean and variance of $U_{\theta j}$ are, respectively, obtained as follows:

$$E\left[U_{\theta j}\right]=i^{-1}\left[\frac{d}{dt}\phi_{U_{\theta j}}(t)\right]_{t=0}=\sum_{k=1}^{K_j}w_{jk}p_{\theta jk},$$

$$V\left[U_{\theta j}\right]=i^{-2}\left[\frac{d^{2}}{dt^{2}}\phi_{U_{\theta j}}(t)\right]_{t=0}-E\left[U_{\theta j}\right]^{2}=\sum_{k=1}^{K_j}w_{jk}^{2}p_{\theta jk}-\left(\sum_{k=1}^{K_j}w_{jk}p_{\theta jk}\right)^{2}.$$

2.3 Distribution of Sum of Dice Roll

The sum of the responses for J items by the respondent with trait θ is expressed as follows:

$$X_\theta=\sum_{j=1}^{J}U_{\theta j}\quad(X_\theta=X|\theta).$$

Based on the assumption of local independence, the response to each item is mutually independent when θ is given. Thus, it can be said that X_θ follows the distribution of the sum of rolling J dice in which the jth die has K_j faces each of whose probability of appearing is not even.

Let us roll J dice and consider the distribution of the sum of the pips. The J dice may be similar or nonsimilar (S^+/S^-), be fair or unfair (F^+/F^-), and has the same M faces or different number of faces (M/D). We denote such a set of J dice as $\{S^+/S^-\}_J\{F^+/F^-\}_{M/D}$. For example, the distribution followed by the sum of ten regular (fair) hexahedral dice is expressed as $S_{10}^{+}F_{6}^{+}$, or the distribution followed by the sum of eight dodecahedral dice, each of which is unfair and mutually nonsimilar, is expressed as $S_8^- F_{12}^-$. Note that when J dice are similar, the number of faces of each die is necessarily the same; thus, the combination of {similar, distinct} is impossible. In addition, if all dice are fair and have the same number of faces, then all dice must be mutually similar. Additionally, note that "similar" is employed in Fig. 1 when there is only one die. Following this notation, the distribution followed by X_θ is hereafter written as follows:

$$X_\theta\sim S_J^-F_D^-(\boldsymbol{W},\boldsymbol{P}_\theta),$$

where $\boldsymbol{W} = \{\boldsymbol{w}_1, \cdots, \boldsymbol{w}_J\}$ and $\boldsymbol{P}_\theta = \left\{\boldsymbol{p}_{\theta 1}, \cdots, \boldsymbol{p}_{\theta J}\right\}$ represent the weights and probabilities for all item categories, respectively.

The distribution of the sum of rolling J regular hexahedral dice, $S_J^+ F_6^+$, was previously unraveled by [45, 49, 56], as described in the center right of Fig. 1. In particular, [45] determined the MGF of $S_J^+ F_6^+$ and provided the PMF of the distribution, and [56] used the DFT to derive the PMF of the distribution. Moreover, [54] provided the PMF of the distribution followed by the sum of J dice, each of which is a regular M-faced polyhedral die, that is, $S_J^+ F_M^+$ (center left of Fig. 1). There are also several websites [7, 44] that calculate the distribution of the sum of regular (and thus similar) polyhedral dice $S_J^+ F_M^+$. However, the distribution followed by the score of a test is precisely the distribution of the sum of rolling heterogeneous dice, and the theoretical property of the distribution is still uncharted, as far as the authors know. The uninvestigated distributions are shaded in Fig. 1. Among them, the $S_J^- F_D^-$ surrounded by the double line in the upper left corner of Fig. 1 is the distribution to be explored in this study. Our discussion begins with the following definition.

Definition 1 (*Conditional PMF of dice roll sum distribution*)
The conditional PMF for $S_J^- F_D^-$ of the respondent with trait θ is expressed as follows:

$$\Pr(X_\theta = x) = \sum_{h=1}^{|\Omega|} I(\mathbf{1}'_J \boldsymbol{u}_h, x) \prod_{j=1}^{J} \prod_{k=1}^{K_j} p_{\theta jk}^{z_{hjk}} \quad (x \in \Omega_{surj}). \tag{3}$$

Remark Let Ω be the J-fold Cartesian product of $\Omega_1, \cdots, \Omega_J$. That is,

$$\Omega = \prod_{j=1}^{J} \Omega_j = \{(\omega_1, \cdots, \omega_J) | \omega_1 \in \Omega_1 \wedge \cdots \wedge \omega_J \in \Omega_J\},$$

where the number of elements (J-tuples) in Ω is $|\Omega| = \prod_{j=1}^{J} K_j$. In addition, let the function summing an element in Ω be denoted as $surj(\cdot)$, the set of all possible sums is expressed as follows:

$$\Omega_{surj} = \{x \in \mathbb{Q} | x = surj(\omega) \wedge \omega \in \Omega\}. \tag{4}$$

This finite set is the sample space (support) of X_θ. The number of elements in Ω_{surj} becomes much smaller than that in Ω because $surj(\cdot)$ is a surjection (not a bijection). For convenience, but without loss of generality, we assume that the elements of Ω_{surj} are arranged in an increasing order. In general, Ω_{surj} becomes a set of discrete and non-equispaced rational numbers. Furthermore, $\boldsymbol{u}_h = [u_{h1} \cdots u_{hJ}]' (J \times 1)$ in Eq. (3) is the hth response pattern among all $|\Omega|$ patterns, and z_{hjk} in the equation represents a dichotomous indicator coded 1 when $u_{hj} = k$, and 0 otherwise. In addition,

$$I(a, b) = \begin{cases} 1, & \text{if } a = b \\ 0, & \text{otherwise} \end{cases}$$

represents an indicator function that returns 1 if $a = b$ and 0 otherwise. Equation (3) sums up the likelihoods of all response patterns whose total is x. However, this calculation is almost infeasible if the number of items J is large. For example, $|\Omega|$ exceeds 95 trillion when $J = 20$ and the number of categories for each item is five. Hereafter, Eq. (3) is referred to as the brute force (BF) method.

2.4 Need for Dice Roll Sum Distribution

The sum of the rolling heterogeneous dice distribution, $S_J^- F_D^-$, applies to more than just the total score distribution of a test. To enhance the motivation for investigating this dice-roll-sum distribution, certain situations in which the distribution can be applied are listed below.

(i) Horse racing

Although there are many variations in betting, let us simply consider betting on the winner of a race with eight horses, H_1–H_8. When one bets 10 dollars each on H_1, H_2, and H_3, whose payouts are 10 times, 5 times, and 3 times the bet, respectively. Then, this race can be regarded as an unfair die with eight faces ($S_1^+ F_8^-$) each of whose pip (profit) is +70, +20, 0, −30, −30, −30, −30, and −30 dollars. If there are 10 races in a day, the total profit of the day follows the distribution of $S_{10}^- F_D^-$, because each of the ten dice (race) has a different number of horses (faces) and is non-similar to the other as each horse's chance of winning (each face's probability of showing on the top) varies from race to race.

(ii) Actuarial science

To briefly describe the insurance system, given m accidents (or diseases) A_1–A_m, a subscriber chooses one or more accidents and periodically pays a small premium to the insurance company instead of receiving a payout when the subscriber suffers the selected accidents. From the insurance company's perspective, each subscriber is an $(m + 1)$-faced die containing A_0 (no accidents), that is, $S_1^+ F_{m+1}^-$. The company gains the subscriber's premium if A_0 shows on top; however, if other faces (A_1–A_m) appear, the company disburses the money based on the contract. Subtraction of the payout from (the amount of) the premium gives the pip (score) of each face of the die. If there are n subscribers, the company's total profit follows that of $S_n^- F_D^-$.

(iii) Psephology

In score voting, which is an election system, voters rate all candidates on a scale of {0, 1, 2, 3, 4, 5}, {−1, 0, 1}, and [0, 1], and the candidate with the highest total score wins. Suppose there are five candidates and 1,000 voters, and each voter assigns a score of {0, 1, 2, 3, 4, 5} to each of the five candidates. Because each voter is considered an unfair five-faced die, the total points polled by each candidate follow $S^-_{1000}F^-_5$. A similar voting system has been adopted for sports. For example, in the election for the winner of the Major League Baseball MVP Award at the end of each season, each voter (30 sports journalists belonging to the Baseball Writers' Association of America) enumerates 10 players and ranks them, which are converted into their respective scores, and the player with the largest total points wins the award. This election method is referred to as the Borda count, and the total score for each candidate is distributed as the sum of the heterogeneous dice.

(iv) Artistic sports

In artistic sports, such as gymnastics, artistic swimming (synchronized swimming), skateboarding, figure skating, snowboarding, and aerial (freestyle skiing), the judges' evaluations can be regarded as score voting for competitors, and the one who receives the highest total point wins the gold medal. The total score of each competitor follows the distribution of the sum of heterogeneous dice.

(v) Darts

Although there are many variations in the rules, the simplest one is that each of the 62 partitions with different areas in the dartboard (20 single-, 20 double-, and 20 triple-areas and inner and outer bulls) is assigned a score, and the player receives the score of the area where the dart is hit. If players attempt eight rounds and compete to obtain the highest total score, the total score of each player follows the distribution of $S^+_8F^-_{62}$. Note that the eight 62-faced dice are considered similar when focusing on a particular player because it can be assumed that the probability of hitting each of the 62 partitions does not change significantly throughout the eight rounds.

2.5 *Some Theoretical Properties of Dice Roll Sum Distribution*

This section explores the properties of $S^-_JF^-_D$ in detail. First, the CF of the $S^-_JF^-_D$ can be obtained from the CF of the categorical distribution.

Theorem 2 (*Conditional CF of dice roll sum distribution*) The total score of the respondent with trait θ, $X_\theta = \sum_{j=1}^{J} U_{\theta j}$, follows that of $S^-_JF^-_D$. Because $U_{1\theta}, U_{2\theta}, \cdots, U_{J\theta}$ are (locally) independent of each other, the CF of X_θ is obtained from Eq. (2), as follows:

$$\phi_{X_\theta}(t) = E\left[e^{itX_\theta}\right] = E\left[e^{it\sum_{j=1}^{J} U_{\theta j}}\right] = \prod_{j=1}^{J} E\left[e^{itU_{\theta j}}\right] = \prod_{j=1}^{J}\sum_{k=1}^{K_j} p_{\theta jk}e^{itw_{jk}}. \quad (5)$$

From this theorem, the moments of $S^-_JF^-_D$ are introduced immediately.

Corollary 3 (*Conditional moment of dice roll sum distribution*) The mth moment of $X_\theta \sim S_J^- F_D^-$ about the origin is expressed as the sum of the mth moment for each item. That is,

$$E\left[X_\theta^m\right] = i^{-m}\left[\frac{d^m}{dt^m}\phi_{X_\theta}(t)\right]_{t=0} = \sum_{j=1}^{J}\sum_{k=1}^{K_j} w_{jk}^m p_{\theta jk}.$$

Note that this is the conditional moment of the total score for the respondent whose trait was θ. From this, the mean and variance of X_θ are obtained as follows:

$$E[X_\theta] = i^{-1}\left[\frac{d}{dt}\phi_{X_\theta}(t)\right]_{t=0} = \sum_{j=1}^{J}\sum_{k=1}^{K_j} w_{jk} p_{\theta jk},$$

$$V[X_\theta] = i^{-2}\left[\frac{d^2}{dt^2}\phi_{X_\theta}(t)\right]_{t=0} - E[X_\theta]^2 = \sum_{j=1}^{J}\left\{\sum_{k=1}^{K_j} w_{jk}^2 p_{\theta jk} - \left(\sum_{k=1}^{K_j} w_{jk} p_{\theta jk}\right)^2\right\}.$$

If all J dice are similar, that is, $X_\theta \sim S_J^+ F_M^-$ or $X_\theta \sim S_J^+ F_M^+$, the random variables $U_{\theta 1}, \cdots, U_{\theta J}$, thereby, become independent and identically distributed. Then, according to the central limit theorem, the distribution of X_θ converges in law to a normal distribution as J increases. In this case, the probability of X_θ being larger and smaller than x is approximated as follows:

$$\Pr(X_\theta \geq x) \approx 1 - N(x|E[X_\theta], V[X_\theta]),$$
$$\Pr(X_\theta \leq x) \approx N(x|E[X_\theta], V[X_\theta]),$$

where $N(\cdot|a, b)$ represents the cumulative distribution function of the normal distribution with mean *a and variance b*. However, when all dice are not similar, that is, $X_\theta \sim S_J^- F_D^-$ *or* $X_\theta \sim S_J^- F_D^+$, from the Chernoff inequality (e.g., [4]) and following MGF of X_θ,

$$M_{X_\theta}(t) = E\left[e^{tX_\theta}\right] = \prod_{j=1}^{J}\sum_{k=1}^{K_j} p_{\theta jk} e^{tw_{jk}},$$

we can evaluate $\Pr(X_\theta \geq x)$ and $\Pr(X_\theta \leq x)$ as follows:

Corollary 4 (*Conditional Chernoff bound for dice roll sum distribution*) For an arbitrary $x \in \mathbb{R}$ and $t \geq 0$, we have

$$\Pr(X_\theta \geq x) \leq \inf_{t\geq 0} e^{-tx} M_{X_\theta}(t) = \inf_{t\geq 0} e^{-tx} \prod_{j=1}^{J} \sum_{k=1}^{K_j} p_{\theta jk} e^{tw_{jk}},$$

$$\Pr(X_\theta \leq x) \leq \inf_{t\geq 0} e^{tx} M_{X_\theta}(-t) = \inf_{t\geq 0} e^{tx} \prod_{j=1}^{J} \sum_{k=1}^{K_j} p_{\theta jk} e^{tw_{jk}}.$$

These bounds can be used to evaluate the likelihood of observing a particular value of the total score, conditioned at a specific level of θ. The practical range of X_θ is limited for each level of θ, and the range of the respondent with a higher θ generally lies further in the positive direction. Thus, if a respondent gets an extremely high or low score compared to his or her θ, the test may contain one or more items that are advantageous or disadvantageous to people sharing one or more of the respondent's demographic variables. When the distribution of an item score differs significantly among respondent subgroups, this situation is traditionally referred to as Simpson's paradox or differential item functioning [20, 40], particularly from the perspective of item writers and psychometricians.

2.6 *Closed Form of Conditional Dice Roll Sum Distribution*

As described in Sect. 2.3, it is practically impossible to calculate the PMF of $S_J^- F_D^-$ using the BF method shown in Eq. (3). However, when the weights of all items are of the Likert-type, as in Example A, a recursive method for obtaining the PMF of $S_J^- F_D^-$ was developed by [17, 48, 52] by extending the original LW algorithm. Moreover, [46] generalized the algorithm to be applicable to the condition that the weights within an item are not integers (but are equally spaced), as in Example B. This study does not consider whether the weights are positive or negative, whole or rational, and equispaced or non-equispaced. Additionally, the same weights can be assigned to multiple categories within an item. In other words, different item scoring schemes, such as Examples A–E, can be used in a test.

In addition, the LW algorithm is a recursive method and, thus, is not expressed in closed form. This section shows a closed form of the PMF of $S_J^- F_D^-$ using the method of applying DFT to CF (DFT-CF) [13, 21, 57]. The DFT-CF method was originally developed to determine the closed form of the Poisson binomial distribution (PB) [19, 26, 36, 53], which is the distribution of the number of successes in J trials when the probability of success in each trial is different. The PB can be denoted as $S_J^- F_2^-$ based on the notation used in this study (see Fig. 1).

Theorem 5 (*Closed form of conditional PMF of dice roll sum distribution*) Let the support of the distribution of $S_J^- F_D^-$ *be* $\Xi = \{\xi_g | g = 0, 1, \cdots, G\}$ and the PMF of ξ_g conditional on θ be $p_{\theta g}$. Then, the closed form of $p_{\theta g}$ is expressed as follows:

$$p_{\theta_g} = \frac{1}{G+1}\sum_{\gamma=0}^{G} C^{-\gamma\xi_g}\prod_{j=1}^{J}\sum_{k=1}^{K_j} p_{\theta jk}C^{\gamma w_{jk}} (g = 0, 1, \ldots, G), \tag{6}$$

where $C = \exp\{2\pi il/(G+1)\}$, l, and G are explained in the following proof.

Proof When all weights in the test are $\boldsymbol{W} = \{\boldsymbol{w}_1, \cdots, \boldsymbol{w}_J\}$, the minimum and maximum possible scores of the test are

$$\xi_{min} = \sum_{j=1}^{J} \min \boldsymbol{w}_j \quad \text{and} \quad \xi_{max} = \sum_{j=1}^{J} \max \boldsymbol{w}_j,$$

respectively. In addition, let D represents the set of all denominators of the rational numbers in $\boldsymbol{W}$, it is noted from Eq. (1) *that all the elements in D* are positive. Let the least common multiple (LCM) of the elements in D be $l = LCM(D)$. The following equispaced sequence

$$\Xi = \left\{\xi_g = \xi_{min} + \frac{g}{l}\middle| g = 0, 1, \ldots, G; G = l(\xi_{max} - \xi_{min})\right\},$$

contains all the elements of the sample space (support) of $S_J^- F_D^-$, Ω_{surj}, that is, $\Omega_{surj} \subseteq \Xi$, because the denominator of every element in Ω_{surj} is $\leq l$. Note that the number of elements in Ξ is $|\Xi| = l(\xi_{max} - \xi_{min}) + 1 = G + 1$.

Denoting the probability (or PMF) of observing ξ_g in Ξ as $p_{\theta g}$ (s.t. $p_{\theta g} = 0, \forall g \in \Xi \setminus \Omega_{surj}$), the CF of $S_J^- F_D^-$ conditional on θ is by definition expressed as follows:

$$\phi_{X_\theta}(t) = E\left[e^{itX_\theta}\right] = \sum_{g=0}^{G} p_{\theta_g} e^{it\xi_g}.$$

Then, dividing both sides of the equation by $(G + 1)$ and substituting $t = 2\pi l\gamma/(G + 1)$, we obtain the following:

$$\phi_{X_\theta}\left(\frac{2\pi l\gamma}{G+1}\right) = \frac{1}{G+1}\sum_{g=0}^{G} p_{\theta g}C^{\gamma\xi_g} \quad (\gamma = 0, 1, \cdots, G), \tag{7}$$

where $C = \exp\{2\pi il/(G + 1)\}$. Equation (7) can be regarded as the inverse DFT when

$$\boldsymbol{p}_\theta = [p_{\theta 0}\ p_{\theta 1}\ \cdots\ p_{\theta G}]' \quad ((G + 1) \times 1)$$

and

$$\boldsymbol{\phi}_{X_\theta} = \left[\phi_{X_\theta}\left(\frac{2\pi l0}{G+1}\right)\phi_{X_\theta}\left(\frac{2\pi l1}{G+1}\right)\cdots\ \phi_{X_\theta}\left(\frac{2\pi lG}{G+1}\right)\right]' \quad ((G + 1) \times 1)$$

are considered as discrete spectra in the frequency domain and discrete analog signals in the time domain, respectively, that is, $\mathcal{IF} : \boldsymbol{p} \rightarrow \boldsymbol{\phi}_{X_\theta}$ Thus, by applying DFT to both sides of Eq. (7), we obtain the following:

$$p_{\theta_g} = \frac{1}{G+1} \sum_{\gamma=0}^{G} C^{-\gamma \xi_g} \phi_{X_\theta} \left(\frac{2\pi l \gamma}{G+1} \right).$$

Finally, by substituting Eq. (5) of Theorem 2 into the above equation, Eq. (6) was obtained. Thus, the proof is complete. ■

The method employed in the proof is based on the DFT-CF method developed by [21], which can be applied when the sample points on the sample space are equispaced. However, the sample space of $S_J^- F_D^-$, Ω_{surj}, is generally non-equispaced. Thus, the DFT was performed after upsampling (interpolating) the sample points of Ω_{surj} and introducing the equispaced sample space of Ξ. This DFT is a simple technique for developing non-equispaced DFT methods (e.g., [11, 41]) and can thus be referred to as the NDFT-CF method.

It should also be noted that Eq. (6) computed using the NDFT-CF method is obtained as a complex number, and the real part of the complex number represents the PMF. In addition, Eq. (6) is not an approximation of the PMF of $S_J^- F_D^-$ but an exact expression while explicitly parameterizing the weights. For example, we can directly examine the effect of changing a particular weight(s) on the shape of the score distribution. In other words, this closed-form can be used to mathematically examine the properties of $S_J^- F_D^-$. Moreover, although the BF method in Definition 1 is infeasible when J is large, this closed form is computable within a realistic time frame and runs in the quadratic time of J, that is, $O\left(G \prod_{j=1}^{J} K_j\right) \approx O(J^2)$.

Example Assume that there is a small test (Test A) consisting of three items, and the weights of the items are

$$\boldsymbol{w}_1 = \left[\frac{1}{4} \, \frac{1}{2} \, \frac{3}{4} \, 1 \right], \quad \boldsymbol{w}_2 = \left[\frac{1}{3} \, 0 \, 2 \, 0 \, 0 \right], \quad \boldsymbol{w}_3 = \left[-\frac{1}{2} \quad 0 \;\; 1 \;\; 0 \, 0 \right],$$

and PMF of the score distribution in Test A were obtained using the NDFT-CF method. First, the minimum and maximum scores of Test A were $\xi_{min} = 1/4 + 0 - 1/2 = -1/4$ and $\xi_{max} = 1 + 2 + 1 = 4$, respectively. Next, the number of possible response patterns for Test A was $|\Omega| = 4 \times 5 \times 5 = 100$, and the set of possible total scores was given as

$$\Omega_{surj} = \left\{ -\frac{1}{4}, 0, \frac{1}{12}, \frac{1}{4}, \frac{1}{3}, \frac{1}{2}, \frac{7}{12}, \frac{3}{4}, \frac{5}{6}, 1, \frac{13}{12}, \frac{5}{4}, \frac{4}{3} \right.$$
$$\left. \frac{3}{2}, \frac{19}{12}, \frac{7}{4}, \frac{11}{6}, 2, \frac{25}{12}, \frac{9}{4}, \frac{7}{3}, \frac{5}{2}, \frac{11}{4}, 3, \frac{13}{4}, \frac{7}{2}, \frac{15}{4}, 4 \right\} \tag{8}$$

whose number of elements was $\left|\Omega_{surj}\right| = 28$. As shown in Eq. (8), Ω_{surj} becomes a set of non-equispaced sequence of rational numbers. Then, because $l = 12$ from $D = \{4, 3, 2\}$, the support of the score distribution of Test A is obtained as follows:

$$\Xi = \left\{ \xi_g = -\frac{1}{4} + \frac{g}{12} \middle| g = 0, 1, \ldots, 52 \right\}.$$

Evidently, Ξ becomes an equispaced sequence of rational numbers and includes all elements in Ω_{surj}. We also assume that the selection rates for each category of the three items at a given level of θ are as follows:

$$\boldsymbol{p}_{\theta 1} = [.20 \;\; .25 \;\; .15 \;\; .40] \text{ (four categories)},$$

$$\boldsymbol{p}_{\theta 2} = [.20 \;\; .30 \;\; .10 \;\; .15 \;\; .25] \text{ (five categories)},$$

$$\boldsymbol{p}_{\theta 3} = [.10 \;\; .20 \;\; .30 \;\; .20 \;\; .20] \text{ (five categories)}.$$

Then, the total score of Test A follows $S_3^- F_D^-(\{w_1, w_2, w_3\}, \{\boldsymbol{p}_{\theta 1}, \boldsymbol{p}_{\theta 2}, \boldsymbol{p}_{\theta 3}\})$.

Table 1 lists the PMF of $S_3^- F_D^-$ computed by the NDFT-CF method and verifies that the results evaluated by the BF and NDFT-CF methods match 52 score patterns from $-1/4 (= -3/12)$ to $48/12 (= 4)$. It should be noted that the BF method evaluates only the probabilities whose score is occurable, whereas those of impossible scores (the shaded cells in Table 1) are not evaluated and are set to 0.

Figure 1 also shows the closed-form function of the PMF for Test A obtained using the NDFT-CF method. The discrete points indicated by the markers are the PMF of Test A, while the solid line passing through all the markers represents the resultant wave created by the NDFT-CF method. To verify the difference in PMF evaluated using the BF and NDFT-CF methods, the mean absolute difference (MAD) and root mean squared difference (RMSD) were used and evaluated as follows:

$$MAD = \frac{1}{53} \sum_{g=0}^{52} \left| p_{\theta_g}^{(BF)} - p_{\theta_g}^{(NDFT)} \right| = 4.7872 \times 10^{-18},$$

$$RMSD = \left\{ \frac{1}{53} \sum_{g=0}^{52} \left(p_{\theta_g}^{(BF)} - p_{\theta_g}^{(NDFT)} \right)^2 \right\}^{\frac{1}{2}} = 7.1737 \times 10^{-18},$$

where $p_{\theta g}^{(BF)}$ and $p_{\theta g}^{(NDFT)}$ represent the PMFs of ξ_g obtained using the BF and NDFT-CF methods, respectively. As these indices illustrate, the NDFT-CF method accurately reproduces the PMF evaluated using the BF method. However, the MAD and RMSD were not perfectly zero. This is because the DFT ticks the discrete real number series on the continuous composite wave of complex sine and cosine curves,

Table 1 PMF of Test A ($S_3^- F_D^-$)

Ξ	BF	NDFT-CF	LW	Ξ	BF	NDFT-CF	LW
−3/12	0.0140	0.0140	0.0140	23/12	0.0000	0.0000	0.0000
−2/12	0.0000	0.0000	0.0000	24/12 (2.0)	0.0865	0.0865	0.0865
−1/12	0.0000	0.0000	0.0000	25/12	0.0090	0.0090	0.0090
0/12 (0.0)	0.0175	0.0175	0.0175	26/12	0.0000	0.0000	0.0000
1/12	0.0040	0.0040	0.0040	27/12	0.0135	0.0135	0.0135
2/12	0.0000	0.0000	0.0000	28/12	0.0240	0.0240	0.0240
3/12	0.0945	0.0945	0.0945	29/12	0.0000	0.0000	0.0000
4/12	0.0050	0.0050	0.0050	30/12 (2.5)	0.0190	0.0190	0.0190
5/12	0.0000	0.0000	0.0000	31/12	0.0000	0.0000	0.0000
6/12 (0.5)	0.1330	0.1330	0.1330	32/12	0.0000	0.0000	0.0000
7/12	0.0270	0.0270	0.0270	33/12	0.0090	0.0090	0.0090
8/12	0.0000	0.0000	0.0000	34/12	0.0000	0.0000	0.0000
9/12	0.0630	0.0630	0.0630	35/12	0.0000	0.0000	0.0000
10/12	0.0380	0.0380	0.0380	36/12 (3.0)	0.0240	0.0240	0.0240
11/12	0.0000	0.0000	0.0000	37/12	0.0000	0.0000	0.0000
12/12 (1.0)	0.1680	0.1680	0.1680	38/12	0.0000	0.0000	0.0000
13/12	0.0180	0.0180	0.0180	39/12	0.0060	0.0060	0.0060
14/12	0.0000	0.0000	0.0000	40/12	0.0000	0.0000	0.0000
15/12	0.0420	0.0420	0.0420	41/12	0.0000	0.0000	0.0000
16/12	0.0480	0.0480	0.0480	42/12 (3.5)	0.0075	0.0075	0.0075
17/12	0.0000	0.0000	0.0000	43/12	0.0000	0.0000	0.0000
18/12 (1.5)	0.0525	0.0525	0.0525	44/12	0.0000	0.0000	0.0000
19/12	0.0120	0.0120	0.0120	45/12	0.0045	0.0045	0.0045
20/12	0.0000	0.0000	0.0000	46/12	0.0000	0.0000	0.0000
21/12	0.0335	0.0335	0.0335	47/12	0.0000	0.0000	0.0000
22/12	0.0150	0.0150	0.0150	48/12 (4.0)	0.0120	0.0120	0.0120

which in practice depends on the setting for the computation accuracy of the software program used (Fig. 2).

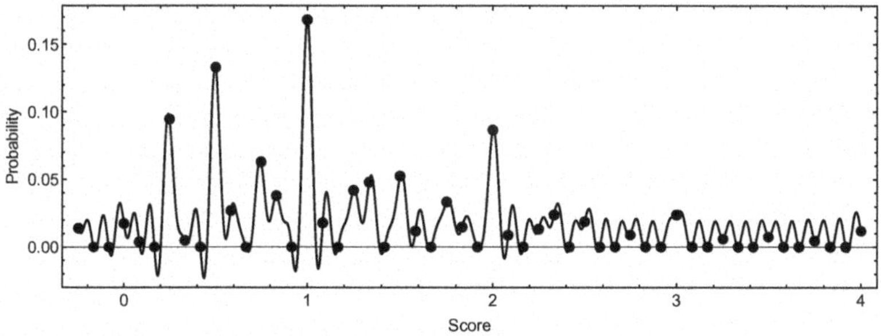

Fig. 2 PMF of Test A by NDFT-CF method

2.7 LW Algorithm for Dice Roll Sum Distribution

Although the NDFT-CF method can compute the PMF of $S_J^- F_D^-$ within a realistic time frame, the method requires more computation time when the support length, $|\Xi| = G + 1$, increases. To further speed up the computation, it is necessary to implement the fast Fourier transform (FFT). Hong (2003) succeeded in shortening the computation time slightly compared to the LW algorithm in calculating the PMF of the PB distribution (or $S_J^- F_2^-$ in Fig. 1), although the difference was not significant. The LW algorithm is sufficiently fast to compute and, in particular, is much easier to implement than the FFT. This section describes the LW algorithm for obtaining the PMF of the $S_J^- F_D^-$.

First, let the minimum and maximum weights in item j be

$$w_{min,j} = \min \boldsymbol{w}_j \text{ and } w_{max,j} = \max \boldsymbol{w}_j,$$

respectively, and the following equispaced sequence

$$\Xi_j = \left\{\xi_{jg} = w_{min,j} + \frac{g}{l} \middle| g = 0, 1, \cdots, G_j; G_j = l\left(w_{max,j} - w_{min,j}\right)\right\}$$

be provided as the support of the categorical distribution for item j. Then, the PMF of the categorical distribution for item j whose $G + 1$ elements are arranged in accordance with Ξ_j is created in a vector form as follows:

$$\boldsymbol{\pi}_{\theta j} = \left[p_{\theta j1} \cdots p_{\theta jG_j}\right]' = \left\{p_{\theta jg} = \sum_{k=1}^{K_j} I\left(\xi_{jg}, w_{jk}\right) p_{\theta jk}\right\} \quad ((G_j + 1) \times 1).$$

This vector can be regarded as the score distribution of the respondent with trait θ for item j. Algorithm 6 shows the pseudocode of the LW algorithm to obtain the total score distribution of $S_J^- F_D^-$ conditioned on θ.

Algorithm 6 LW Algorithm for Conditional Score Distribution

Pseudocode
[1] Specify $\theta \in \mathbb{R}$
[2] Compute $\boldsymbol{\pi}_{\theta 1}$ $\left((G_1+1)\times 1\right)$
[3] Set $\boldsymbol{q}_{\theta 1} = \boldsymbol{\pi}_{\theta 1}$
[4] Set $r_1 = \lvert \boldsymbol{q}_{\theta 1} \rvert = G_1 + 1$
[5] for $j = 1$ to J
[5-1] Compute $\boldsymbol{\pi}_{\theta,j+1}$ $\left((G_{j+1}+1)\times 1\right)$
[5-2] Compute $\boldsymbol{Q}_{\theta,j+1} = \boldsymbol{q}_{\theta j}\boldsymbol{\pi}'_{\theta,j+1}$ $\left(r_j \times (G_{j+1}+1)\right)$
[5-3][*1] Compute $\boldsymbol{R}_{\theta,j+1} = \boldsymbol{Q}_{\theta,j+1}\boldsymbol{E}_{G_{j+1}}$ $\left(r_j \times (G_{j+1}+1)\right)$
[5-4] Compute $r_{j+1} = r_j + G_{j+1} - 1$
[5-5] For $r = 0$ to $r_{j+1} - 1$
[5-5-1][*2] Compute $q_{\theta,j+1,r} = \mathrm{tr}\left(\boldsymbol{R}_{\theta,j+1}, G_{j+1} - 1 - r\right)$
[5-6] Set $\boldsymbol{q}_{\theta,j+1} = \left[q_{\theta,j+1,0}\ q_{\theta,j+1,1} \cdots q_{\theta,j+1,r_{j+1}-1}\right]$ $(r_{j+1}\times 1)$
[6] Obtain $\boldsymbol{p}_\theta^{(LW)} = \boldsymbol{q}_{\theta J}$ as the score distribution for $N_j U_D$ conditioned on θ

*1 $\boldsymbol{E}_a$ is the exchange matrix (backward identity matrix) with size a.

$$*2\ \mathrm{tr}(\boldsymbol{A}, b) = \begin{cases} \sum_{k=1}^{\mathrm{row}(\boldsymbol{A})-b} a_{k,k+b} & b > 0 \\ \sum_{k=1}^{\min(\mathrm{row}(\boldsymbol{A}),\mathrm{col}(\boldsymbol{A}))} a_{k,k} & b = 0 \\ \sum_{k=1}^{\mathrm{col}(\boldsymbol{A})+b} a_{k-b,k} & b < 0 \end{cases}$$

The column labeled LW in Table 1 shows the results of Algorithm 6. As the table shows, the LW algorithm can also produce the PMF made using the BF method. It should be noted that, as the following MAD and RMSD between the BF method and LW algorithm show,

$$MAD = \frac{1}{53}\sum_{g=0}^{52}\left|p_{\theta g}^{(BF)} - p_{\theta g}^{(LW)}\right| = 0,$$

the LW algorithm does not even have the smallest calculation error derived from the computation accuracy of the software program.

2.8 Simulation Study

The NDFT-CF method (Theorem 5) is a closed form, and the LW algorithm (Algorithm 6) is a recursive method, but both can reproduce the PMF of $S_J^- F_D^-$ obtained by the BF method (Definition 1). In this section, we describe a simulation study conducted to confirm the accuracy and efficiency of the NDFT-CF method and LW algorithm. Note that the BF method was not examined in the numerical experiment because it is not computationally feasible when the number of items is large. However, as shown in Sect. 2.7, the LW algorithm has no small error compared with the BF method; thus, the results obtained by the LW algorithm can be considered to be those of the BF method.

First, three levels {10, 20, 40} were provided for the condition of the number of items, and the number of repetitions in each level was set to 100. Next, the number of categories for item j were randomly selected from natural numbers between 2 and 6. That is,

$$K_j \sim U_D[2, 6] \quad (j = 1, \cdots, J;\ J \in \{10, 20, 40\}),$$

where $U_D[a, b]$ $(a \leq b;\ a, b \in \mathbb{N})$ represents a discrete uniform distribution. The weight and probability for category k of item j are then randomly determined as follows:

$$w_{jk} = \frac{a}{b}\big(a \sim U\big[-1, K_j\big];\ b \sim U[1, 6];\ k = 1, \ldots, K_j;\ j = 1, \ldots, J\big),$$

$$p_{\theta jk} = \frac{p^*_{\theta jk}}{\sum_{k'=1}^{K_j} p^*_{\theta jk'}}\big(p^*_{\theta jk} \sim U_C[0, 1];\ k = 1, \ldots, K_j;\ j = 1, \ldots, J\big),$$

where $U_C[a, b]$ $(a \leq b;\ a, b \in \mathbb{R})$ indicates a *continuous* uniform distribution. For $S_J^- F_D^-(\boldsymbol{W}, \boldsymbol{P}_\theta)$ created through this process, the PMF of the distribution was calculated using the NDFT-CF method and LW algorithm, and the accuracy between them was evaluated as follows:

$$MAD = \frac{1}{G+1}\sum_{g=0}^{G}\left|p_{\theta g}^{(LW)} - p_{\theta g}^{(NDFT)}\right|,$$

$$RMSD = \left\{\frac{1}{G+1}\sum_{g=0}^{G}\left(p_{\theta g}^{(LW)} - p_{\theta g}^{(NDFT)}\right)^2\right\}^{\frac{1}{2}}.$$

In addition, the efficiency of each method was evaluated based on the time (s) required for calculation. The numerical experiment was performed using Mathematica 12.0, in a workstation with an Intel Core i7-8565U CPU (3.79 GHz) and a 20.0 GB RAM installed.

Table 2 lists the results of the numerical experiment, and it was observed from the MAD and RMSD that there are practically no differences in the PMF produced by the NDFT-CF method and LW algorithm. However, comparing the computation time, the NDFT-CF method requires 17 s on average when the number of items was

Table 2 Simulation results

No. of items		10	20	40
MAD	Mean SD	1.3746×10^{-16} 5.5741×10^{-17}	1.5165×10^{-16} 4.1550×10^{-17}	1.6919×10^{-16} 3.3549×10^{-17}
RMSD	Mean SD	2.1472×10^{-16} 8.7468×10^{-17}	2.3166×10^{-16} 6.1421×10^{-17}	2.5880×10^{-16} 4.9394×10^{-17}
Time (NDFT)	Mean SD	0.7709 (s) 0.5086 (s)	4.2042 (s) 1.6055 (s)	17.2094 (s) 04.4808 (s)
Time (LW)	Mean SD	0.0559 (s) 0.0456 (s)	0.3230 (s) 0.1349 (s)	01.3484 (s) 00.4001 (s)

40. Additionally, if the number of items is doubled, the computation time increases by approximately four times because the computation time required for the method is $O(J^2)$, as described in Theorem 5. Therefore, the presumed time required to obtain the PMF by running the NDFT-CF is approximately 70 s when the number of items was 80.

Meanwhile, the LW algorithm was approximately 10 times faster than the NDFT-CF method, with an average computation time of 1.3 s when the number of items was 40. Although the NDFT-CF method has the potential to be slightly faster than the LW algorithm if FFT is implemented, as demonstrated by [21] for the PB distribution, the fact that the LW algorithm has exactly zero error with respect to the BF method is a remarkably reliable property, and the superiority of the LW algorithm will, thus, remain unchallenged in practical use. It is, therefore, recommended that the closed-form obtained by the NDFT-CF method be useful for examining, modeling, or predicting the behavior of test scores when the weight scheme of an item(s) is changed, while the LW algorithm is preferable for implementation in software that must routinely calculate test score distributions as in linking two or more tests [38] and estimating the multidimensional traits of each respondent [5, 23].

2.9 *Properties of Populational Dice Roll Sum Distribution*

The one definition, two theorems, two corollaries, and one algorithm presented in Sects. 2.4–2.7 express the properties of the conditional dice-roll-sum distribution at a particular level on the θ scale, that is, $X_\theta \sim S_J^- F_D^-$. In this section, we summarize the population counterparts of the conditional distribution when the density of the trait distribution is $f(\theta|\boldsymbol{\mu})$, where $\boldsymbol{\mu}$ represents the parameter(s) of the density distribution. That is, the distribution of $X_\mu \sim S_J^- F_D^-$. The proofs for the following equations are omitted because each of them can be obtained by marginalizing the corresponding equation with respect to θ over f and the integral calculation can be approximated by considering an adequate number of quadrature points on θ [3].

Definition 1* Population PMF of dice roll sum distribution.

$$\Pr(X = x) = \int_{\mathbb{R}} \Pr(X = x) f(\theta|\mu) d\theta$$
$$= \int_{\mathbb{R}} f(\theta|\mu) \sum_{h=1}^{|\Omega|} I\left(1'_J u_h, x\right) \prod_{j=1}^{J} p_{\theta jk}^{zrjk} d\theta \left(x \in \Omega_{surj}\right)$$

Theorem 2* Populational CF of dice roll sum distribution.

$$\phi_X(t) = \int_{\mathbb{R}} \phi_{X_\theta}(t) f(\theta|\boldsymbol{\mu}) d\theta = \int_{\mathbb{R}} f(\theta|\boldsymbol{\mu}) \prod_{j=1}^{J} \sum_{k=1}^{K_j} p_{\theta jk} e^{itw_{jk}} d\theta$$

Corollary 3* Populational moment of dice roll sum distribution.

$$E\left[X_\theta^m\right] = \int_{\mathbb{R}} E\left[X_\theta^m\right] f(\theta|\boldsymbol{\mu}) d\theta = \int_{\mathbb{R}} f(\theta|\boldsymbol{\mu}) \sum_{j=1}^{J} \sum_{k=1}^{K_j} w_{jk}^m p_{\theta jk} d\theta$$

Corollary 4* Populational Chernoff bound for dice roll sum distribution.
For an arbitrary $x \in \mathbb{R}$ and $t \geq 0$,

$$\Pr(X \geq x) \leq \inf_{t \geq 0} e^{-ta} \int_{\mathbb{R}} M_{X_\theta}(t) f(\theta|\boldsymbol{\mu}) d\theta$$
$$= \inf_{t \geq 0} e^{-ta} \int_{\mathbb{R}} f(\theta|\mu) \prod_{j=1}^{J} \sum_{k=1}^{K_j} p_{\theta jk} e^{tw_{jk}} d\theta,$$
$$\Pr(X \leq x) \leq \inf_{t \geq 0} e^{ta} \int_{\mathbb{R}} M_{X_\theta}(-t) f(\theta|\boldsymbol{\mu}) d\theta$$
$$= \inf_{t \geq 0} e^{ta} \int_{\mathbb{R}} f(\theta|\mu) \prod_{j=1}^{J} \sum_{k=1}^{K_j} p_{\theta jk} e^{-tw_{jk}} d\theta.$$

Theorem 5* Closed form of populational PMF of dice roll sum distribution.

$$p_g = \int_{\mathbb{R}} \frac{f(\theta|\mu)}{G+1} \sum_{\gamma=0}^{G} C^{-\gamma \xi_g} \prod_{j=1}^{J} \sum_{k=1}^{K_j} p_{\theta jk} C^{\gamma w_{jk}} d\theta \left(g = 0, 1, \ldots, G\right)$$

Algorithm 6* LW Algorithm for Populational Score Distribution

Pseudocode
[1*] Set Q quadrature points on θ scale $(\theta_1, \cdots, \theta_Q)$
[2*] for $q = 1$ to Q
[2*-1] Execute commands [2]–[5] in Algorithm 6 and obtain $\boldsymbol{p}_{\theta_q}^{(LW)}$
[2*-2] Compute $\boldsymbol{p}_q^{(LW)} += f(\theta_q \| \boldsymbol{\mu})\boldsymbol{p}_{\theta_q}^{(LW)}$
[3*] Obtain $\boldsymbol{p}_Q^{(LW)}$ as the score distribution for $N_j U_D$ when the trait distribution is $f(\theta \| \boldsymbol{\mu})$

3 Conclusion

This study explored the total score distribution of a test (including academic tests, psychological and medical scales, and social and health survey questionnaires) under the general condition that the numbers of categories and weight scheme of each item are allowed to be distinct and that the weight for each category in an item may be a positive or negative rational number, and then modeled the PMF of the distribution of such a test ($S_{\bar{J}} F_{\bar{D}}$) using the IRT. We first show that the distribution of the test scores conditioned on θ can be considered identical to the distribution of the sum of the rolling heterogeneous dice (Sect. 2.3). This distribution has the potential to be applied to horse racing, actuarial science, psephology, and sports sciences (Sect. 2.4).

In addition, after obtaining the CF of $S_{\bar{J}} F_{\bar{D}}$ (Theorem 2 in Sect. 2.5), the closed form of the PMF of $S_{\bar{J}} F_{\bar{D}}$ (Theorem 5 in Sect. 2.6) was derived by extending the DFT-CF method to be applicable to the case in which the support is non-equispaced, referred to as the NDFT-CF method, which is calculable within a practical time frame even when the number of items is large. The closed-form can be used to theoretically examine the properties of $S_{\bar{J}} F_{\bar{D}}$ and model the total score distribution for a test of concern, because it explicitly parameterizes the weights. This study also used the LW algorithm (Algorithm 6 in Sect. 2.7) to calculate the PMF of the $S_{\bar{J}} F_{\bar{D}}$. Because this recursive algorithm can run very quickly and has no error with respect to the PMF produced by the BF method, the algorithm is recommended to use in actual analysis and implement in software in cases where the total score distribution of $S_{\bar{J}} F_{\bar{D}}$ is required to be calculated frequently or routinely.

Finally, Sect. 2.8 presented some properties of the distribution of $S_{\bar{J}} F_{\bar{D}}$ when the trait of each respondent follows a population trait distribution, $f(\theta|\boldsymbol{\mu})$. This population total score distribution of $S_{\bar{J}} F_{\bar{D}}$ may sometimes be of more interest than the conditional distribution at a particular level of θ, however mathematically, it is simple to obtain by marginalizing the conditional distribution on θ with respect to f.

A limitation of this study is that the weights of the item categories were limited to rational numbers. Although there are not many situations in which the item weights are set to irrational numbers, if there are irrational weights, it is not always possible to create the equispaced support Ξ from the non-equispaced support Ω_{surj} by upsampling. In such cases, the NDFT-CF method must be modified to be directly applicable to the non-equispaced support Ω_{surj}. However, it may be possible to build an LW algorithm under such conditions by considering a possible discrete sequence as the support.

References

1. Bock, R. (1972). Estimating item parameters and latent ability when responses are scored in two or more nominal categories. *Psychometrika, 37*, 29–51.
2. Bock, R. D., Gibbons, R., & Muraki, E. (1988). Full-information item factor analysis. *Applied Psychological Measurement., 12*, 261–280.
3. Bock, R. D., & Lieberman, M. (1970). Fitting a response model for n dichotomously scored items. *Psychometrika, 35*, 179–197.
4. Boucheron, S., Lugosi, G., & Massart, P. (2013). *Concentration inequalities: A nonasymptotic theory of independence*. Oxford University Press.
5. Cai, L. (2015). Lord-Wingersky algorithm version 2.0 for hierarchical item factor models with applications in test scoring, scale alignment, and model fit testing. *Psychometrika, 80*, 535–559.
6. Cai, L., Yang, J. S., & Hansen, M. (2011). Generalized full-information item bifactor analysis. *Psychological Methods, 16*, 221–248.
7. Calc Site. (2021). Probability of the sum of the dice. Retrieved from https://en.calc-site.com/probabilities/dice_total
8. Chen, W. H., & Thissen, D. (1999). Estimation of item parameters for the three-parameter logistic model using the marginal likelihood of summed scores. *British Journal of Mathematical and Statistical Psychology, 52*, 19–37.
9. Deng, S., McCarthy, D. E., Piper, M. E., Baker, T. B., & Bolt, D. M. (2018). Extreme response style and the measurement of intra-individual variability in affect. *Multivariate Behavioral Research, 53*, 199–218.
10. Dimitrov, D. M. (2003). Marginal true-score measures and reliability for binary items as a function of their IRT parameters. *Applied Psychological Measurement, 27*, 440–458.
11. Dutt, A., & Rokhlin, V. (1993). Fast Fourier transforms for nonequispaced data. *SIAM Journal on Scientific Computing, 14*, 1368–1393.
12. Embretson, S. E., & Reise, S. P. (2000). *Item response theory for psychologists*. Psychology Press.
13. Fernandez, M., & Williams, S. (2010). Closed-form expression for the Poisson-binomial probability density function. *IEEE Transactions on Aerospace Electronic Systems, 46*, 803–817.
14. Fraley, R. C., Waller, N. G., & Brennan, K. A. (2000). An item response theory of self-report measures of adult attachment. *Journal of Personality and Social Psychology, 78*, 350–365.
15. Glas, C. A. W., & Verhelst, N. D. (1989). Extensions of the partial credit model. *Psychometrika, 54*, 635–659.
16. González, J., Wiberg, M., & von Davier, A. A. (2016). A note on the Poisson's binomial distribution in item response theory. *Applied Psychological Measurement, 40*, 302–310.
17. Hanson B. A. (1994). Extension of Lord-Wingersky algorithm to computing test score distributions for polytomous items. Retrieved from http://www.openirt.com/b-a-h/papers/note9401.pdf
18. Hambleton, R. K., Swaminathan, H., & Rogers, H. J. (1991). *Fundamentals of item response theory*. Sage.

19. Hoeffding, W. (1956). On the distribution of the number of successes in independent trials. *The Annals of Mathematical Statistics, 27*, 713–721.
20. Holland, P. W., & Wainer, H. (1993). *Differential item functioning*. Lawrence Erlbaum.
21. Hong, Y. (2013). On computing the distribution function for the Poisson binomial distribution. *Computational Statistics and Data Analysis, 59*, 41–51.
22. Hua, C., & Wind, S. A. (2019). Exploring the psychometric properties of the mind-map scoring rubric. *Behaviormetrika, 46*, 73–99.
23. Huang, S., & Cai, L. (2021). Lord-Wingersky algorithm version 2.5 with applications. *Psychometrika, 86*, 973–993.
24. Kaliski, P. K., Wind, S. A., Engelhard, G., Morgan, D. L., Plake, B. S., & Reshetar, R. A. (2013). Using the many-faceted Rasch model to evaluate standard setting judgments. *Educational and Psychological Measurement, 73*, 386–411.
25. Kim, S. (2013). Generalization of the Lord-Wingersky algorithm to computing the distribution of summed test scores based on real-number item scores. *Journal of Educational Measurement, 50*, 381–389.
26. Le Cam, L. (1960). An approximation theorem for the Poisson binomial distribution. *Pacific Journal of Mathematics, 10*, 1181–1197.
27. Lord, F. M. (1980). *Applications of item response theory to practical testing problems*. Lawrence Erlbaum.
28. Lord, F. M., & Novick, M. R. (1968). *Statistical theories of mental test scores*. Addison-Wesley.
29. Lord, F. M., & Wingersky, M. (1984). Comparison of IRT true-score and equipercentile observed-score "equatings." *Applied Psychological Measurement, 8*, 453–461.
30. Masters, G. N. (1982). A Rasch model for partial credit scoring. *Psychometrika, 47*, 149–174.
31. May, K., & Nicewander, W. A. (1994). Reliability and information functions for percentile ranks. *Journal of Educational Measurement, 31*, 313–325.
32. Mayekawa, S., & Arai, S. (2008). Distribution of the sum of scored multinomial random variables and its application to the item response theory. In K. Shigemasu, A. Okada, T. Imaizumi, & T. Hoshino (Eds.), *New trends in psychometrics* (pp. 263–272).
33. Muraki, E. (1992). A generalized partial credit model: Application of an EM algorithm. *Applied Psychological Measurement, 16*, 159–176.
34. Muraki, E., & Carlson, J. E. (1995). Full-information factor analysis for polytomous item responses. *Applied Psychological Measurement, 19*, 73–90.
35. Murphy, K. P. (2012). *Machine learning: A probabilistic perspective*. MIT Press.
36. Nedelman, J., & Wallenius, T. (1986). Bernoulli trials, poisson trials, surprising variances, and Jensen's inequality. *The American Statistician, 40*, 286–289.
37. Ogasawara, H. (1998). A factor analysis model for a mixture of various types of variables. *Behaviormetrika, 25*, 1–12.
38. Orlando, M., Sherbourne, C. D., & Thissen, D. (2000). Summed-score linking using item response theory: Application to depression measurement. *Psychological Assessment, 12*, 354–359.
39. Orlando, M., & Thissen, D. (2000). Likelihood-based item-fit indices for dichotomous item response theory models. *Applied Psychological Measurement, 24*, 50–64.
40. Osterlind, S. J. & Everson, H. T. (2009). *Differential item functioning*. Sage Publishing.
41. Plonka G., Potts D., Steidl G., Tasche M. (2018). Fast Fourier transforms for nonequispaced data. In G. Plonka, D. Potts, G. Steidl, & M. Tasche (Eds.), *Numerical fourier analysis*. Birkhäuser.
42. Reckase, M. D. (2009). *Multidimensional item response theory: Statistics for social and behavioral sciences*. Springer.
43. Samejima, F. (1969). Estimation of latent ability using a response pattern of graded scores. *Psychometrika Monograph, 17*.
44. Sas, W. (2021). *Dice probability calculator*. Retrieved from https://www.omnicalculator.com/statistics/dice
45. Singh, A. K., Dalpatadu, R. J., & Lucas, A. F. (2011). The probability distribution of the sum of several dice: Slot applications. *UNLV Gaming Research and Review Journal, 15*. Retrieved from https://digitalscholarship.unlv.edu/grrj/vol15/iss2/10

46. Stucky, B. D. (2009). Item response theory for weighted summed scores, Doctoral dissertation. The University of North Carolina at Chapel Hill. Retrieved from https://core.ac.uk/download/pdf/210596126.pdf
47. Thissen, D., Chen, W.-H., & Bock, R. D. (2003). *MULTILOG 7 for windows: Multiple category item analysis and test scoring using item response theory*. Scientific Software International, Inc.
48. Thissen, D., Pommerich, M., Billeaud, K., & Williams, V. S. L. (1995). Item response theory for scores on tests including polytomous items with ordered responses. *Applied Psychological Measurement, 19*, 39–49.
49. Uspensky, J. V. (1937). *Introduction to mathematical probability*. McGraw-Hill.
50. Uto, M., & Ueno, M. (2020). A generalized many-facet Rasch model and its Bayesian estimation using Hamiltonian Monte Carlo. *Behaviormetrika, 47*, 469–496.
51. van der Linden, W. J. & Hambleton, R. K. (Eds.). (1996). *Handbook of modern item response theory*. Springer.
52. von Davier, M., & Rost, J. (1995). Polytomous mixed Rasch models. In G. H. Fischer & I. W. Molenaar (Eds.), *Rasch models: Foundations, recent developments, and applications* (pp. 371–379). Springer.
53. Wang, Y. H. (1993). On the number of successes in independent trials. *Statistica Sinica, 3*, 295–312.
54. Weisstein, E. W. (2021) Dice. *MathWorld–A Wolfram Web Resource*. Retrieved from https://mathworld.wolfram.com/Dice.html
55. Whitely, S. E. (1980). Multicomponent latent trait models for ability tests. *Psychometrika, 45*, 479–494.
56. Willmart, J. (2020). Basic D&D statistics: Sums of dice rolls. Retrieved from https://justinwillmert.com/articles/2020/basic-dd-statistics-sums-of-dice-rolls/
57. Zhang, M., Hong, Y., & Balakrishnan, N. (2018). The generalized Poisson-binomial distribution and the computation of its distribution function. *Journal of Statistical Computation and Simulation, 88*, 1515–1527.

The manufacturer's authorised representative in the EU is Springer
Nature Customer Service Centre GmbH, Europaplatz 3, 69115 Heidelberg,
Germany. If you have any concerns regarding our products, please
contact ProductSafety@springernature.com

Printed and bound by CPI Group (UK) Ltd, Croydon, CR0 4YY
01/10/2026
02225892-0011